T0214713

Lecture Notes in Mathematics

Edited by J.-M. Morel, F. Takens and B. Teissier

Editorial Policy
for the publication of monographs

1. Lecture Notes aim to report new developments in all areas of mathematics – quickly, informally and at a high level. Monograph manuscripts should be reasonably self-contained and rounded off. Thus they may, and often will, present not only results of the author but also related work by other people. They may be based on specialized lecture courses. Furthermore, the manuscripts should provide sufficient motivation, examples and applications. This clearly distinguishes Lecture Notes from journal articles or technical reports which normally are very concise. Articles intended for a journal but too long to be accepted by most journals, usually do not have this "lecture notes" character. For similar reasons it is unusual for doctoral theses to be accepted for the Lecture Notes series.

2. Manuscripts should be submitted (preferably in duplicate) either to one of the series editors or to Springer-Verlag, Heidelberg. In general, manuscripts will be sent out to 2 external referees for evaluation. If a decision cannot yet be reached on the basis of the first 2 reports, further referees may be contacted: the author will be informed of this. A final decision to publish can be made only on the basis of the complete manuscript, however a refereeing process leading to a preliminary decision can be based on a pre-final or incomplete manuscript. The strict minimum amount of material that will be considered should include a detailed outline describing the planned contents of each chapter, a bibliography and several sample chapters.
Authors should be aware that incomplete or insufficiently close to final manuscripts almost always result in longer refereeing times and nevertheless unclear referees' recommendations, making further refereeing of a final draft necessary.
Authors should also be aware that parallel submission of their manuscript to another publisher while under consideration for LNM will in general lead to immediate rejection.

3. Manuscripts should in general be submitted in English.
Final manuscripts should contain at least 100 pages of mathematical text and should include
– a table of contents;
– an informative introduction, with adequate motivation and perhaps some historical remarks: it should be accessible to a reader not intimately familiar with the topic treated;
– a subject index: as a rule this is genuinely helpful for the reader.

Continued on inside back-cover

Lecture Notes in Mathematics 1786

Editors:
J.-M. Morel, Cachan
F. Takens, Groningen
B. Teissier, Paris

Springer
Berlin
Heidelberg
New York
Barcelona
Hong Kong
London
Milan
Paris
Tokyo

Steven Dale Cutkosky

Monomialization
of Morphisms
from 3-Folds to Surfaces

Springer

Author

Steven Dale Cutkosky

Department of Mathematics
University of Missouri
Columbia
MO 65211
U.S.A.

e-mail: cutkoskys@missouri.edu
http://www.math.missouri.edu/~cutkosky

Cataloging-in-Publication Data applied for

Die Deutsche Bibliothek - CIP-Einheitsaufnahme

Cutkosky, Steven Dale:
Monomialization of morphisms from 3-folds to surfaces / Steven Dale
Cutkosky. - Berlin ; Heidelberg ; New York ; Barcelona ; Hong Kong ; London
; Milan ; Paris ; Tokyo : Springer, 2002
    (Lecture notes in mathematics ; 1786)
    ISBN 3-540-43780-0

Mathematics Subject Classification (2000): 14E15, 14D06

ISSN 0075-8434
ISBN 3-540-43780-0 Springer-Verlag Berlin Heidelberg New York

This work is subject to copyright. All rights are reserved, whether the whole or part of the material is
concerned, specifically the rights of translation, reprinting, reuse of illustrations, recitation, broadcasting,
reproduction on microfilm or in any other way, and storage in data banks. Duplication of this publication
or parts thereof is permitted only under the provisions of the German Copyright Law of September 9, 1965,
in its current version, and permission for use must always be obtained from Springer-Verlag. Violations are
liable for prosecution under the German Copyright Law.

Springer-Verlag Berlin Heidelberg New York a member of BertelsmannSpringer
Science + Business Media GmbH

http://www.springer.de

© Springer-Verlag Berlin Heidelberg 2002
Printed in Germany

The use of general descriptive names, registered names, trademarks, etc. in this publication does not imply,
even in the absence of a specific statement, that such names are exempt from the relevant protective laws
and regulations and therefore free for general use.

Typesetting: Camera-ready TEX output by the author

SPIN: 10878502   41/3142/du - 543210 - Printed on acid-free paper

CONTENTS

1. INTRODUCTION

These lecture notes give a systematic development of the theory of monomialization of morphisms in algebraic geometry.

Suppose that $f_1, \ldots, f_m$ are polynomials in n variables $x_1, \ldots, x_n$ over a field k (such as $\mathbf{R}$ or $\mathbf{C}$).

$$y_i = f_i(x_1, \ldots, x_n) \text{ for } 1 \le i \le m$$

defines a polynomial mapping (a morphism of affine varieties) $\Phi : k^n \to k^m$. The morphism Φ is dominant if the image of Φ in k^m is dense. This is equivalent to the statement that the $m \times m$ minors of the $m \times n$ matrix $\left(\frac{\partial f_i}{\partial x_j} \right)$ do not all vanish identically.

The morphism Φ is monomial if each of the f_i are monomials in coordinate functions $x_1, \ldots, x_n$ of k^n. That is, there exists an $m \times n$ matrix (a_{ij}) of natural numbers such that

$$y_i = x_1^{a_{i1}} \cdots x_n^{a_{in}} \text{ for } 1 \le i \le m.$$

For such a monomial mapping, Φ is a dominant morphism if and only if (a_{ij}) has rank m. Monomial morphisms are of course much easier to analyze. For instance, it is easy to compute solutions of the x variables in terms of the y variables, and it is easier to compute integrals and determine convergence of integrals of monomial morphisms.

We consider the condition that there is a neighborhood U of the origin in k^n and coordinate functions $\overline{x}_1, \ldots, \overline{x}_n$ on U such that the mapping Φ is monomial (in U) with respect to the coordinates $\overline{x}_1, \ldots, \overline{x}_n$.

This notion works well when we allow U to be an open set in the Euclidean topology. In algebraic geometry, open sets (in the Zariski topology) are the complements of the vanishing locus of a set of polynomials. This topology is too coarse for our notion of monomial on U. The correct notion to make this work is to take U to be an étale neighborhood. This is an algebraic neighborhood which lies between the concept of open in the Zariski topology and open in the Euclidean topology.

Using the concept of monomial in the étale topology, we can define monomial mappings for dominant morphisms of algebraic varieties. A morphism of algebraic varieties is a mapping that can be represented locally by polynomial mappings. A precise definition of a monomial morphism is given in Definition 1.1.

Most morphisms $\Phi : X \to Y$ are not monomial. Hence one can seek to monomialize a morphism, by performing morphisms $\alpha : X_1 \to X$ and $\beta : Y_1 \to Y$, so that there is an induced morphism $X_1 \to Y_1$ which is monomial. Of course, one can always remove from X the closed set where Φ is not monomial to produce a monomial morphism, but such a solution is useless for understanding the mapping, and for applications. To be meaningful, we must insist that α and β are birational (isomorphic on an open set) and are proper. The properness condition rules out the trivial solution of removing from X the closed set where Φ is not monomial. A precise definition of a monomialization of a morphism is given in Definition 1.2.

Consider the morphism $\mathbf{C}^2 \to \mathbf{C}$ given by

$$y = x_1(x_1 - x_2)x_2.$$

This morphism is not monomial in any sense at the origin of $\mathbf{C}$.

If we make a substitution

$$x_1 = \overline{x}_1\overline{x}_2, x_2 = \overline{x}_2$$

then

$$y = \overline{x}_1\overline{x}_2^3(\overline{x}_1 - 1)$$

which is monomial (in the etale topology) at all points of the domain of definition $U_1 = \mathbf{C}^2$ of $(\overline{x}_1, \overline{x}_2)$. However the mapping $U_1 \to \mathbf{C}^2$ is not proper as the only point of $x_2 = 0$ in the image of this mapping is the origin.

If we make the substitution

$$x_1 = \tilde{x}_1, x_2 = \tilde{x}_1\tilde{x}_2$$

we have

$$y = \tilde{x}_1^3(1 - \tilde{x}_2)\tilde{x}_2$$

This mapping is monomial on the domain of definition $V_2 = \mathbf{C}^2$ of $(\tilde{x}_1, \tilde{x}_2)$. We have a natural identification $\overline{x}_1 = \frac{1}{\tilde{x}_2}$ and $\overline{x}_2 = \tilde{x}_1\tilde{x}_2$, which allows us to patch $V_1 - \{\overline{x}_1 = 0\}$ to $V_2 - \{\tilde{x}_2 = 0\}$ to obtain a variety X_1 with a proper morphism $\alpha : X_1 \to \mathbf{C}^2$ such that the resulting morphism $X_1 \to \mathbf{C}$ is monomial. This morphism α is an isomorphism away from the origin of $\mathbf{C}^2$, and $\alpha^{-1}(0)$ is a 1 dimensional projective space (a Riemann sphere). In the language of algebraic geometry, $X_1 \to \mathbf{C}^2$ is the blow-up of the origin.

The definition of monomialization (Definition 1.2) requires the mappings $X_1 \to X$ and $Y_1 \to Y$ to be products of blow-ups of nonsingular subvarieties. Blow-ups are very special proper morphisms. On a nonsingular surface we can only blow up points (algebraically replace the point with a 1 dimensional projective space). On a nonsingular variety of dimension 3, we can blow up points to replace a point with a 2 dimensional projective space, and blow up nonsingular curves to replace them with a ruled surface, a family of 1 dimensional projective spaces over the curve. A single blow-up is a particularly simple example of a monomial morphism. In suitable local coordinates, it will have the form

$$y_1 = x_1, y_2 = x_1x_2, \ldots , y_r = x_1x_r, y_{r+1} = x_{r+1}, \ldots , y_n = x_n \qquad (1)$$

where $y_1 = y_2 = \cdots = y_r = 0$ are local equations of the nonsingular subvariety which is blown up in the n-dimensional variety Y. A product of blow-ups can be represented locally by a series of changes of variables and tranformations of the form (1). A product of blow-ups will thus in general be far from a monomial morphism.

We will now state the problem of monomialization precisely.

Suppose that $\Phi : X \to Y$ is a dominant morphism from a nonsingular k-variety X to a nonsingular k-variety Y, where k is a field of characteristic zero.

The structure of Φ is extremely complicated. However, we can hope to construct a commutative diagram

$$
\begin{array}{ccc}
X_1 & \overset{\Psi}{\to} & Y_1 \\
\downarrow & & \downarrow \\
X & \overset{\Phi}{\to} & Y
\end{array}
\tag{2}
$$

where the vertical maps are products of blow-ups of nonsingular subvarieties, to obtain a morphism $\Psi : X_1 \to Y_1$ which has a relatively simple structure.

The most optimistic conclusion we can hope for is to construct a diagram (2) such that Ψ is monomial.

Definition 1.1. *(Definition 18.20) Suppose that $\Phi : X \to Y$ is a dominant morphism of nonsingular k-varieties (where k is a field of characteristic zero). Φ is **monomial** if for all $p \in X$ there exists an étale neighborhood U of p, uniformizing parameters $(x_1, \ldots, x_n)$ on U, regular parameters $(y_1, \ldots, y_m)$ in $\mathcal{O}_{Y,\Phi(p)}$, and a matrix (a_{ij}) of nonnegative integers (which necessarily has rank m) such that*

$$
y_1 = x_1^{a_{11}} \cdots x_n^{a_{1n}}
$$

$$
\vdots
$$

$$
y_m = x_1^{a_{m1}} \cdots x_n^{a_{mn}}
$$

Definition 1.2. *Suppose that $\Phi : X \to Y$ is a dominant morphism of k-varieties. A morphism $\Psi : X_1 \to Y_1$ is a **monomialization** of Φ if there are sequences of blow-ups of nonsingular subvarieties $\alpha : X_1 \to X$ and $\beta : Y_1 \to Y$, and a morphism $\Psi : X_1 \to Y_1$ such that the diagram*

$$
\begin{array}{ccc}
X_1 & \overset{\Psi}{\to} & Y_1 \\
\downarrow & & \downarrow \\
X & \overset{\Phi}{\to} & Y
\end{array}
$$

commutes, and Ψ is a monomial morphism.

It is natural to ask the following question.

Question *Suppose that $\Phi : X \to Y$ is a dominant morphism of k-varieties (over a field k of characteristic zero). Does there exist a monomialization of Φ?*

By resolution of singularities and resolution of indeterminacy, we easily reduce to the case where X and Y are nonsingular.

The characteristic of k must be zero in the question. If char $k = p > 0$, a monomialization may not exist even for curves.

$$
t = x^p + x^{p+1}
$$

gives a simple example of a mapping of curves which cannot be monomialized, since $\sqrt[p]{1+x}$ is inseparable over $k[x]$. The obstruction to monomialization in positive characteristic is thus wild ramification.

In [14], we prove that a local analog of the question (if k has characteristic 0) has a positive answer for generically finite morphisms. A discussion of these results is given in chapter 2.

An extension of monomialization to morphisms of surfaces in characteristic $p \geq 0$ is given in [18] when wild ramification is not present. Further local extensions are given in [19].

In these lecture notes, we give a positive answer (Theorem 1.3) to the question in the case of a dominant morphism from a 3 fold to a surface. Precise definitions of a 3 fold and a surface are given in chapter 5, Notations.

Theorem 1.3. *(Theorem 18.21) Suppose that $\Phi : X \to S$ is a dominant morphism from a 3 fold X to a surface S (over an algebraically closed field k of characteristic zero). Then there exist sequences of blow-ups of nonsingular subvarieties $X_1 \to X$ and $S_1 \to S$ such that the induced map $\Phi_1 : X_1 \to S_1$ is a monomial morphism.*

From this we deduce that it is possible to toroidalize a dominant morphism from a 3 fold to a surface. A toroidal morphism $X \to Y$ is a morphism which is monomial with respect to fixed SNC divisors on X and Y ([24], [6], Definition 19.1).

Theorem 1.4. *(Theorem 19.11) Suppose that $\Phi : X \to S$ is a dominant morphism from a 3 fold X to a surface S (over an algebraically closed field k of characteristic zero) and D_S is a reduced 1 cycle on S such that $E_X = \Phi^{-1}(D_S)_{red}$ contains $sing(X)$ and $sing(\Phi)$. Then there exist sequences of blow-ups of nonsingular subvarieties $\pi_1 : X_1 \to X$ and $\pi_2 : S_1 \to S$ such that the induced morphism $X_1 \to S_1$ is a toroidal morphism with respect to $\pi_2^{-1}(D_S)_{red}$ and $\pi_1^{-1}(E_X)_{red}$.*

Suppose that $\Phi : X \to Y$ is a dominant morphism of nonsingular k-varieties, and $\dim(Y) > 1$.

To begin with, we point out that monomialization is not a direct consequence of embedded resolution of singularities and principalization of ideals.

Suppose that $p \in X$ is a point where Φ is not smooth, and $q = \Phi(p)$. Let $(y_1, \ldots, y_m)$ be regular parameters in $\mathcal{O}_{Y,q}$. By standard theorems on resolution, we have a sequence of blow-ups of nonsingular subvarieties $\pi : X_1 \to X$ such that if $p_1 \in \pi^{-1}(p)$, then there exist regular parameters $(x_1, \ldots, x_n)$ in $\mathcal{O}_{X_1,p_1}$, a matrix (a_{ij}) with nonnegative coefficients and units $\delta_1, \ldots, \delta_m \in \mathcal{O}_{X_1,p_1}$ such that

$$
\begin{aligned}
y_1 &= x_1^{a_{11}} \cdots x_n^{a_{1n}} \delta_1 \\
&\vdots \\
y_m &= x_1^{a_{m1}} \cdots x_n^{a_{mn}} \delta_n
\end{aligned}
\tag{3}
$$

In general, p_1 will lie on a single exceptional component of π, and p_1 will be disjoint from the strict transforms of codimension 1 subschemes of X determined by $y_i = 0$, $1 \leq i \leq m$, on a neighborhood of $\Phi^{-1}(q)$. In this case we will have

$a_{ij} = 0$ if $j > 1$, since the $x_i = 0$ are either local equations of exceptional components of $X_1 \to X$ or are local equations of the strict transforms of irreducible components of $y_i = 0$. Thus (a_{ij}) will have rank 1.

There thus cannot exist regular parameters $(\overline{x}_1, \ldots, \overline{x}_n)$ in $\hat{\mathcal{O}}_{X_1, p_1}$ such that

$$y_1 = \overline{x}_1^{a_{11}} \cdots \overline{x}_n^{a_{1n}}$$
$$\vdots$$
$$y_m = \overline{x}_1^{a_{m1}} \cdots \overline{x}_n^{a_{mn}}$$

since this would imply that $\text{rank}(a_{ij}) = m > 1$. $\text{rank}(a_{ij}) < m$ would imply, by Zariski's subspace theorem [3], that Φ is not dominant.

In fact, in general it is necessary to blow up in both X and Y to construct a monomialization. For instance, if we blow up a point p on a nonsingular surface S, blow up a point on the exceptional curve E_1, blow up the intersection point of the new exceptional curve E_2 with the strict transform of E_1, then blow up a general point on the new exceptional curve E_3 with exceptional curve E_4, we get a birational map $\pi : S_1 \to S$ such that if $p_1 \in E_4$ is a general point we have regular parameters (u, v) in $\mathcal{O}_{S,p}$ and regular parameters (x, y) in $\hat{\mathcal{O}}_{S_1, p_1}$ such that

$$u = x^2, v = \alpha x^3 + x^4 y.$$

π is not monomial at p_1 and further blow ups over S_1 will produce a morphism which is further from being monomial.

Suppose that Y is a nonsingular surface. If $\pi_2 : Y_1 \to Y$ is a sequence of blow-ups of points over $q \in Y$, and $q_1 \in \pi_2^{-1}(q)$ is a point which only lies on a single exceptional component E of π_2, then there exist regular parameters (u, v) in $\mathcal{O}_{Y,q}$ and $(\overline{x}, \overline{y})$ in $\hat{\mathcal{O}}_{Y_1, q_1}$ such that

$$u = \overline{x}^a$$
$$v = P(\overline{x}) + \overline{x}^b \overline{y} \tag{4}$$

where $a, b \in \mathbf{N}$ and $P(\overline{x})$ is a polynomial of degree $\leq b$. This follows since $Y_1 \to Y$ can be factored near p_1 by sequences of blow-ups of the form

$$x_i = x_{i+1}^{a_{i+1}}, y_i = x_{i+1}^{b_{i+1}}(x_{i+1} + \alpha_{i+1}).$$

If we perform a sequence of blow-ups of nonsingular subvarieties $\pi_1 : X_1 \to X$, and if $p_1 \in (\Phi \circ \pi_1)^{-1}(q)$ is such that $\hat{\mathcal{O}}_{X_1, p_1}$ has regular parameters

$$(\overline{x}_1, \overline{x}_2, \overline{x}_3, \cdots, \overline{x}_n)$$

such that

$$u = \overline{x}_1^a$$
$$v = P(\overline{x}_1) + \overline{x}_1^b \overline{x}_2 \tag{5}$$

of the form of (4), we will have a factorization $X_1 \to S_1$ which is a morphism in a neighborhood of p_1, and $X_1 \to S_1$ will be monomial at p_1.

A strategy for monomializing a dominant morphism from a nonsingular variety X to a nonsingular surface S is thus to first perform a sequence of blow-ups of nonsingular subvarieties $\pi_1 : X_1 \to X$ so that for all points p of X_1, appropriate regular parameters (u, v) in $\mathcal{O}_{S_1, q}$ where $q = \Phi \circ \pi_1(p)$ will have simple forms which we will call prepared, which include the form of (5). This is accomplished (easily) in Lemma 3.3 when X is a surface, and (not so easily) in Chapter 17 of this book when X has dimension 3.

An interesting case when the existence of a global monomialization is still open is for birational morphisms of nonsingular, characteristic 0 varieties of dimension ≥ 3. Such birational maps are known to have a simple structure, since they can be factored by alternating sequences of blow-ups and blow-downs ("weak factorization") [7]. A local form of "weak factorization" along a valuation was proven earlier by us in Theorem 1.6 [14]. Our local proof of "strong factorization" in dimension 3 appears in [13].

The results of these notes can be stated in terms of logarithmic differential forms. The related problems of resolution of differential forms and vector fields are still open in dimension 3, although there is a local proof, which implies resolution of vector fields locally along a valuation in dimension 3 by Cano [10].

Suppose that $\Phi : X \to Y$ is a dominant morphism from a nonsingular k-variety X to a nonsingular k-variety Y, where k is an algebraically closed field of characteristic zero.

We will give an interpretation, in terms of logarithmic differential forms, of the concept of toroidal morphisms in Lemma 1.5. We will only give an outline of the proof of this lemma, as it is included for motivation, and does not play a role in the proofs of the results of these notes.

Suppose that there are simple normal crossing divisors D_Y on Y and D_X on X such that $\Phi^{-1}(D_Y)_{red} = D_X$ and $\Phi(\text{sing}(\Phi)) \subset D_Y$.

$\Omega^1_X(\log D_X)$ is the locally free sheaf on X which has the basis

$$\frac{dx_1}{x_1}, \ldots, \frac{dx_r}{x_r}, dx_{r+1}, \ldots, dx_n$$

at a point $p \in X$ if $(x_1, \ldots, x_n)$ are regular parameters at p such that $x_1 \cdots x_r = 0$ is a local equation of D_X at p.

$$\Omega^t_X(\log D_X) = \wedge^t \Omega^1_X(\log D_X)$$

for $t \geq 1$. We have a natural inclusion

$$\Phi^*(\Omega^t_Y(\log D_Y)) \subset \Omega^t_X(\log D_X).$$

Lemma 1.5. Φ *is toroidal if and only if* $\Phi^*(\Omega^1_Y(\log D_Y))$ *is a subbundle of* $\Omega^1_X(\log D_X)$.

The proof follows easily from the implicit function theorem and formal properties of differentiation. We obtain from this the following interpretation of embedded resolution of hypersurface singularities.

Theorem 1.6. *Suppose that C is a nonsingular curve. Then there exists a sequence of blow-ups of nonsingular subvarieties $X_1 \to X$ such that $X_1 \to C$ is a toroidal morphism.*

We give an outline of the proof here, and give a more detailed proof later in Theorem 3.1. We need only consider the finite set of points $\Phi(\text{sing } \Phi) = \{p_i\}$ in C, and blow up to make each $\Phi^{-1}(p_i)$ a simple normal crossings divisor.

We observe that if $\dim Y > 1$, it is in general necessary to blow up in Y as well as in X to construct a monomialization or toroidalization. For instance, if we blow up a point p on a nonsingular surface S, blow up a point on the exceptional curve E_1, blow up the intersection point of the new exceptional curve E_2 with the strict transform of E_1, then blow up a general point on the new exceptional curve E_3 with exceptional curve E_4, we get a birational map $\pi : S_1 \to S$ such that if $p_1 \in E_4$ is a general point we have regular parameters (u, v) in $\mathcal{O}_{S,p}$ and regular parameters (x, y) in $\hat{\mathcal{O}}_{S_1,p_1}$ such that

$$u = x^2, v = \alpha x^3 + x^4 y.$$

π is not monomial at p_1 and further blow-ups over S_1 will produce a morphism which is further from being monomial.

Now we impose the further condition that Y is a surface S, but X has arbitrary dimension ≥ 2.

If $q \in D_S$, we will say that $u, v \in \mathcal{O}_{S,q}$ are permissible parameters at q if $u = 0$ or $uv = 0$ is a local equation of D_S at q.

We will say that Φ is strongly prepared at $p \in D_X$ if there exist permissible parameters (u, v) at $q = \Phi(p)$ and regular parameters $(x_1, \ldots, x_n)$ in $\hat{\mathcal{O}}_{X,p}$ such that $x_1 \cdots x_l = 0$ is a local equation of D_X at p and one of the following forms holds:

1.

$$\begin{aligned} u &= (x_1^{a_1} \cdots x_l^{a_l})^m \\ v &= P(x_1^{a_1} \cdots x_l^{a_l}) + x_1^{b_1} \cdots x_l^{b_l} \end{aligned}$$

or

2.

$$\begin{aligned} u &= (x_1^{a_1} \cdots x_l^{a_l})^m \\ v &= P(x_1^{a_1} \cdots x_l^{a_l}) + x_1^{b_1} \cdots x_l^{b_l} x_{l+1}. \end{aligned}$$

or

3.

$$\begin{aligned} u &= x_1^{a_1} \cdots x_{l-1}^{a_{l-1}} \\ v &= x_2^{b_2} \cdots x_l^{b_l} \end{aligned}$$

where $(a_1, \ldots, a_l) = 1$ and P is a series. In Case (1) we must have

$$\text{rank} \begin{pmatrix} a_1 & \cdots & a_l \\ b_1 & \cdots & b_l \end{pmatrix} = 2$$

by Zariski's subspace theorem [3]. Note that if $uv = 0$ is a local equation of D_S, then we must have that $uv = x_1^{c_1} \cdots x_l^{c_l}$ unit for some $c_1, \ldots, c_l$.

The concept of strong preparation is connected to logarithmic differentiation by the statement of Lemma 1.7. We will only give an outline of the proof of this lemma, as it is included for motivation, and does not play a role in the proofs of the results of these notes.

Lemma 1.7. $\Phi : X \to S$ *is strongly prepared if and only if*

$$\Phi^* \Omega_S^2 (log\ D_S) = \mathcal{I}\mathcal{M}$$

where $\mathcal{I} \subset \mathcal{O}_X$ *is an ideal sheaf, and* $\mathcal{M}$ *is a subbundle of* $\Omega_X^2 (log\ D_X)$.

The proof is a straight forward calculation in formal differentiation.

The problem of toroidalization naturally breaks up into two steps.

1. Construct a sequence of blow-ups with nonsingular centers $X_1 \to X$ so that $X_1 \to S$ is strongly prepared.

2. Construct a sequence of blow-ups with nonsingular centers $X_2 \to X_1$ and $S_1 \to S$ so that $X_2 \to S_1$ is a toroidal morphism.

When X is itself a surface, $\Omega_X^2 (log\ D_X)$ is an invertible sheaf. Thus $\Phi : X \to S$ is itself strongly prepared and step (1) is vacuous. Step (2) can then be accomplished without tremendous difficulty, by considering the special local forms we have for the equations of the mapping.

If dim $X \geq 3$, $\Phi : X \to S$ is in general far from being strongly prepared. A simple example is given in Example 6.3.

The main result of these notes is the realization of Step 1., the construction of a strongly prepared morphism, if X has dimension 3. The bulk of these notes is devoted to this proof.

The algorithms in these notes should extend with some work to the case where X has arbitrary dimension and S has dimension 2.

The Author would like to thank Professor Seshadri and the Chennai Mathematical Institute for their hospitality during the fall of 1999, and Professors Tony Iarrobino, Mark Levine and Northeastern University for their hospitality during the Spring of 2000, while this manuscript was being prepared. This research was partially supported by NSF.

2. LOCAL MONOMIALIZATION

A local version of the problem of monomialization considered in these lecture notes is completely proven in [14] for the case of generically finite morphisms. These theorems are stated below without proof (Theorems 2.1 and 2.2). These results show that the obstruction to monomialization (at least for generically finite morphisms) is a global problem, coming from patching local solutions.

Suppose that $R \subset S$ is a local homomorphism of local rings essentially of finite type over a field k and that V is a valuation ring of the quotient field K of S, such that V dominates S. Then we can ask if there are sequences of monoidal transforms $R \to R'$ and $S \to S'$ such that V dominates S', S' dominates R', and $R' \to S'$ is a monomial mapping.

$$
\begin{array}{ccc}
R' & \to & S' \subset V \\
\uparrow & & \uparrow \\
R & \to & S
\end{array}
\tag{6}
$$

A monoidal transform of a local ring R is the local ring R' of a point in the blow up of a nonsingular subvariety of $\mathrm{spec}(R)$ such that R' dominates R. If R is a regular local ring, then R' is a regular local ring.

Theorem 2.1. *(Monomialization)(Theorem 1.1 [14]) Suppose that $R \subset S$ are regular local rings, essentially of finite type over a field k of characteristic zero, such that the quotient field K of S is a finite extension of the quotient field J of R.*

Let V be a valuation ring of K which dominates S. Then there exist sequences of monoidal transforms $R \to R'$ and $S \to S'$ such that V dominates S', S' dominates R' and there are regular parameters $(x_1, ..., x_n)$ in R', $(y_1, ..., y_n)$ in S', units $\delta_1, \ldots, \delta_n \in S'$ and a matrix (a_{ij}) of nonnegative integers such that $Det(a_{ij}) \neq 0$ and

$$
\begin{aligned}
x_1 &= y_1^{a_{11}} ..... y_n^{a_{1n}} \delta_1 \\
&\;\;\vdots \\
x_n &= y_1^{a_{n1}} ..... y_n^{a_{nn}} \delta_n.
\end{aligned}
\tag{7}
$$

Thus (since $\mathrm{char}(k) = 0$) there exists an etale extension $S' \to S''$ where S'' has regular parameters $\bar{y}_1, \ldots, \bar{y}_n$ such that $x_1, \ldots, x_n$ are pure monomials in $\bar{y}_1, \ldots, \bar{y}_n$.

The standard theorems on resolution of singularities allow one to easily find R' and S' such that (7) holds, but, in general, we will not have the essential condition $Det(a_{ij}) \neq 0$. The difficulty of the problem is to achieve this condition.

It is an interesting open problem to prove Theorem 2.1 in positive characteristic, even in dimension 2. Theorem 2.1 implies local simultaneous resolution from above in all dimensions (and characteristic 0) [15], which has been conjectured to be true by Abhyankar [5]. Global simultaneous resolution is false even for surfaces [16].

A quasi-complete variety over a field k is an integral finite type k-scheme which satisfies the existence part of the valuative criterion for properness (c.f. Chapter

0 [23] where the notion is called complete). Quasi-complete and separated is equivalent to proper.

The construction of a monomialization by quasi-complete varieties follows from Theorem 2.1.

Theorem 2.2. *(Theorem 1.2 [14]) Let k be a field of characteristic zero, Φ : $X \to Y$ a generically finite morphism of nonsingular proper k-varieties. Then there are birational morphisms of nonsingular quasi-complete k-varieties α : $X_1 \to X$ and $\beta : Y_1 \to Y$, and a locally monomial morphism $\Psi : X_1 \to Y_1$ such that the diagram*

$$\begin{array}{ccc} X_1 & \overset{\Psi}{\to} & Y_1 \\ \downarrow & & \downarrow \\ X & \overset{\Phi}{\to} & Y \end{array}$$

commutes and α and β are locally products of blow-ups of nonsingular subvarieties. That is, for every $z \in X_1$, there exist affine neighborhoods V_1 of z, V of $x = \alpha(z)$, such that $\alpha : V_1 \to V$ is a finite product of monoidal transforms, and there exist affine neighborhoods W_1 of $\Psi(z)$, W of $y = \alpha(\Psi(z))$, such that $\beta : W_1 \to W$ is a finite product of monoidal transforms.

In this theorem, a monoidal transform of a nonsingular k-scheme S is the map $T \to S$ induced by an open subset T of $\mathrm{Proj}(\oplus \mathcal{I}^n)$, where $\mathcal{I}$ is the ideal sheaf of a nonsingular subvariety of S.

Theorems 1.1 and 1.2 of [14] are analogs for morphisms of the theorems on local uniformization and local resolution of singularities of varieties of Zariski [36], [37].

3. Monomialization of Morphisms in Low Dimensions

We will outline proofs of monomialization in the previously known cases. Suppose that k is an algebraically closed field of characteristic zero and $\Phi : X \to Y$ is a dominant morphism of nonsingular k varieties.

Let sing(Φ) be the closed subset of X where Φ is not smooth.

If Φ is a dominant morphism from a variety to a curve, the existence of a global monomialization follows immediately from resolution of singularities. In fact, it is really a restatement of embedded resolution of hypersurface singularities.

Theorem 3.1. *Suppose that* $\Phi : X \to C$ *is a dominant morphism from a k-variety to a curve. Then* Φ *has a monomialization.*

Proof. Suppose that $\Phi : X \to C$ where C is a nonsingular curve, X is a nonsingular n fold. $\Phi(\text{sing}(\Phi))$ is a finite number of points of C, so we may fix a regular parameter t at a point $q \in \Phi(\text{sing}(\Phi))$, and monomialize the mapping above q.

By embedded resolution of hypersurfaces, there exists a sequence of blow-ups of nonsingular subvarieties which dominate subvarieties of $\Phi^{-1}(q)$, $\pi : X_1 \to X$ such that for all $p \in (\Phi \circ \pi)^{-1}(q)$, there exist regular parameters $(x_1, \ldots, x_n)$ at p such that

$$t = u x_1^{a_1} \cdots x_n^{a_n}$$

where $a_1 > 0$, $u \in \mathcal{O}_{X_1,p}$ is a unit. If $x_1 = \overline{x}_1 u^{-\frac{1}{a_1}}$, we have

$$t = \overline{x}_1^{a_1} \cdots x_n^{a_n}.$$

□

If $\Phi : T \to S$ is a dominant morphism of surfaces, monomialization is not a direct corollary of resolution of singularities. One proof of monomialization in this case (over $\mathbf{C}$) is given by Akbulut and King in [8].

In our paper [18] with Olivier Piltant, we show that if L is a perfect field and $\Phi : T \to S$ is a dominant morphism of L-surfaces, then Φ can be monomialized if Φ is unramified. That is, no wild ramification occurs with respect to any divisorial valuation of $L(T)$ over $L(S)$. This condition occurs, for instance, if $p \nmid [K : L(S)]$ where K is a Galois closure of $L(T)$ over $L(S)$.

We will now outline a simple proof of monomialization for morphisms of surfaces (when k is algebraically closed of characteristic zero).

Theorem 3.2. *Suppose that* $\Phi : T \to S$ *is a dominant morphism of surfaces over k. Then* Φ *has a monomialization.*

If Φ is a monomial mapping, then Φ comes from an expression

$$\begin{aligned} u &= x^a y^b \\ v &= x^c y^d \end{aligned} \tag{8}$$

where $ad - bc \neq 0$.

sing(Φ) must be contained in $xy = 0$. At a point p on $x = 0$ we have regular parameters $(\hat{x}, \hat{y})$ in $\hat{\mathcal{O}}_{T,p}$ such that

$$\hat{x} = x, \hat{y} = y - \alpha$$

for some $\alpha \in k$. If $a > 0$ and $c > 0$ we have

$$
\begin{aligned}
u &= \hat{x}^a(\hat{y} + \alpha)^b = \overline{x}^a \\
v &= \hat{x}^c(\hat{y} + \alpha)^d = \beta\overline{x}^c + \overline{x}^c\overline{y}
\end{aligned}
\tag{9}
$$

where

$$
\hat{x} = \overline{x}(\hat{y} + \alpha)^{-\frac{b}{a}}, \overline{y} = (\hat{y} + \alpha)^{d-\frac{cb}{a}} - \beta
$$

with $\beta = \alpha^{d-\frac{cb}{a}}$.

If $a = 0$ or $c = 0$ we also obtain a form (9) with respect to regular parameters (u_1, v_1) in $\mathcal{O}_{S,\Phi(p)}$.

Thus Φ is monomial at a point p if and only if there exist regular parameters in $\hat{\mathcal{O}}_{T,p}$ such that one of the forms (8) or (9) hold.

We will say that Φ is prepared at $p \in T$ if there exist regular parameters (u, v) in $\mathcal{O}_{S,\Phi(p)}$, regular parameters (x, y) in $\hat{\mathcal{O}}_{T,p}$, and a power series P such that one of the following forms holds at p.

$$
\begin{aligned}
u &= x^a \\
v &= P(x) + x^c y
\end{aligned}
\tag{10}
$$

or

$$
\begin{aligned}
u &= (x^a y^b)^m \\
v &= P(x^a y^b) + x^c y^d
\end{aligned}
\tag{11}
$$

where $(a, b) = 1$ and $ad - bc \neq 0$.

We first observe that by resolution of singularities and elimination of indeterminacy, there exists a commutative diagram

$$
\begin{array}{ccc}
T_1 & \overset{\Phi_1}{\to} & S_1 \\
\downarrow & & \downarrow \\
T & \overset{\Phi}{\to} & S
\end{array}
$$

where the vertical maps are products of blow-ups of points, $\mathrm{sing}(\Phi_1)$ is a simple normal crossings (SNC) divisor, and for all $p \in \mathrm{sing}(\Phi_1)$, there exist regular parameters (u, v) at $\Phi_1(p)$ such that $u = 0$ is a local equation of $\mathrm{sing}(\Phi_1)$ at p.

The essential observation is that Φ_1 is now prepared. We give a simple proof that appears in [8].

Lemma 3.3. Φ_1 *is prepared.*

Proof. Suppose that $p \in T_1$. With our assumptions, one of the following must hold at p.

$$
\begin{aligned}
u &= x^a \\
u_x v_y - u_y v_x &= \delta x^e
\end{aligned}
\tag{12}
$$

where δ is a unit or

$$
\begin{aligned}
u &= (x^a y^b)^m \\
u_x v_y - u_y v_x &= \delta x^e y^f
\end{aligned}
\tag{13}
$$

where $a, b, e, f > 0$, $(a, b) = 1$ and δ is a unit.

Write $v = \sum a_{ij}x^i y^j$ with $a_{ij} \in k$. First suppose that (12) holds. Then $ax^{a-1}v_y = \delta x^e$ implies we have the form (10) (after making a change of parameters in $\hat{\mathcal{O}}_{T_1,p}$). Now suppose that (13) holds.

$$u_x v_y - u_y v_x = \sum m(aj - bi)a_{ij}x^{am+i-1}y^{bm+j-1} = \delta x^e y^f.$$

Thus

$$v = \sum_{aj-bi=0} a_{ij}x^i y^j + \epsilon x^{e+1-am}y^{f+1-bm}$$

where ϵ is a unit. After making a change of parameters, multiplying x by a unit, and multiplying y by a unit, we get the form (11). $\qquad\square$

It is now not difficult to construct a monomialization. We must blow up points q on S_1 over which the map is not monomial at some point over q, and blow up points on T_1 to make $m_q\mathcal{O}_{T_1}$ principal. If we iterate this procedure, it can be shown that we construct a commutative diagram

$$
\begin{array}{ccc}
T_2 & \overset{\Phi_2}{\rightarrow} & S_2 \\
\downarrow & & \downarrow \\
T_1 & \overset{\Phi_1}{\rightarrow} & S_1
\end{array}
$$

such that Φ_2 is monomial.

4. An Overview of the Proof of
Monomialization of Morphisms From 3 Folds to Surfaces

Suppose that k is an algebraically closed field of characteristic zero, and Φ : $X \to Y$ is a dominant morphism of nonsingular k-varieties.

A natural first step in monomializing a morphism $\Phi : X \to Y$ is to use resolution of singularities and resolution of indeterminacy to construct a commutative diagram

$$
\begin{array}{ccc}
X_1 & \overset{\Phi_1}{\to} & Y_1 \\
\downarrow & & \downarrow \\
X & \overset{\Phi}{\to} & Y
\end{array}
$$

where the vertical maps are products of blow-ups of nonsingular subvarieties, $\text{sing}(\Phi_1)$ is a simple normal crossings (SNC) divisor, and for all $p \in \text{sing}(\Phi_1)$, there exist regular parameters $(u_1, \dots, u_n)$ at $\Phi_1(p)$ such that $u_1 = 0$ is a local equation of $\text{sing}(\Phi_1)$ at p.

We observed that if X and Y are surfaces, then Φ_1 is prepared (Lemma 3.3). Unfortunately, even for morphisms from a 3 fold to a surface, Φ_1 may be quite complicated (Example 6.3).

A key step in the local proof of monomialization, Theorem 2.1, is to define a new invariant, which measures how far the situation is from a specific form which is close to being monomial. In this local valuation theoretic proof we make use of special products of monoidal transforms defined by Zariski called Perron transforms [37]. Under appropriate application of Perron transforms our invariant does not increase, and we can in fact make the invariant decrease, by an appropriate algorithm.

An essential difficulty globally is that our invariant can increase after a permissible monoidal transform (Example 7.2). This is a significant difference from resolution of singularities, where a foundational result is that the multiplicity of an ideal does not go up under permissible blow-ups.

We will give a brief overview of the proof of Theorem 18.21 (Monomialization of morphisms from 3 folds to surfaces).

Step 1. (Preparation) First construct a diagram

$$
\begin{array}{ccc}
X' & \overset{\Phi'}{\to} & S' \\
\downarrow & & \downarrow \\
X & \overset{\Phi}{\to} & S
\end{array}
$$

where the vertical maps are products of blow-ups of nonsingular subvarieties such that X', S' are nonsingular, there exist reduced SNCS divisors $D_{S'}$ on S', $E_{X'} = (\Phi')^{-1}(D_{S'})_{red}$ on X' such that $\text{sing}(\Phi') \subset E_{X'}$ and components of $E_{X'}$ on X' dominating distinct components of $D_{S'}$ are disjoint. Such a morphism Φ' will be called weakly prepared (Definition 6.1 and Lemma 6.2). For the duration of this section, we will suppose that $\Phi : X \to S$ is weakly prepared.

For all $p \in X$ there exist regular parameters (u, v) in $\mathcal{O}_{S,q}$ ($q = \Phi(p)$) and regular parameters (x, y, z) in $\hat{\mathcal{O}}_{X,p}$ such that $u = 0$ is a local equation of E_X,

$u = 0$ (or $uv = 0$) is a local equation of D_S and exactly one of the following cases hold (Definition 6.4):

1. p is a 1 point:
$$u = x^a, v = P(x) + x^b F_p$$
where $x \nmid F$, F has no terms which are monomials in x.

2. p is a 2 point:
$$u = (x^a y^b)^m, v = P(x^a y^b) + x^c y^d F_p$$
where $(a, b) = 1$, $x \nmid F_p$, $y \nmid F_p$, $x^c y^d F_p$ has no terms which are monomials in $x^a y^b$.

3. p is a 3 point:
$$u = (x^a y^b z^c)^m, v = P(x^a y^b z^c) + x^d y^e z^f F_p$$
where $(a, b, c) = 1$, $x \nmid F_p$, $y \nmid F_p$, $z \nmid F_p$, $x^d y^e z^f F_p$ has no terms which are monomials in $x^a y^b z^c$.

(u, v) are called permissible parameters at q and (x, y, z) are called permissible parameters at p for (u, v) (Definition 6.4).

The structure of the singularities of F_p can be very complicated (Example 6.3). This is in sharp contrast to the case of a morphism of surfaces (Lemma 3.3).

We say that Φ is prepared at $p \in E_X$ if there exist permissible parameters (u, v) at $\Phi(p)$ and permissible parameters (x, y, z) at p such that one of the following four forms hold (Definition 6.5)

1. $u = x^a, v = P(x) + x^b y$,
2. $u = (x^a y^b)^m, v = P(x^a y^b) + x^c y^d$,
3. $u = (x^a y^b)^m, v = P(x^a y^b) + x^c y^d z$,
4. $u = (x^a y^b z^c)^m, v = P(x^a y^b z^c) + x^d y^e z^f$ with

$$\text{rank} \left\{ \begin{array}{ccc} a & b & c \\ d & e & f \end{array} \right\} = 2.$$

The main theorem of this article is Theorem 17.3, whose statement we repeat here.

Theorem 4.1. *(Theorem 17.3) Suppose that $\Phi : X \to S$ is a dominant morphism from a 3 fold to a surface and $D_S \subset S$ is a reduced 1 cycle such that $E_X = \Phi^{-1}(D_S)_{red}$ contains sing(X) and sing(Φ). Then there exist sequences of monoidal transforms with nonsingular centers $\pi_1 : S_1 \to S$ and $\pi_2 : X_1 \to X$ such that $\Phi_{X_1} : X_1 \to S_1$ is prepared with respect to $D_{S_1} = \pi_2^{-1}(D_S)_{red}$.*

Sections 6 through 17 are devoted to the proof of Theorem 17.3. Sections 6 through 9 prove foundational results, and sections 10 through 17 achieve the proof of the theorem.

Our main invariant is
$$\nu(p) = \text{mult}(F_p).$$
This invariant is independent of permissible parameters in the forms above (Definition 6.8).
$$S_r(X) = \{p \in X \mid \nu(p) \geq r\}$$

is a constructible (but not Zariski closed) subset of X (Proposition 6.21 and Example 6.12).

We construct a preparation by blowing up points and curves in the Zariski closure $\overline{S}_r(X)$ of $S_r(X)$. Much of the complexity of the proof arises from the analysis at closure points of $S_r(X)$. The difficulty occurs when passing from 3 points to 1 points or 2 points (or from 3 points to 2 points).

Other important local invariants are $\tau(p)$ and $\gamma(p)$ (Definition 6.8). We observe that (by Definition 6.9 and Remark 6.10) we have preparation at $p \in E_X$ if

1. If p is a 1 point then $\nu(p) \leq 1$.
2. If p is a 2 point then $\gamma(p) \leq 1$.
3. If p is a 3 point then $\nu(p) = 0$.

The method of proof of the main theorem, Theorem 17.3, is to construct a series of reductions involving the invariants ν, γ and τ to eventually get a reduction, after enough permissible monoidal transforms.

Let $B_2(X)$ be the (possibly not closed) curve of 2 points in X, $B_3(X)$ be the (finite) set of 3 points in X, $\overline{B}_2(X) = B_2(X) \cup B_3(X)$ be the Zariski closure of $B_2(X)$ in X.

Suppose that $C \subset \overline{S}_r(X)$ is a curve and $p \in C$. The relationship between the series F_p (of Definition 6.4) and the ideal $\hat{\mathcal{I}}_{C,p}$ is very subtle (c.f. Examples 6.15 and 6.12). The general idea is that if C makes simple normal crossings (SNCs) with $\overline{B}_2(X)$, then it is possible to give a reasonable local description, if we restrict to permissible parameters for C at p, which are defined after Lemma 6.16.

Section 7 analyzes the effect of the blow-up of a point (a quadratic transform) $X_1 \to X$. In Theorem 7.1 it is shown that the multiplicity can go up by at most 1 after the blow-up of a point. An example illustrating the case where the multiplicity does increase is given in Example 7.2. The multiplicity can only go up if we blow up a 2 (or 3) point p, and consider the multiplicity at a 1 (or a 1 or 2) point q above p. We also analyze the change of the invariants τ and γ under the blow-up of a point.

The most difficult part of this chapter is to give a partial analysis of the structure of curves C in $\overline{S}_r(X_1) \cap \pi^{-1}(p)$. This takes place in Theorem 7.8 and Lemma 7.9.

Chapter 8 analyzes the effect of the blow-up of a curve $C \subset \overline{S}_r(X)$ which makes SNCs with $\overline{B}_2(X)$. The analysis depends on whether C is r-small or r-big, and if C is a 2 curve or if C contains a 1 point. There are thus 4 basic cases. To make an effective analysis, we may be required to perform a series of quadratic transforms at some points.

If $C \subset \overline{S}_r(X)$ makes SNCs with $\overline{B}_2(X)$, then either $F_p \in \hat{\mathcal{I}}_{C,p}^r$ for all $p \in C$ (with respect to permissible parameters for C at p) and we say that C is r-big, or $F_p \in \hat{\mathcal{I}}_{C,p}^{r-1}$, $F_p \notin \hat{\mathcal{I}}_{C,p}^r$ for all $p \in C$ (with respect to permissible parameters for C at p) and we say that C is r-small.

We have seen (sections 7 and 8) that multiplicity can go up after blowing up, and that we have an analysis of how τ and γ vary if the multiplicity stays the same or goes up. This allows us to formulate a condition $\overline{A}_r(X)$ which will not go up under permissible monoidal transforms (Definition 10.1).

We also consider a refined condition $A_r(X)$, which is stable under permissible monoidal transforms, and has the property that $\overline{S}_r(X) \cup \overline{B}_2(X)$ is close to having SNCs (Definition 10.2). The 3 points are the most difficult to control, as $\overline{S}_r(S) \cup \overline{B}_2(X)$ cannot be required to have SNCs at 3 points p with $\nu(p) \geq r - 1$. We obtain the improvement of $\overline{A}_r(X)$ to the condition $A_r(X)$ by performing permissible monoidal transforms in Lemma 10.3.

The inductive theorem which must be proven is Theorem 17.1, which shows that if $X \to S$ satisfies $\overline{A}_r(X)$ with $r \geq 2$, then there exists a permissible sequence of monoidal transforms $Y \to X$ such that $\overline{A}_{r-1}(Y)$ holds.

There are two basic ingredients in the proof. The first is to use the information on permissible monoidal transforms centered at points and nonsingular curves in $\overline{S}_r(X)$ obtained in sections 7 and 8 to obtain a reduction at all but a finite number of difficult points which are in 4 special forms, that are analyzed in sections 11 and 12. These special forms (at a point p) are locally reduced by blowing up a generic curve C through p such that C is resolved at a generic point of C. This process is iterated by blowing up sections over C. The idea is that in these special forms, blowing up sections over C, (which are not in $\overline{S}_r$) allow one to reduce to a two dimensional situation, and use Pusieux series type arguments which are developed in section 9.

The analyses of all of these special cases are similar. A representative case is the proof of Theorem 11.5. In Lemma 11.1 and Theorem 11.2, we construct a formal sequence of blow-ups of sections over C, so that conclusions similar to those of a "good point" (c.f. [4], [26], [17]) hold everywhere above p . In Theorem 11.4, we approximate this sequence by blow-ups of algebraic sections. This is similar to a method used in [13] and [14]. Finally, in Theorem 11.5, we perform a sequence of blow-ups of r-big curves in $\overline{S}_r$, followed by a sequence of blow-ups of r-small curves in $\overline{S}_r$, and some quadratic transforms so that a reduction is obtained everywhere above p.

The main step accomplished in section 13 is Theorem 13.8. We eliminate 2 curves C such that C is r-small or $r - 1$ big, reduce all 3 points to multiplicity $\leq r - 2$, and reduce the 2 points with $\nu(p) = r$ and $\tau(p) = 1$ to a special form (162). We require our local reduction of the special form Theorem 12.4 in this proof.

In section 14 we obtain a further reduction (in Theorem 14.7) so that $C_r(X)$ (of Definition 14.1) holds. In the proof, we must patch in the local analysis of Theorem 12.4 into the proof of Theorem 14.6, and the construction of Theorem 12.2 into the proof of Theorem 14.7. This requires a sequence of quadratic transforms to be performed at special points of generic curves (Theorem 14.5) before we can globalize the local blow-ups of sections over the generic curve considered in the proof of Theorem 12.2.

In section 15 we reduce all 1 and 2 points to multiplicity $\leq r - 1$ and reduce the 2 points with $\nu(p) = r - 1$ to $\tau(p) > 0$ or the special case (163). The proof makes a careful patching of the local solutions from sections 11 and 12.

Theorem 17.1 (via Theorem 16.1) proves the final reduction to $\overline{A}_{r-1}$.

Step 2.(Monomialization) In order to achieve monomialization, we make an inductive argument, for which we must consider a slightly weaker condition than being prepared, which we call strongly prepared (Definition 18.1). Suppose that $\Phi : X \to S$ is strongly prepared. We construct a commutative diagram

$$
\begin{array}{ccc}
X' & \overset{\Phi'}{\to} & S' \\
\downarrow & & \downarrow \\
X & \overset{\Phi}{\to} & S
\end{array}
$$

such that $X' \to X$ is a product of blow-ups of nonsingular curves, $S' \to S$ is a product of blow-ups of points and Φ' is monomial. This is accomplished in Theorem 18.19. Theorem 18.21 follows.

We give a brief idea of how monomialization is obtained. $\pi : S' \to S$ is a sequence of blow-ups of points. If $q \in S$ and $q_1 \in \pi^{-1}(q)$ then there exist regular parameters (u, v) in $\mathcal{O}_{S,q}$ and (u_1, v_1) in $\hat{\mathcal{O}}_{S',q_1}$ such that

$$
\begin{aligned}
u &= u_1^a \\
v &= P(u_1) + u_1^b v_1
\end{aligned}
$$

or

$$
\begin{aligned}
u &= (u_1^a v_1^b)^m \\
v &= P(u_1^a v_1^b) + u_1^c v_1^d
\end{aligned}
$$

with $ad - bc \neq 0$ and $(a, b) = 1$.

If $p \in X$ is a point of the form 1. of Step 1. (Definition 6.5), then there exists $\overline{\pi} : S_1 \to S$ and $q_1 \in \overline{\pi}^{-1}(q)$ with regular parameters (u_1, v_1) in $\mathcal{O}_{S_1,q_1}$, $(\overline{x}, \overline{y}, \overline{z})$ in $\hat{\mathcal{O}}_{X,p}$ such that

$$
\begin{aligned}
u_1 &= \overline{x}^{\overline{a}} \\
v_1 &= \overline{x}^b(\alpha + \overline{y}),
\end{aligned}
$$

so that (in these cases) we have attained monomialization.

To make this work in all cases, we have an essentially canonical procedure for achieving Step 2. We blow up on S the (finitely many) images of all non monomial points of X, then blow up nonsingular curves on X to resolve the indeterminacy of the resulting rational map. An invariant improves. By induction we eventually construct Φ'.

Toroidalization (Theorem 19.11) is proven in section 19.

5. NOTATIONS

We will suppose that k is an algebraically closed field of characteristic zero. By a variety we will mean a separated, integral finite type k-scheme.

Suppose that Z is a variety and $p \in Z$. Then m_p will denote the maximal ideal of $\mathcal{O}_{Z,p}$.

Definition 5.1. *A reduced divisor D on a nonsingular variety Z of dimension n is a simple normal crossing divisor (SNC divisor) if*

1. *All components of D are nonsingular.*
2. *Suppose that $p \in X$. Let $D_1, \ldots, D_s$ be the components of D containing p. Then $s \leq n$ and there exist regular parameters $(x_1, \ldots, x_n)$ in $\mathcal{O}_{X,p}$ such that $x_i = 0$ are local equations of D_i at p for $1 \leq i \leq s$.*

A curve is a 1 dimensional k variety. A surface is a 2 dimensional k variety. A 3 fold is a 3 dimensional k variety. A point of a variety will mean a closed point.

By a generic point or a generic curve on a variety Z, we will mean a point or a curve which satisfies a good condition which holds on an open set (in some parameterizing space) of points or curves.

Suppose that Z is a variety and $p \in Z$. the blow-up of p or the quadratic transform of p will denote $Z_1 = \mathrm{Proj}(\oplus_{n \geq 0} m_p^n)$. If $V \subset Z$ is a nonsingular subvariety then the blow-up of V or the monoidal transform of Z centered at V will denote $Z_1 = \mathrm{Proj}(\oplus_{n \geq 0} \mathcal{I}_V^n)$.

If R is a regular local ring with maximal ideal m, then a quadratic transform of R is $R_1 = R[\frac{m}{x}]_{m_1}$ where $0 \neq x \in m$ and m_1 is a maximal ideal of R_1.

Suppose that $P(x) = \sum_{i=0}^{\infty} b_i x^i \in k[[x]]$ is a series. Given $t \in \mathbf{N}$, $P_t(x)$ will denote the polynomial

$$P_t(x) = \sum_{i=0}^{t} b_i x^i.$$

Given a series $f(x_1, \ldots, x_n) \in k[[x_1, \ldots, x_n]]$ $\nu(f)$, $\mathrm{mult}(f)$ or $\mathrm{ord}(f)$ will denote the order of f.

If $x \in \mathbf{Q}$, $[x] = n$ if $n \in \mathbf{N}$, $n \leq x < n + 1$. $\{x\} = [x] - x$. The greatest common divisor of $a_1, \ldots, a_n \in \mathbf{N}$ will be denoted by $(a_1, \ldots, a_n)$.

Definition 6.1. *Suppose that* $\Phi_X : X \to S$ *is a dominant morphism from a nonsingular 3 fold* X *to a nonsingular surface* S, *with reduced SNC divisors* D_S *on* S *and* E_X *on* X *such that* $\Phi_X^{-1}(D_S)_{red} = E_X$. *Let* $sing(\Phi_X)$ *be the locus of singular points of* Φ_X.

We will say that Φ_X *is weakly prepared if*

1. $sing(\Phi_X) \subset E_X$ *and*
2. *If* $p \in S$ *is a singular point of* D_S, C_1 *and* C_2 *are the components of* D_S *containing* p, T_1 *is a component of* E_X *dominating* C_1 *and* T_2 *is a component of* E_X *dominating* C_2 *then* T_1 *and* T_2 *are disjoint.*

Lemma 6.2. *Suppose that* $\Phi : X \to S$ *is a dominant morphism from a 3 fold* X *to a surface* S, D_S *is a reduced Weil divisor on* S *such that* $sing(\Phi) \subset \Phi^{-1}(D_S)$ *and the singular locus of* X, $sing(X) \subset \Phi^{-1}(D_S)$. *Then there exists a commutative diagram*

$$
\begin{array}{ccc}
X_1 & \overset{\Phi_1}{\to} & S_1 \\
\downarrow \pi_1 & & \downarrow \pi_2 \\
X & \overset{\Phi}{\to} & S
\end{array}
$$

such that π_1 *and* π_2 *are products of blow-ups of nonsingular subvarieties, and if* $D_{S_1} = \pi_2^{-1}(D_S)_{red}$, $E_{X_1} = (\Phi \circ \pi_1)^{-1}(D_S)_{red}$, *then* ϕ_1 *is weakly prepared.*

Proof. By resolution of singularities and resolution of indeterminacy of mappings [23], there exists

$$
\begin{array}{ccc}
\overline{X} & \overset{\overline{\Phi}}{\to} & \overline{S} \\
\downarrow & & \downarrow \\
X & \to & S
\end{array}
$$

such that $\overline{X}$ and $\overline{S}$ are nonsingular, $\pi^{-1}(D_S)_{red} = D_{\overline{S}}$ and $E_{\overline{X}} = \overline{\Phi}^{-1}(D_{\overline{S}})_{red}$ are SNC divisors.

Suppose that E_1 and E_2 are components of $E_{\overline{X}}$ which dominate distinct components C_1 and C_2 of $\overline{S}$. If $E_1 \cap E_2 \neq \emptyset$ then there exists a sequence of blow-ups $\overline{X}_1 \to \overline{X}$ with nonsingular centers which map into $C_1 \cap C_2$ with induced map $\overline{\Phi}_1 : \overline{X}_1 \to \overline{S}$ such that the strict transform of E_1 and E_2 are disjoint on $\overline{X}_1$, and $E_{\overline{X}_1} = \overline{\Phi}_1^*(D_{\overline{S}})_{red}$ is a SNC divisor.

One way to construct this is to blow up the conductor of $E_1 \cup E_2$ to separate the strict transforms of E_1 and E_2 (c.f. section 2 of [20]), and then resolve the singularities of the resulting variety.

Iterating this procedure, we construct a weakly prepared morphism. $\qquad\square$

Example 6.3. *The structure of weakly prepared morphisms can be quite complicated.*

Consider the germ of maps

$$
u = x^a, v = x^c F
$$

with $a \geq 2$, $c \geq 0$ *where*

$$
F = x^r z + h(x, y)
$$

where h is arbitrary. The singular locus of this map germ is the variety defined by the ideal where the Jacobian has rank < 2. That is, the variety with ideal $J = \sqrt{(x^{a+c-1}F_y, x^{a+c-1}F_z)}$. Since $F_z = x^r$, we have that $\sqrt{J} = (x)$.

Examples of this kind can be used to construct weakly prepared projective morphisms satisfying the assumptions of Φ_1, by resolving the indeterminacy of the induced rational map $\mathbf{P}^3 \to \mathbf{P}^2$. A reasonably easy example to calculate is

$$\begin{aligned} u &= x^2 \\ v &= y^2 + xz. \end{aligned}$$

Throughout this section we will suppose that $\Phi_X : X \to S$ is weakly prepared.

We define permissible parameters (u, v) at points $q \in D_S$ by the following rules

1. If q is a nonsingular point of D_S, then regular parameters (u, v) in $\mathcal{O}_{S,q}$ are permissible parameters at q if $u = 0$ is a local equation for D_S. Necessarily, $u = 0$ is a local equation for E_X in $\mathcal{O}_{X,p}$ for all $p \in \Phi_X^{-1}(q)$.

2. If q is a singular point of D_S, then regular parameters (u, v) in $\mathcal{O}_{S,q}$ are permissible parameters at q if $uv = 0$ is a local equation for D_S at q. Necessarily, $uv = 0$ is a local equation for E_X at p for all $p \in \Phi_X^{-1}(q)$ and either $u = 0$ or $v = 0$ is a local equation of E_X at p.

Definition 6.4. *Suppose that (u, v) are permissible parameters at $q \in D_S$, $p \in \Phi_X^{-1}(q)$ and that $u = 0$ is a local equation of E_X at p. Regular parameters (x, y, z) in $\hat{\mathcal{O}}_{X,p}$ are called permissible parameters at p for (u, v) if $u = x^a y^b z^c$ with $a \geq b \geq c \geq 0$.*

If (x, y, z) are permissible parameters at p for (u, v), then one of the following forms holds for v at p.

1. *p is a 1 point:*

$$\begin{aligned} u &= x^a \\ v &= P_p(x) + x^b F_p \end{aligned}$$

   where $a > 0$, $x \nmid F_p$ and $x^b F_p$ has no terms which are powers of x,
2. *p is a 2 point:*

$$\begin{aligned} u &= (x^a y^b)^m \\ v &= P_p(x^a y^b) + x^c y^d F_p \end{aligned}$$

   where $a, b > 0$ $(a, b) = 1$, $x, y \nmid F_p$ and $x^c y^d F_p$ has no terms which are powers of $x^a y^b$,
3. *p is a 3 point:*

$$\begin{aligned} u &= (x^a y^b z^c)^m \\ v &= P_p(x^a y^b z^c) + x^d y^e z^f F_p \end{aligned}$$

   where $a, b, c > 0$, $(a, b, c) = 1$, $x, y, z \nmid F_p$ and $x^d y^e z^f F_p$ has no terms which are powers of $x^a y^b z^c$.

We will say that (x, y, z) are permissible parameters at p and that the above expression of V is the normalized form of v with respect to these parameters. We will also say that F_p is normalized with respect to (x, y, z).

The leading form of F_p will be denoted by L_p.

With the notation of Definition 6.4, we see that if p is a 1 point, then

$$\hat{\mathcal{I}}_{\text{sing}(\Phi_X), p} = \sqrt{x^{a+b-1}\left(\frac{\partial F}{\partial y}, \frac{\partial F}{\partial z}\right)}. \tag{14}$$

If p is a 2 point, then

$$\overline{\hat{\mathcal{I}}_{\text{sing}(\Phi_X), p}} = \sqrt{x^{ma+c-1}y^{mb+d-1}\left((ad - bc)F + ay\frac{\partial F}{\partial y} - bx\frac{\partial F}{\partial x}, y\frac{\partial F}{\partial z}, x\frac{\partial F}{\partial z}\right)} \tag{15}$$

If p is a 3 point then

$$\overline{\hat{\mathcal{I}}_{\text{sing}(\Phi_X), p}} = \sqrt{x^{ma+d-1}y^{mb+e-1}z^{mc+f-1}\left(\begin{array}{c} (ae - bd)zF + ayz\frac{\partial F}{\partial y} - bxz\frac{\partial F}{\partial x}, \\ (af - cd)yF + ayz\frac{\partial F}{\partial z} - cxy\frac{\partial F}{\partial x}, \\ (bf - ce)xF + bxz\frac{\partial F}{\partial z} - cxy\frac{\partial F}{\partial y} \end{array}\right)} \tag{16}$$

Definition 6.5. *We will say that permissible parameters (u, v) for $\Phi_X(p) \in D_S$ are prepared at $p \in E_X$ if $u = 0$ is a local equation of E_X at p and there exist permissible parameters (x, y, z) at p such that one of the following forms hold:*

$$\begin{aligned} u &= x^a \\ v &= P(x) + x^b y \end{aligned} \tag{17}$$

or

$$\begin{aligned} u &= (x^a y^b)^m \\ v &= P(x^a y^b) + x^c y^d \end{aligned} \tag{18}$$

with $ad - bc \neq 0$

$$\begin{aligned} u &= (x^a y^b)^m \\ v &= P(x^a y^b) + x^c y^d z \end{aligned} \tag{19}$$

or

$$\begin{aligned} u &= (x^a y^b z^c)^m \\ v &= P(x^a y^b z^c) + x^d y^e z^f \end{aligned} \tag{20}$$

with

$$\text{rank}\begin{pmatrix} a & b & c \\ d & e & f \end{pmatrix} = 2.$$

We will say that Φ_X is prepared with respect to D_S if for every $p \in E_X$ there exist permissible parameters for $\Phi_X(p)$ which are prepared at p.

Lemma 6.6. *Suppose that $p \in E_X$, (u, v) are permissible parameters at $q = \Phi_X(p)$ such that $u = 0$ is a local equation of E_X. Then $r = \nu(F_p)$ is independent of permissible parameters (x, y, z) at p for (u, v).*

If p is a 1 point then $\nu(F(0, y, z))$ is independent of permissible parameters (x, y, z) at p for (u, v). If p is a 1 point and

$$F_p = \sum_{i+j+k \geq r} a_{ijk} x^i y^j z^k,$$

then

$$\tau(F_p) = max\{j + k \mid a_{ijk} \neq 0 \ with \ i + j + k = r\}$$

is independent of permissible parameters (x, y, z) for (u, v) at p.

If p is a 2 point, then $\nu(F(0, 0, z))$ is independent of permissible parameters (x, y, z) at p for (u, v). If p is a 2 point and

$$F_p = \sum_{i+j+k \geq r} a_{ijk} x^i y^j z^k,$$

then

$$\tau(F_p) = max\{k \mid a_{ijk} \neq 0 \ with \ i + j + k = r\}$$

is independent of permissible parameters (x, y, z) for (u, v) at p.

Proof. Suppose that (u, v) are permissible parameters at q such that $u = 0$ is a local equation of E_X at p, (x, y, z), (x_1, y_1, z_1) are permissible parameters at p for (u, v). First suppose that p is a 1 point. We have a (normalized) expression

$$u = x^a, v = P(x) + x^b F$$

Thus

$$
\begin{aligned}
x &= \omega x_1 \\
y &= y(x_1, y_1, z_1) = b_{21} x_1 + b_{22} y_1 + b_{23} z_1 + \cdots \\
z &= z(x_1, y_1, z_1) = b_{31} x_1 + b_{32} y_1 + b_{33} z_1 + \cdots
\end{aligned}
$$

where $\omega^a = 1$ and $b_{22} b_{33} - b_{23} b_{32} \neq 0$, and

$$u = x_1^a, v = P_1(x_1) + x_1^b F_1$$

where

$$
\begin{aligned}
P_1 &= P(\omega x_1) + x_1^b \omega^b F(\omega x_1, y(x_1, 0, 0), z(x_1, 0, 0)) \\
F_1 &= \omega^b [F(\omega x_1, y(x_1, y_1, z_1), z(x_1, y_1, z_1)) - F(\omega x_1, y(x_1, 0, 0), z(x_1, 0, 0))]
\end{aligned}
$$

Substituting into

$$F_p = \sum_{i+j+k \geq r} a_{ijk} x^i y^j z^k$$

we get that $\nu(F) = \nu(F_1)$, $\nu(F(0, y, z)) = \nu(F_1(0, y_1, z_1))$ so that F_1 is normalized with respect to (x_1, y_1, z_1), and $\tau(F) = \tau(F_1)$.

Now suppose that p is a 2 point. Then

$$u = (x^a y^b)^m, v = P(x^a y^b) + x^c y^d F.$$

Set $r = \nu(F)$. We have one of the following two cases.

case 1

$$
\begin{aligned}
x &= \alpha x_1 \\
y &= \beta y_1 \\
z &= z(x_1, y_1, z_1) = a_1 x_1 + a_2 y_1 + a_3 z_1 + \cdots
\end{aligned}
$$

where $\omega = \alpha^a \beta^b$ satisfies $\omega^m = 1$ and $a_3 \neq 0$, or

case 2

$$x = \alpha y_1$$
$$y = \beta x_1$$
$$z = z(x_1, y_1, z_1) = a_1 x_1 + a_2 y_1 + a_3 z_1 + \cdots$$

where $\omega = \alpha^a \beta^b$ satisfies $\omega^m = 1$, and $a_3 \neq 0$.

In case 1, set $t_0 = \max\{\frac{c}{a}, \frac{d}{b}\}$. For $t \geq t_0$, set

$$
\begin{aligned}
b_t &= \left(\frac{\partial^{t(a+b)-c-d}(\alpha^c \beta^d F)}{\partial x_1^{ta-c} \partial y_1^{tb-d}} \right) \big|_{x_1=y_1=z_1=0} \\
F_1 &= \alpha^c \beta^d F - \sum_{t \geq t_0} b_t x_1^{ta-c} y_1^{tb-d} \\
P_1 &= P(\omega x_1^a y_1^b) + \sum_{t \geq t_0} b_t (x_1^a y_1^b)^t.
\end{aligned}
\tag{21}
$$

Then

$$u = (x_1^a y_1^b)^m, v = P_1(x_1^a y_1^b) + x_1^c y_1^d F_1.$$

F_1 is normalized with respect to (x_1, y_1, z_1) and $\nu(F(0,0,z)) = \nu(F_1(0,0,z_1))$. Set $\alpha(0,0,0) = \alpha_0$. $\beta(0,0,0) = \beta_0$. Let L, L_1 be the respective leading forms of F and F_1. Then

$$L_1 = \alpha_0 \beta_0 L(\alpha_0 x_1, \beta_0 y_1, a_1 x_1 + a_2 y_2 + a_3 z_1)$$

if there does not exist natural numbers i_0, j_0 such that $(c + i_0)b - (d + j_0)a = 0$ and $i_0 + j_0 = r$,

$$L_1 = \alpha_0 \beta_0 L(\alpha_0 x_1, \beta_0 y_1, a_1 x_1 + a_2 y_2 + a_3 z_1) - \bar{c} x_1^{i_0} y_1^{j_0}$$

for some $\bar{c} \in k$, if there exist natural numbers i_0, j_0 such that $(c+i_0)b-(d+j_0)a = 0$ and $i_0 + j_0 = r$. Thus $\nu(F) = \nu(F_1)$ and $\tau(F) = \tau(F_1)$.

To verify Case 2, we now need only consider the effect of a substitution

$$x = y_1, y = x_1.$$

Finally, suppose that p is a 3 point. We have

$$u = (x^a y^b z^c)^m, v = P(x^a y^b z^c) + x^d y^e z^f F$$

There exists $\sigma \in S_3$, and unit series α, β, γ with constant terms $\alpha_0, \beta_0, \gamma_0$ respectively, such that

$$x = \alpha w_{\sigma(1)}, y = \beta w_{\sigma(2)}, z = \gamma w_{\sigma(3)}$$

where

$$w_1 = x_1, w_2 = y_1, w_3 = z_1,$$

and if $\omega = \alpha^a \beta^b \gamma^c$, then $\omega^m = 1$. Set $t_0 = \max\{\frac{d}{a}, \frac{e}{b}, \frac{f}{c}\}$. For $t \geq t_0$, set

$$
\begin{aligned}
b_t &= \frac{\partial^{t(a+b+c)-d-e-f}(\alpha^d \beta^e \gamma^f F)}{\partial w_{\sigma(1)}^{ta-d} \partial w_{\sigma(2)}^{tb-e} \partial w_{\sigma(3)}^{tc-f}} \big|_{w_{\sigma(1)}=w_{\sigma(2)}=w_{\sigma(3)}=0} \\
F_1 &= \alpha^d \beta^e \gamma^f F(\alpha w_{\sigma(1)}, \beta w_{\sigma(2)}, \gamma w_{\sigma(3)}) - \sum_{t \geq t_0} b_t w_{\sigma(1)}^{ta-d} w_{\sigma(2)}^{tb-e} w_{\sigma(3)}^{tc-f} \\
P_1 &= P(\omega w_{\sigma(1)}^a w_{\sigma(2)}^b w_{\sigma(3)}^c) + \sum_{t \geq t_0} b_t w_{\sigma(1)}^{ta} w_{\sigma(2)}^{tb} w_{\sigma(3)}^{tc}.
\end{aligned}
\tag{22}
$$

Thus

$$
\begin{aligned}
u &= (w_{\sigma(1)}^a w_{\sigma(2)}^b w_{\sigma(3)}^c)^m, \\
v &= P_1(w_{\sigma(1)}^a w_{\sigma(2)}^b w_{\sigma(3)}^c) + w_{\sigma(1)}^d w_{\sigma(2)}^e w_{\sigma(3)}^f F_1(w_{\sigma(1)}, w_{\sigma(2)}, w_{\sigma(3)})
\end{aligned}
$$

where the leading form of F_1 is

$$L_1 = \alpha_0^d \beta_0^e \gamma_0^f F(\alpha_0 w_{\sigma(1)}, \beta_0 w_{\sigma(2)}, \gamma_0 w_{\sigma(3)})$$

if there does not exist natural numbers i_0, j_0, k_0 such that

$$(d+i_0)b - (e+j_0)a = 0, (d+i_0)c - (f+k_0)a = 0, \text{ and } i_0 + j + 0 + k_0 = r,$$

$$L_1 = \alpha_0^d \beta_0^e \gamma_0^f F(\alpha_0 w_{\sigma(1)}, \beta_0 w_{\sigma(2)}, \gamma_0 w_{\sigma(3)}) - \bar{c} x_1^{i_0} y_1^{j_0} z_1^{k_0}$$

for some $\bar{c} \in k$, if there exist natural numbers i_0, j_0, k_0 such that

$$(d+i_0)b - (e+j_0)a = 0, (d+i_0)c - (f+k_0)a = 0 \text{ and } i_0 + j_0 + k_0 = r.$$

Thus F_1 is normalized with respect to (x_1, y_1, z_1) and $\nu(L_1) = \nu(L)$. $\qquad\square$

Lemma 6.7. *Suppose that $p \in E_X$, $q = \Phi_X(p)$. Then $r = \nu(F_p)$ is independent of permissible parameters (u, v) at q such that $u = 0$ is a local equation of E_X at p.*

If p is a 1 point then $\nu(F_p(0, y, z))$ is independent of permissible parameters (u, v) at q such that $u = 0$ is a local equation of E_X at p. If p is a 1 point, and

$$F_p = \sum_{i+j+k \geq r} a_{ijk} x^i y^j z^k,$$

then

$$\tau(F_p) = max\{j + k \mid \text{ there exists } a_{ijk} \neq 0 \text{ with } i + j + k = r\}$$

is independent of permissible parameters (x, y, z) at p for (u, v) such that $u = 0$ is a local equation of E_X at p.

If p is a 2 point, then $\nu(F_p(0, 0, z))$ is independent of permissible parameters (u, v) at q such that $u = 0$ is a local equation of E_X at p. If p is a 2 point, and

$$F_p = \sum_{i+j+k \geq r} a_{ijk} x^i y^j z^k,$$

then

$$\tau(F_p) = max\{k \mid \text{ there exists } a_{ijk} \neq 0 \text{ with } i + j + k = r\}$$

is independent of permissible parameters (x, y, z) at p for (u, v) such that $u = 0$ is a local equation of E_X at p.

Proof. Let m be the maximal ideal of $\hat{\mathcal{O}}_{X,p}$. Suppose that (u, v) and (u_1, v_1) are permissible parameters at q such that $u = 0$ is a local equation of E_X at p and $u_1 = 0$ is a local equation of E_X at p. We will show that the multiplicities of the Lemma are the same for these two sets of permissible parameters.

Case 1 Suppose that p is a 1 point. Then (u_1, v_1) and (u, v) are related by a composition of changes of parameters of the types of Cases 1.1, 1.2 and 1.3 below. It thus suffices to prove the Lemma in each of these 3 cases.

Case 1.1 Suppose that $v_1 = u$, $u_1 = v$. We have

$$u = x^a, v = P(x) + x^c F$$

with $r = \nu(F) > 0$. In this case we must have $v = $ unit u in $\mathcal{O}_{X,p}$. $v = $ unit u is equivalent to $p = \bar{u}(x)x^d$ where $\bar{u}$ is a unit and $0 < d \leq c$. Set

$$x = \bar{x}(\bar{u}(x) + x^{c-d} F)^{\frac{-1}{d}}.$$

Then $v = \bar{x}^d$. Set $\tau = \frac{-1}{d}$. Write $\bar{u}(x) = a_0 + a_1 x + \cdots$.

$$(\bar{u}(x) + x^{c-d} F)^{\frac{-1}{d}} = \bar{u}(x)^{\tau} + \tau \bar{u}(x)^{\tau-1} x^{c-d} F + \frac{\tau(\tau-1)}{2} \bar{u}(x)^{\tau-2} x^{2(c-d)} F^2 + \cdots$$
$$\equiv \bar{u}(x)^{\tau} \bmod \bar{x}^{c-d} m^r$$

We thus have

$$x \equiv \overline{xu}(x)^{\tau} \bmod \bar{x}^{c-d+1} m^r. \tag{23}$$

Now suppose that $P_0(x, \bar{x})$ is a series. By substitution of (23), we see that there exist series A_1 and P_1 such that

$$P_0(x, \bar{x}) \equiv A_1(\bar{x}) + \bar{x} P_1(x, \bar{x}) \bmod \bar{x}^{c-d+1} m^r$$

By iteration, we get that there is a polynomial $\overline{P}(\bar{x})$, such that $\overline{P}(0) = P_0(0, 0)$,

$$P_0(x, \bar{x}) \equiv \overline{P}(\bar{x}) \bmod \bar{x}^{c-d+1} m^r. \tag{24}$$

We get from (24) that

$$\bar{u}(x) \equiv Q(\bar{x}) \bmod \bar{x}^{c-d+1} m^r$$

where $Q(0) = \bar{u}(0)$. Set $u_0 = \bar{u}(0)$. We also see that

$$x \equiv \bar{x} Q(\bar{x})^{\tau} \bmod \bar{x}^{c-d+1} m^r$$

Set $\lambda = \frac{-a}{d}$.

$$\begin{aligned} u &= \bar{x}^a (\bar{u}(x) + x^{c-d} F)^{\lambda} \\ &= \bar{x}^a [\bar{u}(x)^{\lambda} + \lambda \bar{u}(x)^{\lambda-1} x^{c-d} F + \frac{\lambda(\lambda-1)}{2} \bar{u}^{\lambda-2} x^{2(c-d)} F^2 + \cdots] \\ &\equiv \bar{x}^a [Q(\bar{x})^{\lambda} + \lambda Q(\bar{x})^{\lambda-1+\tau(c-d)} \bar{x}^{c-d} F \\ &\quad + \frac{\lambda(\lambda-1)}{2} Q(\bar{x})^{\lambda-2+2\tau(c-d)} \bar{x}^{2(c-d)} F^2 + \cdots] \\ &\quad \bmod \bar{x}^{a+c-d+1} m^r \end{aligned}$$

Thus

$$v = \bar{x}^d$$
$$u = P_1(\bar{x}) + \bar{x}^{a+c-d} F_1(\bar{x}, y, z)$$

where $\nu(p_1) = a$, and

$$\begin{aligned} F_1 &\equiv \lambda u_0^{\lambda-1+\tau(c-d)} F(u_0 \bar{x}, y, z) + \frac{\lambda(\lambda-1)}{2} u_0^{\lambda-2+2\tau(c-d)} \bar{x}^{c-d} F(u_0 \bar{x}, y, z)^2 \\ &\quad + \cdots \bmod \bar{x} m^r. \end{aligned}$$

Thus $\nu(F(x, y, z)) = \nu(F_1(\bar{x}, y, z))$, $\nu(F(0, y, z)) = \nu(F_1(0, y, z))$ and $\tau(F) = \tau(F_1)$.

Case 1.2 Suppose that $u_1 = \alpha u$, $v_1 = v$, where $\alpha(u, v)$ is a unit series. We have

$$u = x^a, v = p(x) + x^b F$$

with $r = \nu(F) > 0$. Set $\lambda = \frac{-1}{a}$. Define $x = \bar{x} \alpha^{\lambda}$, so that $u_1 = \bar{x}^a$. Write

$$\alpha = \alpha_0(u_1) + \alpha_1(u_1) v + \cdots$$

$$\begin{aligned}
\alpha^\lambda &= \alpha_0(u_1)^\lambda + \lambda\alpha_0(u_1)^{\lambda-1}(\alpha_1(u_1)v + \alpha_2(u_1)v^2 + \cdots) \\
&\quad + \tfrac{\lambda(\lambda-1)}{2}\alpha_0(u_1)^{\lambda-2}(\alpha_1(u_1)v + \alpha_2(u_1)v^2 + \cdots)^2 + \cdots \\
&\equiv \alpha_0(u_1)^\lambda + \lambda\alpha_0(u_1)^{\lambda-1}(\alpha_1(u_1)P(x) + \alpha_2(u_1)P(x)^2 + \cdots) \\
&\quad + \tfrac{\lambda(\lambda-1)}{2}\alpha_0(u_1)^{\lambda-2}(\alpha_1(u_1)P(x) + \alpha_2(u_1)P(x)^2 + \cdots)^2 + \cdots \bmod \overline{x}^b m^r
\end{aligned} \tag{25}$$

We have an expression

$$\alpha^\lambda \equiv A_0(\overline{x}) + \overline{x}B_0(\alpha^\lambda, \overline{x}) \bmod \overline{x}^b m^r. \tag{26}$$

Substitute (25) into (26) to get

$$\alpha^\lambda \equiv A_0(\overline{x}) + \overline{x}A_1(\overline{x}) + \overline{x}^2 B_1(\alpha^\lambda, \overline{x}) \bmod \overline{x}^b m^r$$

By iteration, we get that there is a series $S(\overline{x})$ with $S(0) = \alpha_0(0)^\lambda = \overline{\alpha}$, such that

$$\alpha^\lambda \equiv S(\overline{x}) \bmod \overline{x}^b m^r.$$

and

$$x = \alpha^\lambda \overline{x} \equiv S(\overline{x})\overline{x} \bmod \overline{x}^{b+1} m^r.$$

$$v = P(x) + x^b F \equiv P(\overline{x}S(\overline{x})) + \overline{x}^b S(\overline{x})^b F(\overline{x}S(\overline{x}), y, z)] \bmod \overline{x}^{b+1} m^r$$

Thus

$$u_1 = \overline{x}^a$$
$$v = P_1(\overline{x}) + \overline{x}^b F_1(\overline{x}, y, z)$$

where

$$\begin{aligned}
F_1 &\equiv S(\overline{x})^b F(\overline{x}S(\overline{x}), y, z) \bmod \overline{x}m^r \\
&\equiv \overline{\alpha}^b F(\overline{\alpha}\overline{x}, y, z) \bmod \overline{x}m^r
\end{aligned}$$

Thus $\nu(F) = \nu(F_1)$, $\nu(F(0, y, z)) = \nu(F_1(0, y, z))$ and $\tau(F) = \tau(F_1)$.

Case 1.3 Suppose that $u_1 = u$, $v_1 = \alpha u + \beta v$. Write

$$\begin{aligned}
\alpha &= \sum \alpha_{ij} u^i v^j \\
\beta &= \sum \beta_{ij} u^i v^j
\end{aligned}$$

with $\beta_{00} \neq 0$.

We have

$$u = x^a, v = P(x) + x^b F$$

$r = \nu(F) > 0$.

$$\begin{aligned}
v_1 &= \sum \alpha_{ij} u^{i+1} v^j + \sum \beta_{ij} u^i v^{j+1} \\
&= \sum \alpha_{ij} x^{a(i+1)}(P(x) + x^b F)^j + \sum \beta_{ij} x^{ai}(P(x) + x^b F)^{j+1} \\
&= \sum \alpha_{ij} x^{a(i+1)}(P(x)^j + jx^b P(x)^{j-1} F \\
&\quad + \tfrac{j(j-1)}{2} x^{2b} P(x)^{j-2} F^2 + \cdots + x^{bj} F^j) \\
&\quad + \sum \beta_{ij} x^{ai}(P(x)^{j+1} + (j+1)x^b P(x)^j F \\
&\quad + \tfrac{(j+1)j}{2} x^{2b} P(x)^{j-1} F^2 + \cdots + x^{b(j+1)} F^{j+1}) \\
&= Q(x) + H(x)F + x^{2b} F^2 G(x, y, z)
\end{aligned}$$

where

$$H(x) = \sum \alpha_{ij} x^{a(i+1)} jx^b P(x)^{j-1} + \sum \beta_{ij} x^{ai}(j+1)P(x)^j x^b.$$

We can further write

$$H(x) = x^b(\beta_{00} + x\Omega(x)).$$

$$v_1 = Q(x) + x^b(\beta_{00} + x\Omega(x))F + x^{2b}F^2G$$
$$= P_1(x) + x^bF_1$$

where $P_1(x) = Q(x)$, and

$$F_1 = (\beta_{00} + x\Omega(x))F + x^bF^2G.$$

Thus $\nu(F) = \nu(F_1)$, $\nu(F(0, y, z)) = \nu(F_1(0, y, z))$ and $\tau(F) = \tau(F_1)$.

Case 2 Suppose that p is a 2 point. It suffices to prove the Lemma in the three subcases 2.1, 2.2 and 2.3.

Case 2.1 Suppose that $u_1 = v$, $v_1 = u$. We have an expression

$$u = (x^a y^b)^k, v = P(x^a y^b) + x^c y^d F.$$

Set $r = \nu(F)$. If

$$r = 0, c \leq \operatorname{ord}(P)a \text{ and } d \leq \operatorname{ord}(P)b \tag{27}$$

then the multiplicities of the Lemma are the same for the two sets of parameters, so suppose that (27) doesn't hold. $v = x^e y^f$ unit (for some e, f) implies that there exists $t > 0$ such that

$$v = (x^a y^b)^t(\overline{u}(x^a y^b) + x^{c-at} y^{d-bt} F)$$

where $\overline{u}$ is a unit power series. Set $\tau = \frac{-1}{at}$,

$$x = \overline{x}(\overline{u}(x^a y^b) + x^{c-at} y^{d-bt} F)^\tau.$$

$$\begin{aligned}(\overline{u}(x^a y^b) + x^{c-at} y^{d-bt} F)^\tau &= \overline{u}(x^a y^b)^\tau + \tau\overline{u}(x^a y^b)^{\tau-1} x^{c-at} y^{d-bt} F \\ &+ \frac{\tau(\tau-1)}{2}\overline{u}(x^a y^b)^{\tau-2} x^{2(c-at)} y^{2(d-bt)} F^2 + \cdots \\ &\equiv \overline{u}(x^a y^b)^\tau \bmod \overline{x}^{c-at} y^{d-bt} m^r\end{aligned}$$

$$\begin{aligned}x^a y^b &= \overline{x}^a y^b(\overline{u}(x^a y^b) + x^{c-at} y^{d-bt} F)^{a\tau} \\ &\equiv \overline{x}^a y^b \overline{u}(x^a y^b)^{a\tau} \bmod \overline{x}^{c-at+a} y^{d-bt+b} m^r.\end{aligned} \tag{28}$$

Now suppose that $P_0(x^a y^b, \overline{x}^a y^b)$ is a series. By substitution of (28), we see that

$$P_0(x^a y^b, \overline{x}^a y^b) \equiv A_1(\overline{x}^a y^b) + \overline{x}^a y^b P_1(x^a y^b, \overline{x}^a y^b) \bmod \overline{x}^{c-at+a} y^{d-bt+b} m^r$$

By iteration, we get that there is a polynomial $Q(\overline{x}^a y^b)$, such that $u_0 = Q(0) = \overline{u}(0)$,

$$\overline{u}(x^a y^b) \equiv Q(\overline{x}^a y^b) \bmod \overline{x}^{c-at+a} y^{d-bt+b} m^r. \tag{29}$$

we get from (29) that

$$\begin{aligned}x &\equiv \overline{x}\overline{u}(x^a y^b) \bmod \overline{x}^{c-at+1} y^{d-bt} m^r \\ &\equiv \overline{x}Q(\overline{x}^a y^b) \bmod \overline{x}^{c-at+1} y^{d-bt} m^r\end{aligned}$$

Set $\lambda = \frac{-k}{t}$.

$$
\begin{aligned}
u &= (x^a y^b)^k = (\overline{x}^a y^b)^k [\overline{u}(x^a y^b) + x^{c-at} y^{d-bt} F]^\lambda \\
&= (\overline{x}^a y^b)^k [\overline{u}(x^a y^b)^\lambda + \lambda \overline{u}(x^a y^b)^{\lambda-1} x^{c-at} y^{d-bt} F \\
&\quad + \tfrac{\lambda(\lambda-1)}{2} \overline{u}(x^a y^b)^{\lambda-2} x^{2(c-at)} y^{2(d-bt)} F^2 + \cdots] \\
&\equiv (\overline{x}^a y^b)^k [Q(\overline{x}^a y^b)^\lambda + \lambda Q(\overline{x}^a y^b)^{\lambda-1+c-at} \overline{x}^{c-at} y^{d-bt} F(\overline{x}Q(\overline{x}^a y^b), y, z) \\
&\quad + \tfrac{\lambda(\lambda-1)}{2} Q(\overline{x}^a y^b)^{\lambda-2+2(c-at)} \overline{x}^{2(c-at)} y^{2(d-bt)} F(\overline{x}Q(\overline{x}^a y^b), y, z)^2 \\
&\quad + \cdots] \bmod \overline{x}^{ak+c-at+1} y^{bk+d-bt} m^r
\end{aligned}
$$

Thus

$$
v = (\overline{x}^a y^b)^t
$$
$$
u = P_1(\overline{x}^a y^b) + \overline{x}^{ak+c-at} y^{bk+d-bt} F_1(\overline{x}, y, z)
$$

where

$$
\begin{aligned}
F_1(\overline{x}, y, z) &\equiv \lambda Q(\overline{x}^a y^b)^{\lambda-1+c-at} F(\overline{x}Q(\overline{x}^a y^b), y, z) \\
&\quad + \tfrac{\lambda(\lambda-1)}{2} Q(\overline{x}^a y^b)^{\lambda-2+2(c-at)} \overline{x}^{c-at} y^{d-bt} F(\overline{x}Q(\overline{x}^a y^b), y, z)^2 \\
&\quad + \cdots \bmod \overline{x} m^r \\
&\equiv \lambda u_0^{\lambda-1+c-at} F(\overline{x}u_0, y, z) \\
&\quad + \tfrac{\lambda(\lambda-1)}{2} u_0^{\lambda-2+2(c-at)} \overline{x}^{c-at} y^{d-bt} F(\overline{x}u_0, y, z)^2 \\
&\quad + \cdots \bmod \overline{x} m^r.
\end{aligned}
$$

Thus $\nu(F) = \nu(F_1)$, $\nu(F_1(0,0,z)) = \nu(F(0,0,z))$ and $\tau(F) = \tau(F_1)$.

Case 2.2 Suppose that p is a 2 point and that $u_1 = \alpha u$, $v_1 = v$. We have an expression

$$
u = (x^a y^b)^k, v = P(x^a y^b) + x^c y^d F
$$

Set $r = \nu(F)$. Write

$$
\alpha = \alpha_0(u_1) + \alpha_1(u_1)v + \cdots
$$

Set $\lambda = \frac{-1}{ak}$,

$$
x = \overline{x}\alpha^\lambda.
$$

We have that

$$
u_1 = (\overline{x}^a y^b)^k.
$$

$$
\begin{aligned}
\alpha^\lambda &= \alpha_0(u_1)^\lambda + \lambda \alpha_0(u_1)^{\lambda-1}(\alpha_1(u_1)v + \alpha_2(u_1)v^2 + \cdots) \\
&\quad + \tfrac{\lambda(\lambda-1)}{2} \alpha_0(u_1)^{\lambda-2}(\alpha_1(u_1)v + \alpha_2(u_1)v^2 + \cdots)^2 + \cdots \\
&\equiv \alpha_0(u_1)^\lambda + \lambda \alpha_0(u_1)^{\lambda-1}(\alpha_1(u_1)P(x^a y^b) \\
&\quad + \alpha_2(u_1)P(x^a y^b)^2 + \cdots) + \tfrac{\lambda(\lambda-1)}{2} \alpha_0(u_1)^{\lambda-2} \\
&\quad (\alpha_1(u_1)P(x^a y^b) + \alpha_2(u_1)P(x^a y^b)^2 + \cdots) + \cdots \bmod \overset{\circ}{\overline{x}}{}^c y^d m^r
\end{aligned}
$$

Now suppose that $P_0(x^a y^b, \overline{x}^a y^b)$ is a series. By substitution of the above equation, we see that

$$
P_0(x^a y^b, \overline{x}^a y^b) \equiv A_1(\overline{x}^a y^b) + \overline{x}^a y^b P_1(x^a y^b, \overline{x}^a y^b) \bmod \overline{x}^c y^d m^r
$$

By iteration, we get that there is a polynomial $S(\overline{x}^a y^b)$, such that $u_0 = S(0) = \overline{u}(0)$,

$$
\alpha^\lambda \equiv S(\overline{x}^a y^b) \bmod \overline{x}^c y^d m^r. \tag{30}
$$

we get from (30) that

$$x = \alpha^\lambda \bar{x} \equiv S(\bar{x}^a y^b)\bar{x} \bmod \bar{x}^{c+1} y^d m^r$$

$$\begin{aligned}
v &= P(x^a y^b) + x^c y^d F \\
&\equiv P(\bar{x}^a y^b S(\bar{x}^a y^b)^a) + \bar{x}^c y^d S(\bar{x}^a y^b)^c F(S(\bar{x}^a y^b)\bar{x}, y, z) \\
&\quad \bmod \bar{x}^{c+1} y^d m^r
\end{aligned}$$

so that

$$\begin{aligned}
u_1 &= (\bar{x}^a y^b)^k \\
v &= P_1(\bar{x}^a y^b) + \bar{x}^c y^d F_1(\bar{x}, y, z)
\end{aligned}$$

where

$$\begin{aligned}
F_1 &\equiv S(\bar{x}^a y^b)^c F(\bar{x} S(\bar{x}^a y^b), y, z) \bmod \bar{x} m^r \\
&\equiv u_0^c F(u_0 \bar{x}, y, z) \bmod \bar{x} m^r
\end{aligned}$$

Thus $\nu(F) = \nu(F_1)$, $\nu(F(0,0,z)) = \nu(F_1(0,0,z))$, and $\tau(F) = \tau(F_1)$.

Case 2.3 Suppose that p is a 2 point and that $u_1 = u$, $v_1 = \alpha u + \beta v$. We have an expression

$$u = (x^a y^b)^k, v = P(x^a y^b) + x^c y^d F$$

where $r = \nu(F)$. Write

$$\begin{aligned}
\alpha &= \sum \alpha_{ij} u^i v^j \\
\beta &= \sum \beta_{ij} u^i v^j
\end{aligned}$$

with $\beta_{00} \neq 0$.

$$\begin{aligned}
v_1 &= \sum \alpha_{ij} u^{i+1} v^j + \sum \beta_{ij} u^i v^{j+1} \\
&= \sum \alpha_{ij} (x^a y^b)^{(i+1)k} (P(x^a y^b) + x^c y^d F)^j \\
&\quad + \sum \beta_{ij} (x^a y^b)^{ik} (P(x^a y^b) + x^c y^d F)^{j+1} \\
&= \sum \alpha_{ij} (x^a y^b)^{(i+1)k} (P(x^a y^b)^j + j(x^c y^d) P(x^a y^b)^{j-1} F + \cdots + (x^c y^d)^j F^j) \\
&\quad + \sum \beta_{ij} (x^a y^b)^{ik} (P(x^a y^b)^{j+1} + (j+1) x^c y^d P(x^a y^b)^j F \\
&\quad + \cdots + (x^c y^d)^{j+1} F^{j+1}) \\
&= Q(x^a y^b) + H(x, y) F + (x^c y^d)^2 F^2 G(x, y, z)
\end{aligned}$$

where

$$H(x, y) = x^c y^d (\beta_{0,0} + x^a y^b \Omega(x, y)).$$

Then

$$\begin{aligned}
u &= (x^a y^b)^k \\
v_1 &= Q_1(x^a y^b) + x^c y^d F_1
\end{aligned}$$

where

$$F_1 \equiv \beta_{0,0} F \bmod (xy m^r + (x^c y^d) m^{2r})$$

Thus $\nu(F) = \nu(F_1)$, $\nu(F(0,y,z)) = \nu(F_1(0,y,z))$, and $\tau(F) = \tau(F_1)$.

Case 3 Suppose that p is a 3 point. It suffices to prove the Lemma in the three subcases 3.1, 3.2 and 3.3.

Case 3.1 Suppose that $u_1 = v$, $v_1 = u_1$.

$$u = (x^a y^b z^c)^k, v = P(x^a y^b z^c) + x^d y^e z^f F$$

where $r = \nu(F)$, If

$$r = 0, d \le \text{ord}(P)a, e \le \text{ord}(P)b \text{ and } f \le \text{ord}(P)c \tag{31}$$

then the multiplicities of the Lemma are the same for the two sets of parameters, so suppose that (31) doesn't hold.

In this case we must have that $v = x^\alpha y^\beta z^\gamma$ unit, so that $P(s) = s^t \overline{u}(s)$ where $\overline{u}$ is a unit power series and

$$v = (x^a y^b z^c)^t [\overline{u}(x^a y^b z^c) + x^{d-ta} y^{e-bt} z^{f-ct} F].$$

Set $\tau = \frac{-1}{at}$,

$$x = \overline{x}(\overline{u}(x^a y^b z^c) + x^{d-ta} y^{e-bt} z^{f-ct} F)^\tau.$$

$$(\overline{u}(x^a y^b z^c) + x^{d-ta} y^{e-bt} z^{f-ct} F)^\tau$$
$$= \overline{u}(x^a y^b z^c)^\tau + \tau \overline{u}(x^a y^b z^c)^{\tau-1} x^{d-ta} y^{e-bt} z^{f-ct} F$$
$$+ \frac{\tau(\tau-1)}{2} \overline{u}(x^a y^b z^c)^{\tau-2} x^{2(d-ta)} y^{2(e-bt)} z^{2(f-ct)} F^2 + \cdots$$
$$\equiv \overline{u}(x^a y^b z^c)^\tau \mod \overline{x}^{d-ta} y^{e-bt} z^{f-ct} m^r$$

$$x^a y^b z^c = \overline{x}^a y^b z^c (\overline{u}(x^a y^b z^c) + x^{d-ta} y^{e-bt} z^{f-ct} F)^{a\tau}$$
$$\equiv \overline{x}^a y^b z^c \overline{u}(x^a y^b z^c)^{a\tau} \mod \overline{x}^{a+d-ta} y^{e-bt+b} z^{f-ct+c} m^r$$

Now suppose that $P_0(x^a y^b z^c, \overline{x}^a y^b z^c)$ is a series. By substitution of the above equation, we see that

$$P_0(x^a y^b z^c, \overline{x}^a y^b z^c)$$
$$\equiv A_1(\overline{x}^a y^b z^c) + \overline{x}^a y^b z^c P_1(x^a y^b z^c, \overline{x}^a y^b z^c) \mod \overline{x}^{a+d-ta} y^{e-bt+b} z^{f-ct+c} m^r$$

By iteration, we get that there is a polynomial $Q(\overline{x}^a y^b z^c)$, such that if $u_0 = \overline{u}(0)$, $Q(0) = \overline{u}(0) = u_0$,

$$\overline{u}(x^a y^b z^c) \equiv Q(\overline{x}^a y^b z^c) \mod \overline{x}^{a+d-ta} y^{e-bt+b} z^{f-ct+c} m^r$$

Thus

$$x \equiv \overline{x}\overline{u}(x^a y^b z^c)^\tau \mod \overline{x}^{d-ta+1} y^{e-bt} z^{f-ct} m^r$$
$$\equiv \overline{x}Q(\overline{x}^a y^b z^c)^\tau \mod \overline{x}^{d-ta+1} y^{e-bt} z^{f-ct} m^r$$

Set $\lambda = \frac{-k}{t}$.

$$u = (x^a y^b z^c)^k = (\overline{x}^a y^b z^c)^k (\overline{u}(x^a y^b z^c) + x^{d-ta} y^{e-bt} z^{f-ct} F)^\lambda$$
$$= (\overline{x}^a y^b z^c)^k [\overline{u}(x^a y^b z^c)^\lambda + \lambda \overline{u}(x^a y^b z^c)^{\lambda-1} x^{d-ta} y^{e-bt} z^{f-ct} F$$
$$+ \frac{\lambda(\lambda-1)}{2} \overline{u}(x^a y^b z^c)^{\lambda-2} x^{2(d-ta)} y^{2(e-bt)} z^{2(f-ct)} F^2 + \cdots]$$
$$\equiv (\overline{x}^a y^b z^c)^k [Q(\overline{x}^a y^b z^c)^a)^\lambda$$
$$+ \lambda Q(\overline{x}^a y^b z^c)^{\lambda-1+\tau(d-ta)} \overline{x}^{d-ta} y^{e-bt} z^{f-ct} F(\overline{x}Q(\overline{x}^a y^b z^c)^\tau, y, z)$$
$$+ \frac{\lambda(\lambda-1)}{2} Q(\overline{x}^a y^b z^c)^{\lambda-2+2\tau(d-ta)} \overline{x}^{2(d-ta)} y^{2(e-bt)} z^{2(f-ct)} F(\overline{x}Q(\overline{x}^a y^b z^c)^\tau, y, z)^2$$
$$+ \cdots] \mod \overline{x}^{ak+d-ta+1} y^{bk+e-bt} z^{ck+f-ct} m^r$$

$$v = (\overline{x}^a y^b z^c)^t$$
$$u = P_1(\overline{x}^a y^b z^c) + \overline{x}^{ak+d-ta} y^{bk+e-bt} z^{ck+f-ct} F_1(\overline{x}, y, z)$$

where

$$
\begin{aligned}
F_1 &\equiv \lambda Q(\overline{x}^a y^b z^c)^{\lambda-1+\tau(d-ta)} F(\overline{x} Q(\overline{x}^a y^b z^c)^\tau, y, z) \\
&+ \tfrac{\lambda(\lambda-1)}{2} Q(\overline{x}^a y^b z^c)^a)^{\lambda-2+2\tau(d-ta)} \overline{x}^{d-ta} y^{e-bt} z^{f-ct} F(\overline{x} Q(\overline{x}^a y^b z^c)^\tau, y, z)^2 \\
&+ \cdots \bmod \overline{x} m^r \\
&\equiv \lambda u_0^{\lambda-1+\tau(d-ta)} F(u_0^\tau \overline{x}, y, z) \\
&+ \tfrac{\lambda(\lambda-1)}{2} u_0^{\lambda-2+\tau(d-ta)} \overline{x}^{d-ta} y^{e-bt} z^{f-ct} F(u_0 \overline{x}, y, z)^2 + \cdots \bmod \overline{x} m^r
\end{aligned}
$$

Thus $\nu(F_1) = \nu(F)$.

Case 3.2 Suppose that $u_1 = \alpha u$, $v_1 = v$.

$$
u = (x^a y^b z^c)^k, \quad v = P(x^a y^b z^c) + x^d y^e z^f F
$$

where $r = \nu(F)$. Set $\lambda = \frac{-1}{ak}$, $x = \overline{x} \alpha^\lambda$. Thus

$$
u_1 = (\overline{x}^a y^b z^c)^k.
$$

Write

$$
\alpha = \alpha_0(u_1) + \alpha_1(u_1) v + \cdots
$$

$$
\begin{aligned}
\alpha^\lambda &\equiv \alpha_0(u_1)^\lambda + \lambda \alpha_0(u_1)^{\lambda-1}(\alpha_1(u_1) v + \alpha_2(u_1) v^2 + \cdots) \\
&+ \tfrac{\lambda(\lambda-1)}{2} \alpha_0(u_1)^{\lambda-2}(\alpha_1(u_1) v + \alpha_2(u_1) v^2 + \cdots)^2 + \cdots \\
&\equiv \alpha_0(u_1)^\lambda + \lambda \alpha_0(u_1)^{\lambda-1}(\alpha_1(u_1) P(x^a y^b z^c) + \alpha_2(u_1) P(x^a y^b z^c)^2 + \cdots) \\
&+ \tfrac{\lambda(\lambda-1)}{2} \alpha_0(u_1)^{\lambda-2}(\alpha_1(u_1) P(x^a y^b z^c) + \alpha_2(u_1) P(x^a y^b z^c)^2 + \cdots)^2 \\
&+ \cdots \bmod \overline{x}^d y^e z^f m^r
\end{aligned}
$$

Thus

$$
\alpha^\lambda \equiv A_0(\overline{x}^a y^b z^c) + \overline{x}^a y^b z^c B_0(\alpha^\lambda, \overline{x}^a y^b z^c) \bmod \overline{x}^d y^e z^f m^r
$$

Substitute the above equation into itself and iterate to get

$$
\alpha^\lambda \equiv S(\overline{x}^a y^b z^c) \bmod \overline{x}^d y^e z^f m^r
$$

Set $\overline{\alpha} = \alpha(0)$. Then $S(0) = \overline{\alpha}^\lambda$.

$$
x = \alpha^\lambda \overline{x} \equiv S(\overline{x}^a y^b z^c) \overline{x} \bmod \overline{x}^{d+1} y^e z^f m^r
$$

$$
\begin{aligned}
v &= P(x^a y^b z^c) + x^d y^e z^f F \\
&\equiv P(\overline{x}^a y^b z^c S(\overline{x}^a y^b z^c)^a) \\
&+ \overline{x}^d y^e z^f S(\overline{x}^a y^b z^c)^d F(S(\overline{x}^a y^b z^c) \overline{x}, y, z) \bmod \overline{x}^{d+1} y^e z^f m^r
\end{aligned}
$$

Thus

$$
\begin{aligned}
u_1 &= (\overline{x}^a y^b z^c)^k \\
v &= P_1(\overline{x}^a y^b z^c) + \overline{x}^d y^e z^f F_1(\overline{x}, y, z)
\end{aligned}
$$

where

$$
\begin{aligned}
F_1 &\equiv S(\overline{x}^a y^b z^c)^d F(S(\overline{x}^a y^b z^c) \overline{x}, y, z) \bmod \overline{x} m^r \\
&\equiv \overline{\alpha}^{\lambda d} F(\overline{\alpha}^\lambda \overline{x}, y, z) \bmod \overline{x} m^r
\end{aligned}
$$

Thus $\nu(F_1) = \nu(F)$.

Case 3.3 Suppose that $u_1 = u$, $v_1 = \alpha u + \beta v$. We have an expression

$$
u = (x^a y^b z^c)^k, \quad v = P(x^a y^b z^c) + x^d y^e z^f F
$$

where $r = \nu(F)$.

Write

$$\begin{aligned}
\alpha &= \sum \alpha_{ij} u^i v^j \\
\beta &= \sum \beta_{ij} u^i v^j
\end{aligned}$$

with $\beta_{00} \neq 0$.

$$\begin{aligned}
v_1 &= \sum \alpha_{ij} u^{i+1} v^j + \sum \beta_{ij} u^i v^{j+1} \\
&= \sum \alpha_{ij} (x^a y^b z^c)^{(i+1)k} (P(x^a y^b z^c) + x^d y^e z^f F)^j \\
&\quad + \sum \beta_{ij} (x^a y^b z^c)^{ik} (P(x^a y^b z^c) + x^d y^e z^f F)^{j+1} \\
&= \sum \alpha_{ij} (x^a y^b z^c)^{(i+1)k} (P(x^a y^b z^c)^j + j(x^d y^e z^f) P(x^a y^b z^c)^{j-1} F \\
&\quad + \cdots + (x^d y^e z^f)^j F^j) \\
&\quad + \sum \beta_{ij} (x^a y^b z^c)^{ik} (P(x^a y^b z^c)^{j+1} + (j+1) x^d y^e z^f P(x^a y^b z^c)^j F \\
&\quad + \cdots + (x^d y^e z^f)^{j+1} F^{j+1}) \\
&= Q(x^a y^b z^c) + HF + (x^d y^e z^f)^2 F^2 G
\end{aligned}$$

where

$$H = x^d y^e z^f (\beta_{0,0} + x^a y^b z^c \Omega)$$

Then

$$\begin{aligned}
u &= (x^a y^b z^c)^k \\
v_1 &= Q_1(x^a y^b z^c) + x^d y^e z^f F_1
\end{aligned}$$

where

$$F_1 \equiv \beta_{0,0} F \mod (xyzm^r + x^d y^e z^f m^{2r})$$

Thus $\nu(F) = \nu(F_1)$. $\qquad\qquad\qquad\qquad\qquad\qquad\qquad\qquad\square$

By Lemmas 6.6 and 6.7, we can make the following definitions, with the notation of Definition 6.4.

Definition 6.8. *Suppose that $p \in E_X$, (u, v) are permissible parameters at $\Phi_X(p)$, (x, y, z) are permissible parameters at p for (u, v) such that $u = 0$ is a local equation of E_X at p. Thus one of the forms of Definition 6.4 holds. Define $\nu(p) = \nu(F_p)$. If p is a 1 point, define $\gamma(p) = \text{mult}(F_p(0, y, z))$. If p is a 2 point, define $\gamma(p) = \text{mult}(F_p(0, 0, z))$.*

Suppose that $p \in X$ is a 1 point such that

$$\begin{aligned}
u &= x^a \\
v &= P(x^a) + x^b F_p \\
F_p &= \sum_{i+j+k \geq r} a_{ijk} x^i y^j z^k
\end{aligned}$$

where $\nu(p) = r$. Define

$$\tau(p) = \max\{j + k \mid \text{ there exits } a_{ijk} \neq 0 \text{ with } i + j + k = r\}.$$

If p is a 1 point, we have $1 \leq \tau(p) \leq \nu(p)$. Suppose that $p \in X$ is a 2 point such that

$$\begin{aligned}
u &= (x^a y^b)^m \\
v &= P(x^a y^b) + x^c y^d F_p \\
F_p &= \sum_{i+j+k \geq r} a_{ijk} x^i y^j z^k
\end{aligned}$$

where $\nu(p) = r$. Define

$$\tau(p) = \max\{k \mid \text{ there exits } a_{ijk} \neq 0 \text{ with } i + j + k = r\}.$$

Define

$$S_r(X) = \{p \in E_X | \nu(p) \geq r\}.$$

Let $\overline{S}_r(X)$ be the Zariski closure of $S_r(X)$ in X.

Definition 6.9. *A point $p \in E_X$ is resolved if the following condition holds.*

1. *If p is a 1 point then $\nu(p) \leq 1$.*
2. *If p is a 2 point then $\gamma(p) \leq 1$.*
3. *If p is a 3 point then $\nu(p) = 0$.*

Remark 6.10. *If $p \in E_X$ is resolved and (u,v) are permissible parameters at $\Phi_X(p)$ such that $u = 0$ is a local equation of E_X at p, then (u,v) are prepared at p.*

Lemma 6.11. *$S_r(X) \subset sing(\Phi_X)$ for $r \geq 2$, and all 3 points are contained in $sing(\Phi_X)$. If $p \in S_1(X)$ is a 2 point then $p \in sing(\Phi_X)$.*

Proof. The Lemma is immediate from (14), (15) and (16). $\square$

Example 6.12. *$S_r(X)$ is in general not Zariski closed. Consider the 2 point p with local equations*

$$\begin{aligned} u &= xy \\ v &= x^2 y. \end{aligned}$$

$\nu(p) = 0$. At 1 points q on the surface $x = 0$ there are regular parameters (x, y_1, z) with $y = y_1 + \alpha$ for some $0 \neq \alpha \in k$. Set $\overline{x} = x(y_1 + \alpha)$. There are permissible parameters $(\overline{x}, \overline{y}, z)$ at q such that

$$\begin{aligned} u &= \overline{x} \\ v &= \alpha^{-1}\overline{x}^2 + \overline{x}^2\overline{y}. \end{aligned}$$

Thus $\nu(q) = 1$.

Lemma 6.13. *Suppose that $p \in E_X$ is a 1 point and that $I \subset \hat{\mathcal{O}}_{X,p}$ is a reduced ideal such that if $x = 0$ is a local equation of E_X at p then $x \in I$. Then the condition $F_p \in I^s$ (with $s \in \mathbf{N}$) and the condition $F_p \in m_p I^s$ (with $s \in \mathbf{N}$) are independent of the choice of permissible parameters (u,v) at $\Phi_X(p)$ such that $u = 0$ is a local equation of E_X at p, and permissible parameters (x,y,z) for (u,v) at p.*

Proof. If $I = m_p\hat{\mathcal{O}}_{X,p}$, the Lemma follows from Lemmas 6.6 and 6.7. So we assume that $I = (x, f)$ for some series $f(y, z)$.

If (x, y, z) and (x_1, y_1, z_1) are permissible parameters at p for (u, v) then with the notation of the proof of Lemma 6.6,

$$F_1 = \omega^b[F - F(\omega x_1, y(x_1, 0, 0), z(x_1, 0, 0))].$$

and

$$x^{\nu(p)} \mid F(\omega x_1, y(x_1(0,0), z(x_1, 0, 0))]$$

implies $F_1 \in I^s$ (or $F_1 \in mI^s$), $x \in I$ and $s \leq \nu(p)$ (or $s \leq \nu(p) - 1$).

Now suppose that (u, v), (u_1, v_1) are permissible parameters at $f(p)$. Suppose that $v_1 = u$, $u_1 = v$. With the notation of Case 1.1 of the proof of Lemma 6.7, $F \in I^s$ implies (23) can be modified to

$$x \equiv \overline{x}\overline{u}(x)^\tau \mod \overline{x}^{c-d+1} I^s$$

and thus

$$x \equiv \overline{x}Q(\overline{x})^\tau \mod \overline{x}^{c-d+1} I^s$$

We thus have

$$\begin{aligned}
F_1 &\equiv \lambda u_0^{\lambda-1+\tau(c-d)} F(u_0\overline{x}, y, z) \\
&\quad + \tfrac{\lambda(\lambda-1)}{2} u_0^{\lambda-2+2\tau(c-d)} \overline{x}^{c-d} F(u_0\overline{x}, y, z)^2 + \cdots \mod \overline{x}I^s
\end{aligned}$$

since $I = (x, f(y, z))$ for some f, we have $F_1 \in I^s$. We have a similar proof when $F \in mI^s$. We can replace m^τ in the formulas of case 1.1 of Lemma 6.7 with mI^s.

In the proofs of cases 1.2 and 1.3, we can also replace m^τ in all the formulas with I^s (or mI^s). Again, since $x \in I$, we get $F_1 \in I^s$ (or $F_1 \in mI^s$). $\qquad\square$

Lemma 6.14. *Suppose that C is a 2 curve and $p \in C$. Then the condition $F_p \in \hat{\mathcal{I}}_{C,p}^s$, (with $s \in \mathbb{N}$) is independent of permissible parameters at $\Phi_X(p)$ and p.*

Proof. Suppose that (u, v) are permissible parameters at $\Phi_X(p)$. We will first show that the condition is independent of permissible parameters for (u, v) at p.

If p is a 2 point, this follows from the proof of Lemma 6.6, with the observation that, in the notation of (21), $F \in \hat{\mathcal{I}}_{C,p}^s$ implies

$$\frac{\partial^{t(a+b)-c-d}(\alpha^c \beta^d F)}{\partial x_1^{ta-c} \partial y_1^{tb-d}} \in \hat{\mathcal{I}}_{C,p}^{s-t(a+b)+c+d},$$

so that $\sum b_t x_1^{ta-c} y_1^{tb-d} \in \hat{\mathcal{I}}_{C,p}^s$, and thus $F_1 \in \hat{\mathcal{I}}_{C,p}^s$.

If p is a 3 point, this also follows from the proof of Lemma 6.6. With the notation of (22), after possibly permuting the parameters $(w_{\sigma(1)}, w_{\sigma(2)}, w_{\sigma(3)})$, we have $\hat{\mathcal{I}}_{C,p} = (w_{\sigma(1)}, w_{\sigma(2)})$.

If $G \in \hat{\mathcal{O}}_{X,p}$ is a series and $G \in \hat{\mathcal{I}}_{C,p}^a$ for some a, we have that

$$\frac{\partial G}{\partial w_{\sigma(1)}}, \frac{\partial G}{\partial w_{\sigma(2)}} \in \hat{\mathcal{I}}_{C,p}^{a-1}$$

and

$$\frac{\partial G}{\partial w_{\sigma(3)}} \in \hat{\mathcal{I}}_{C,p}^a$$

Thus

$$b_t w_{\sigma(1)}^{ta-d} w_{\sigma(2)}^{tb-e} w_{\sigma(3)}^{tc-f} \in \hat{\mathcal{I}}_{C,p}^s$$

for all t, and $F_1 \in \hat{\mathcal{I}}_{C,p}^s$.

The independence of the conditions from permissible parameters (u, v) at $\Phi_X(p)$ follows from cases 2.1 - 3.3 of Lemma 6.7, with m^τ replaced by $\hat{\mathcal{I}}_{C,p}^s = (x, y)^s$ in the formulas of these cases. $\qquad\square$

Example 6.15. *If p is a 2 point, the condition $F_p \in I^s$ where $I \subset \hat{\mathcal{O}}_{X,p}$ is a reduced ideal can depend on the choice of permissible parameters at p.*

Proof. Consider

$$u = xy, v = z^2 + xz$$

the Jacobian is $J = (xz, y(2z + x), x(2z + x))$.

$$x^2 = 2xz + x^2 - 2xz \in J.$$

$\sqrt{J} = (x, yz)$. (x, y, z) are permissible parameters for (u, v) at p. Let $I = (x, z)$. $F \in I^2$.

We have other permissible parameters $(x, y, \bar{z})$ at p, where $\bar{z} = z - y$. Then $I = (\bar{z} + y, x)$. The normalized form of v with respect to these new parameters is

$$u = xy, v = xy + F$$

where

$$F = [(\bar{z} + y)^2 + x\bar{z}] \notin I^2.$$

$\square$

Lemma 6.16. *Suppose that p is a 2 point, and C is a curve, making SNCs with the 2 curve through p. Then the condition $F_p \in \hat{\mathcal{I}}_{C,p}^s$ with $s \in \mathbf{N}$ is independent of permissible parameters (u, v) at $\Phi_X(p)$ and permissible parameters (x, y, z) at p for (u, v) such that $\hat{\mathcal{I}}_{C,p} = (x, z)$.*

We will call parameters as in Lemma 6.16 permissible parameters for C at p.

Proof. Suppose that (u, v) are permissible parameters at $\Phi_X(p)$. We will first show that this is independent of such permissible parameters at p for (u, v). Suppose that (x, y, z) and (x_1, y_1, z_1) are permissible parameters for (u, v) at p such that $\hat{\mathcal{I}}_{C,p} = (x, z) = (x_1, z_1)$ and

$$\begin{aligned} u &= (x^a y^b)^m \\ v &= P(x^a y^b) + x^c y^d F \end{aligned}$$

with $F \in \hat{\mathcal{I}}_{C,p}^s = (x, y)^s$. We have

$$x = \alpha x_1, y = \beta y_1, z = z(x_1, y_1, z_1) = \omega z_1 + \gamma x_1$$

where α, β, ω are units in $\hat{\mathcal{O}}_{X,p}$ and $\gamma \in \hat{\mathcal{O}}_{X,p}$. If $G \in \hat{\mathcal{O}}_{X,p}$ is such that $G \in (x_1, z_1)^a$ then

$$\frac{\partial G}{\partial x_1} \in (x_1, z_1)^{a-1}$$

and

$$\frac{\partial G}{\partial y_1} \in (x_1, z_1)^a.$$

Thus

$$\frac{\partial^{t(a+b)-c-d}(\alpha^c \beta^d F)}{\partial x_1^{ta-c} \partial y_1^{tb-d}} \in (x_1, z_1)^{s-(ta-c)}$$

In (21) of Lemma 6.6, we have $b_t = 0$ if $s > (ta - c)$, so that $F_1 \in (x_1, z_1)^s$.

The independence of the condition $F \in \hat{\mathcal{I}}^s_{C,p}$ from choice of permissible parameters (u, v) at $\Phi_X(p)$ follows from cases 2.1-2.3 of Lemma 6.7, with m^r replaced by $\hat{\mathcal{I}}^s_{C,p} = (x, z)^s$ is the formulas of these cases. $\qquad \square$

Let $B_2(X)$ be the (possibly not closed) curve of 2 points in X, $B_3(X) = \{p_1, \ldots, p_r\}$ the set of 3 points in X. Let $\overline{B}_2(X) = B_2(X) \cup B_3(X)$ be the Zariski closure of $B_2(X)$ in X.

Definition 6.17. *Suppose that $Z \subset E_X$ is a reduced closed subscheme of dimension ≤ 1 and $p \in E_X$. We will say that Z makes SNCs with $\overline{B}_2(X)$ at p if*

1. *All components of Z are nonsingular at p.*
2. *If $C_1, \ldots, C_s$ are the curves of Z containing p and $D_1, \ldots, D_t$ are the components of $\overline{B}_2(X)$ containing p, then $C_1, \ldots, C_s, D_1, \ldots, D_t$ have independent tangent directions at p.*

We will say that Z makes SNCs with $\overline{B}_2(X)$ if Z makes SNCs with $\overline{B}_2(X)$ at p for all $p \in E_X$.

Definition 6.18. *Suppose that $p \in X$, U is an affine neighborhood of p in X, and $\sigma : V \to U$ is an étale cover. Then we will say that V is an étale neighborhood of p. Suppose that $D \subset X$. We will write $D \cap V$ to denote $\sigma^{-1}(D \cap U)$.*

Definition 6.19. *(c.f. Chapter 3, Section 6 [29].) Suppose that V is an affine k-variety. $x_1, \ldots, x_n \in \Gamma(V, \mathcal{O}_V)$ are uniformizing parameters on V if the natural morphism $V \to spec(k[x_1, \ldots, x_n])$ is étale.*

Lemma 6.20. *Suppose that (x, y, z) are permissible parameters at p for (u, v) such that $y, z \in \mathcal{O}_{X,p}$. Then there exists an affine neighborhood U of p and an étale cover V of U such that (x, y, z) are uniformizing parameters on V.*

Proof. With the notations of Definition 6.4, let $\bar{a} = a$ if p is a 1 point, $\bar{a} = ma$ if p is a 2 or 3 point. There exists a unit $\lambda \in \mathcal{O}_{X,p}$ and $\tilde{x} \in \mathcal{O}_{X,p}$ such that $x^{\bar{a}} = \lambda \tilde{x}^{\bar{a}}$. There exists an affine neighborhood U_1 of p such that $\tilde{x}, y, z, \lambda \in R = \Gamma(U_1, \mathcal{O}_X)$ and λ is a unit in R. Set $S = R[\lambda^{\frac{1}{\bar{a}}}]$, $V_1 = \text{spec}(S)$. $f : V_1 \to U_1$ is an étale cover. $k[x, y, z] \to S$ defines a morphism $g : V_1 \to \mathbf{A}^3$. Let a be the origin of $\mathbf{A}^3$. $q \in g^{-1}(a)$ if and only if $x, y, z \in m_q$ which holds if and only if $\tilde{x}, y, z \in m_q$. Thus $g^{-1}(a) = f^{-1}(p)$. $\hat{\mathcal{O}}_{V_1,q} = k[[\tilde{x}, y, z]] = k[[x, y, z]]$ for all $q \in g^{-1}(a)$. Thus g is étale at all points of $g^{-1}(a)$. Since this is an open condition, (Proposition 4.5 [21]) there exists a closed set Z_1 of V_1 which is disjoint from $f^{-1}(p)$ such that $g \mid (V_1 - Z_1)$ is étale. Let U be an affine neighborhood of p in U_1 which is disjoint from the closed set $f(Z_1)$. Let $V = f^{-1}(U)$. Then V is an étale cover of U on which x, y, z are uniformizing parameters. $\qquad \square$

Proposition 6.21. *$S_r(X) \cap (X - \overline{B}_2(X))$ is Zariski closed in $X - \overline{B}_2(X)$ and $S_r(X) \cap B_2(X)$ is Zariski closed in $B_2(X)$. Thus $S_r(X)$ is a constructible set.*

Proof. First suppose that p is a 1 point. Then there are regular parameters $\tilde{x}, y, z$ in $\mathcal{O}_{X,p}$, permissible parameters x, y, z at p, and a unit $\lambda \in \mathcal{O}_{X,p}$ such that

$$u = x^a = \lambda \tilde{x}^a$$
$$v = P(x) + x^b F_p(x, y, z).$$

$\tilde{x}, y, z$ are uniformizing parameters in an affine neighborhood U of p, and there exists an étale neighborhood $\sigma : V = \mathrm{spec}(S) \to U$ of p such that (x, y, z) are uniformizing parameters on V, $x = 0$ is a local equation of $E_X \cap V$ in V. Let

$$I = \left(\frac{\partial^{i+j+k} v}{\partial x^i \partial y^j \partial z^k} \mid j + k > 0, i + j + k \le b + r - 1 \right) \subset S,$$

$Z = V(I) \subset V$.

Suppose that $p' \in E_X \cap V$. Then if $\alpha = y(p'), \beta = z(p')$, we have that

$$u = x^a$$
$$v = \sum \frac{1}{i!j!k!} \frac{\partial^{i+j+k} v}{\partial x^i \partial y^j \partial z^k}(0, \alpha, \beta) x^i (y - \alpha)^j (z - \beta)^k$$

and

$$v - v(\sigma(p')) = P_{p'}(x) + x^b F_{p'}.$$

$\nu(p') \ge r$ if and only if $p' \in V(I)$. Let $Z_1 = \sigma(Z)$. $S_r(X) \cap U = Z_1$ is closed in U.

Now suppose that p is a 2 point. Then there are regular parameters $\tilde{x}, y, z$ in $\mathcal{O}_{X,p}$ and permissible parameters x, y, z at p and a unit λ in $\mathcal{O}_{X,p}$ such that

$$u = (x^a y^b)^m = \lambda(\tilde{x}^a y^b)^m$$
$$v = P(x^a y^b) + x^c y^d F_p(x, y, z)$$

There exists an étale neighborhood $\sigma : V = \mathrm{spec}(S) \to U$ of p such that (x, y, z) are uniformizing parameters on V, $xy = 0$ is a local equation of $E_X \cap V$ in V. Let C be the 2 curve in X containing p. Suppose that $p' \in C \cap V$. Then if $\beta = z(p')$, we have that

$$u = (x^a y^b)^m$$
$$v - v(\sigma(p')) = P_{p'}(x^a y^b) + x^c y^d F_{p'}.$$

$$v = \sum \frac{1}{i!j!k!} \frac{\partial^{i+j+k} v}{\partial x^i \partial y^j \partial z^k}(0, 0, \beta) x^i y^j (z - \beta)^k.$$

Let

$$I = \left(\frac{\partial^{i+j+k} v}{\partial x^i \partial y^j \partial z^k} \mid k > 0 \text{ or } k = 0 \right.$$
$$\left. \text{and } a(d + j) - b(c + i) = 0 \text{ and } i + j + k \le c + d + r - 1 \right) \subset S,$$

$p' \in C$ and $\nu(p') \ge r$ if and only if $p' \in V(I) \cap C$. $Z = V(I) \subset V$. Let $Z_1 = \sigma(Z)$. $S_r(X) \cap C \cap U = C \cap Z_1 \cap U$ is closed in $C \cap U$. $\square$

Lemma 6.22. *Suppose that $p \in E_X$ is a 1 point or a 2 point.*

1. *Suppose that (x, y, z) are permissible parameters at p, $I \subset \hat{\mathcal{O}}_{x,p}$ is a reduced ideal and $F_p \in I^r$ for some $r \ge 2$. Then $\hat{I}_{\overline{S_r(X)},p} \subset I$.*
2. *Suppose that (x, y, z) are permissible parameters at p, $I = (x, f(y, z)) \subset \hat{\mathcal{O}}_{x,p}$ is a reduced ideal and $F_p \in (x) + I^2$. Then $\hat{I}_{\overline{S_2(X)},p} \subset I$.*

Proof. Suppose that $F_p \in I^r$ for some $r \geq 2$. First assume that p is a 1 point. Since $x \in I$ and $r \leq \nu(p)$, we can make a permissible change of parameters, and renormalize to get that $y, z \in \mathcal{O}_{X,p}$ and $F_p \in I^r$.

$$\hat{\mathcal{I}}_{\overline{S_r(X)},p} = \sqrt{\left(\frac{\partial^{i+j+k} F_p}{\partial x^i \partial y^j \partial z^k} \mid i+j+k \leq r-1, j+k > 0\right)}. \tag{32}$$

$F_p \in I^r$ implies

$$\frac{\partial^{i+j+k} F_p}{\partial x^i \partial y^j \partial z^k} \in I$$

for all $i + j + k \leq r - 1$. Thus $\hat{\mathcal{I}}_{\overline{S}_r(X),p} \subset I$.

Suppose that p is a 2 point.

$$u = (x^a y^b)^m$$
$$v = P(x^a y^b) + x^c y^d F_p$$

with $F_p \in I^r$ and $r \geq 2$, $F_p \in I^r$ implies

$$xy \in \hat{\mathcal{I}}_{\mathrm{sing}(\Phi_X),p}$$
$$= \sqrt{x^{ma+c-1} y^{mb+d-1}\left((ad-bc)F_p + ay\frac{\partial F_p}{\partial y} - bx\frac{\partial F_p}{\partial x}, y\frac{\partial F_p}{\partial z}, x\frac{\partial F_p}{\partial z}\right)} \subset I,$$

so that $x \in I$ or $y \in I$. Without loss of generality, $x \in I$.

There exist permissible parameters $(\overline{x}, \overline{y}, \overline{z})$ at p such that $\overline{y}, \overline{z} \in \mathcal{O}_{X,p}$, $\overline{x}^a \overline{y}^b = x^a y^b$ and

$$x = \sigma \overline{x}, y = \tau \overline{y}, z = \overline{z} + h$$

for some series $\sigma, \tau, h \in \hat{\mathcal{O}}_{X,p}$ with

$$\sigma \equiv 1 \bmod m^{\overline{a}},$$
$$\tau \equiv 1 \bmod m^{\overline{a}}$$
$$h \equiv 0 \bmod m^{\overline{a}}$$

where $m = m_p \hat{\mathcal{O}}_{X,p}$,

$$\overline{a} \geq \frac{r+c}{a}(a+b) - (c+d).$$

We have

$$u = (\overline{x}^a \overline{y}^b)^m$$
$$v = P(\overline{x}^a \overline{y}^b) + \overline{x}^c \overline{y}^d [\sigma^c \tau^d F_p(\sigma \overline{x}, \tau \overline{y}, \overline{z} + h)]$$
$$\sigma^c \tau^d F_p(\sigma \overline{x}, \tau \overline{y}, \overline{z} + h) \equiv F_p(\overline{x}, \overline{y}, \overline{z}) \bmod m^{\overline{a}}$$

Let $v = P_1(\overline{x}^a \overline{y}^b) + \overline{x}^c \overline{y}^d F_1$ be the normalized form of v. Since $F_p(\overline{x}, \overline{y}, \overline{z})$ is normalized, we can only remove terms

$$(\overline{x}^a \overline{y}^b)^t / \overline{x}^c \overline{y}^d$$

with $t(a+b) - (c+d) \geq \overline{a}$ from $\sigma^c \tau^d F_p(\sigma \overline{x}, \tau \overline{y}, \overline{z} + h)$ to construct F_1. Since this condition implies

$$at - c \geq r$$

we have $F_1 \in I^r$. We can thus assume that $y, z \in \mathcal{O}_{X,p}$.

Set

$$w = \frac{v - P_t(x^a y^b)}{x^c y^d}$$

with $t > c + d + r$. Thus

$$w = F_p + x^m y^m h(x, y)$$

with $m > r$. $F_p \in I^r$ implies $w \in I^r$ which implies that

$$\frac{\partial^{i+j+k} w}{\partial x^i \partial y^j \partial z^k} \in I$$

if $i + j + k \le r - 1$.

There exists an étale neighborhood $\sigma : V \to U$ of p such that (x, y, z) are uniformizing parameters on V, $xy = 0$ is a local equation of $E_X \cap V$ in V.

Suppose that

$$q \in V\left(x, \frac{\partial^{i+j+k} w}{\partial x^i \partial y^j \partial z^k} \mid i + j + k \le r - 1\right) \subset V$$

is a 1 point in V. q has permissible parameters $(\tilde{x}, \tilde{y}, \tilde{z})$ defined by

$$\tilde{y} = y - \alpha, \tilde{z} = z - \beta, \tilde{x} = x y^{\frac{b}{a}} \tag{33}$$

for some $\alpha, \beta \in k$. Thus

$$
\begin{aligned}
u &= \tilde{x}^{am} \\
v &= P_t(\tilde{x}^a) + \alpha^{d - \frac{cb}{a}} \tilde{x}^c w(\tilde{x} \alpha^{-\frac{b}{a}}, \alpha, \beta) \\
&\quad + \tilde{x}^c \left[(\tilde{y} + \alpha)^{d - \frac{cb}{a}} w(x, y, z) - \alpha^{d - \frac{cb}{a}} w(\tilde{x} \alpha^{-\frac{b}{a}}, \alpha, \beta)\right]
\end{aligned}
$$

$\nu_q(w) \ge r$ implies $q \in S_r(V)$.

Suppose that

$$q \in V\left(x, \frac{\partial^{i+j+k} w}{\partial x^i \partial y^j \partial z^k} \mid i + j + k \le r - 1\right) \subset V$$

is a 2 point in V. q has permissible parameters $(x, y, \tilde{z})$ where

$$\tilde{z} = z - \beta.$$

for some $\beta \in k$.

$$
\begin{aligned}
u &= (x^a y^b)^m \\
v &= P_t(x^a y^b) + x^c y^d \left[\sum_{(c+i)b - a(j+d) = 0} \frac{\partial^{i+j} w}{\partial x^i \partial y^j}(q) x^i y^j\right] \\
&\quad + x^c y^d \left[w - \sum_{(c+i)b - a(j+d) = 0} \frac{\partial^{i+j} w}{\partial x^i \partial y^j}(q) x^i y^j\right]
\end{aligned}
$$

Again, $\nu_q(w) \ge r$ implies $q \in S_r(V)$. So

$$V(I) \subset V\left(x, \frac{\partial^{i+j+k} w}{\partial x^i \partial y^j \partial z^k} \mid i + j + k \le r - 1\right) \subset \overline{S}_r(V)$$

implies

$$\hat{I}_{\overline{S}_r, p} \subset \sqrt{\left(x, \frac{\partial^{i+j+k} w}{\partial x^i \partial y^j \partial z^k} \mid i + j + k \le r - 1\right)} \subset I.$$

We now prove 2. Suppose that the assumptions of 2. hold. If p is a 1 point, then (32) implies $\hat{\mathcal{I}}_{\overline{S}_2(X),p} \subset I$.

If p is a 2 point, then arguing as in the proof of 1., we set

$$w = \frac{v - P_t(x^a y^b)}{x^c y^d}$$

and conclude that $\frac{\partial w}{\partial y}, \frac{\partial w}{\partial z} \in I$. Suppose that $q \in V(x, \frac{\partial w}{\partial y}, \frac{\partial w}{\partial z})$ is a 1 point. Then there exists $\overline{c} \in k$ such that $\text{mult}(w - \overline{c}x) \geq 2$, where the multiplicity is computed at q. q has permissible parameters as in (33).

$$x \equiv \alpha^{-\frac{b}{a}} \tilde{x} \mod m_q^2 \hat{\mathcal{O}}_{X,q}$$

implies $\text{mult}(w - \overline{c}\alpha^{-\frac{b}{a}} \tilde{x}) \geq 2$, so that $q \in \overline{S}_2(X)$. We have a simpler argument if $q \in V(x, \frac{\partial w}{\partial y}, \frac{\partial w}{\partial z})$ is a 2 point. Thus

$$\hat{\mathcal{I}}_{\overline{S}_2(X),p} \subset \sqrt{(x, \frac{\partial w}{\partial y}, \frac{\partial w}{\partial z})} \subset I.$$

$\square$

Lemma 6.23. *Suppose that $C \subset X$ is a curve and there exists $p \in X$ such that $F_p \in \hat{\mathcal{I}}_{C,p}^r$ with $r \geq 2$. Then $C \subset E_X$.*

1. *Suppose that $C \subset E_X$ is a curve and there exists $p \in X$ such that $F_p \in \hat{\mathcal{I}}_{C,p}^r$ with $r \geq 1$. Suppose that $q \in C$ is a 1 point. Then $F_q \in \hat{\mathcal{I}}_{C,q}^r$.*

2. *Suppose that C is a 2 curve and there exists $p \in C$ such that $F_p \in \hat{\mathcal{I}}_{C,p}^r$ with $r \geq 1$. If $q \in C$ is a 2 point, then $F_q \in \hat{\mathcal{I}}_{C,q}^r$.*

Proof. We will first show that 1 or 2 hold for all but finitely many $q \in C$.

First Suppose that p is a 1 point. By Lemma 6.13, we may assume that $y, z \in \mathcal{O}_{X,p}$.

$$u = x^a, v = P(x) + x^b F_p.$$

In an étale neighborhood U of p, (x, y, z) are uniformizing parameters. Let $I = \hat{\mathcal{I}}_{C,p}$. If $F_p \in I^r$ with $r \geq 2$ then

$$\frac{\partial F_p}{\partial y}, \frac{\partial F_p}{\partial z} \in I^{r-1} \subset I.$$

$$u \in \hat{\mathcal{I}}_{\text{sing}(\Phi_X),p} = \sqrt{x^{a-1+b}(\frac{\partial F_p}{\partial y}, \frac{\partial F_p}{\partial z})} \subset I.$$

implies $x \in I$.

Now assume that $F_p \in \hat{\mathcal{I}}_{C,p}^r$ with $r \geq 1$ so that $C \subset E_X$, either by assumption if $r = 1$, or by the above argument if $r \geq 2$. Thus $x \in \hat{\mathcal{I}}_{C,p}$. Set $w = \frac{v - P_{b+r}(x)}{x^b}$.

After possibly replacing U with a smaller étale neighborhood of p, there exists a reduced ideal $J = (x, f) \subset \Gamma(U, \mathcal{O}_U)$ such that $J\hat{\mathcal{O}}_{X,p} = I$. If $q \in V(J) \subset U$, then q has regular parameters $(x, y - \alpha, z - \beta)$ for some $\alpha, \beta \in k$. $w \in J^r \hat{\mathcal{O}}_{X,p}$

implies $w \in J^r$ (since J is a complete intersection implies J^r has no embedded components). Since $\nu(F_q) \geq r$, we have

$$F_q = w - \sum_{i \geq r} \frac{1}{i!} \frac{\partial^i w}{\partial x^i}(0, \alpha, \beta) x^i \in J^r \hat{\mathcal{O}}_{X,q}.$$

Thus for all but finitely many $q \in C$, 1. holds.

Now suppose that p is a 2 point.

$$u = (x^a y^b)^m, v = P(x^a y^b) + x^c y^d F$$

where $F = F_p$.

Suppose that $F \in \hat{\mathcal{I}}^r_{C,p}$ with $r \geq 2$. Then $F, \frac{\partial F}{\partial x}, \frac{\partial F}{\partial y}, \frac{\partial F}{\partial z} \in \hat{\mathcal{I}}_{C,p}$. By (15),

$$u \in \hat{\mathcal{I}}_{\text{sing}(\Phi_X),p} \subset \sqrt{(F, \frac{\partial F}{\partial x}, \frac{\partial F}{\partial y}, \frac{\partial F}{\partial z})} \subset \hat{\mathcal{I}}_{C,p}$$

implies $x \in \hat{\mathcal{I}}_{C,p}$ or $y \in \hat{\mathcal{I}}_{C,p}$.

Now assume that $F_p \in \hat{\mathcal{I}}^r_{C,p}$ with $r \geq 1$. Then $C \subset E_X$, either by assumption if $r = 1$, or by the above argument if $r \geq 2$. Thus we have x or $y \in \hat{\mathcal{I}}_{C,p}$. Suppose that $x \in \hat{\mathcal{I}}_{C,p}$. As in the proof of Lemma 6.22, we may assume that $y, z \in \mathcal{O}_{X,p}$. Set $t = c + d + r$, $w = \frac{v - P_t(x^a y^b)}{x^c y^d}$. There exists an étale neighborhood U of p such that $u = xy = 0$ is a local equation of E_X in U, (x, y, z) are uniformizing parameters in U, and a reduced ideal

$$J = (x, f) \subset \Gamma(U, \mathcal{O}_U)$$

such that $J\hat{\mathcal{O}}_{X,p} = \hat{\mathcal{I}}_{C,p}$. $w \in J^r$ since J is a complete intersection.

Suppose that C is not a 2 curve, so that $\hat{\mathcal{I}}_{C,p} \neq (x, y)$. After possibly replacing U with a smaller étale neighborhood of p, we can assume that $U \cap C \cap \overline{B}_2(X) = p$.

If $q \neq p$, and $q \in V(J) \subset U$, then q has regular parameters $(x, y - \alpha, z - \beta)$ such that $\alpha \neq 0$. $\Phi_X(q)$ has permissible parameters

$$u_1 = u, v_1 = v - v(\Phi_X(q))$$

with permissible parameters $(\overline{x}, \overline{y}, \overline{z})$, defined by

$$x = \overline{x}(\overline{y} + \alpha)^{\frac{-b}{a}}, \overline{y} = y - \alpha, \overline{z} = z - \beta$$

$(\overline{y} + \alpha)^{d - \frac{bc}{a}} w \in \hat{\mathcal{I}}^r_{C,q}$, $\overline{x} \in \hat{\mathcal{I}}_{C,q}$, and

$$F_q = (\overline{y} + \alpha)^{d - \frac{bc}{a}} w - \Omega(\overline{x})$$

with mult$(\Omega) \geq r$, which implies $F_q \in \hat{\mathcal{I}}^r_{C,q}$.

Now suppose that C is a 2 curve, so that $\hat{\mathcal{I}}_{C,p} = (x, y)$. If $q \in V(J)$, then $(u, v - v(\Phi_X(q)))$ are permissible parameters at $\Phi_X(q)$, and q has permissible parameters $(x, y, \overline{z})$ with $\overline{z} = z - \alpha$. $w \in \hat{\mathcal{I}}^r_{C,q}$, $x \in \hat{\mathcal{I}}_{C,q}$ and

$$F_q = w - \frac{\Omega(x^a y^b)}{x^c y^d}$$

for some Ω with mult$(\frac{\Omega(x^a y^b)}{x^c y^d}) \geq r$. Thus $F_q \in \hat{\mathcal{I}}^r_{C,q}$.

Now suppose that p is a 3 point.
$$u = (x^a y^b z^c)^m, v = P(x^a y^b z^c) + x^d y^e z^f F$$
We can assume that $y, z \in \mathcal{O}_{X,p}$. $F \in \hat{\mathcal{I}}^r_{C,p}$ with $r \geq 2$ implies that
$$F, \frac{\partial F_p}{\partial x}, \frac{\partial F_p}{\partial y}, \frac{\partial F_p}{\partial z} \in \hat{\mathcal{I}}_{C,p}$$
By (16),
$$u \in \hat{\mathcal{I}}_{\text{sing}(f),p} \subset \sqrt{(F, \frac{\partial F_p}{\partial x}, \frac{\partial F_p}{\partial y}, \frac{\partial F_p}{\partial z})} \subset \hat{\mathcal{I}}_{C,p}$$
Thus x, y or $z \in \hat{\mathcal{I}}_{C,p}$.

Now suppose that $F \in \hat{\mathcal{I}}^r_{C,p}$ with $r \geq 1$. If $r = 1$, then x, y or $z \in \hat{\mathcal{I}}_{C,p}$ by assumption. If $r \geq 2$, then x, y or $z \in \hat{\mathcal{I}}_{C,p}$ by the above argument. Suppose that $x \in \hat{\mathcal{I}}_{C,p}$. Set $t = d + e + f + r$,
$$w = \frac{v - P_t(x^a y^b z^c)}{x^d y^e z^f}.$$
There exists an étale neighborhood U of p such that (x, y, z) are uniformizing parameters in U, $u = xyz = 0$ is a local equation of E_X in U and $J = \Gamma(U, \mathcal{I}_C) = (x, f)$ is a complete intersection. $w \in J^r$ since J is a complete intersection.

Suppose that C is not a 2 curve. Then we can assume that $U \cap \overline{B}_2(X) \cap C = p$. If $q \in V(J) \subset U$ and $q \neq p$, then $\Phi_X(q)$ has permissible parameters
$$u_1 = u, v_1 = v - v(\Phi_X(q))$$
with permissible parameters $(\overline{x}, \overline{y}, \overline{z})$ at q, with
$$x = \overline{x}(\overline{y} + \alpha)^{\frac{-b}{a}}(\overline{z} + \beta)^{-\frac{c}{a}}, y = \overline{y} + \alpha, z = \overline{z} + \beta$$
with $\alpha, \beta \neq 0$. $w \in \hat{\mathcal{I}}^r_{C,q}, \overline{x} \in \hat{\mathcal{I}}_{C,q}$ and
$$F_q = (\overline{y} + \alpha)^{e - \frac{bd}{a}}(\overline{z} + \beta)^{f - \frac{cd}{a}} w - \Omega(\overline{x}).$$
$\text{mult}(\Omega) \geq r$ implies $F_q \in \hat{\mathcal{I}}^r_{C,q}$.

Suppose that C is a 2 curve, $\hat{\mathcal{I}}_{C,p} = (x, y)$, $q \in V(J)$. Then
$$u_1 = u, v_1 = v - v(\Phi_X(q))$$
are permissible parameters at $\Phi_X(q)$, with permissible parameters $(\overline{x}, y, \overline{z})$ at q,
$$\overline{z} = z - \alpha, x = \overline{x}(\overline{z} + \alpha)^{\frac{-c}{a}}$$
$$u = (\overline{x}^a y^b)^m = (\overline{x}^{\overline{a}} y^{\overline{b}})^{\overline{m}},$$
with $(\overline{a}, \overline{b}) = 1$, $w \in \mathcal{I}^r_{C,q} = (\overline{x}, y)^r$ implies
$$F_q = (z + \alpha)^{f - \frac{dc}{a}} w - \frac{\Omega(\overline{x}^{\overline{a}} y^{\overline{b}})}{\overline{x}^d y^e}$$
with $\text{mult}(\frac{\Omega(\overline{x}^{\overline{a}} y^{\overline{b}})}{\overline{x}^d y^e}) \geq r$. Thus $F_q \in \hat{\mathcal{I}}^r_{C,q}$.

We conclude that 1. or 2. hold for all but finitely many $q \in C$.

Suppose that $q \in C$ is a 1 point. We have at q,

$$u = x^a, v = P(x) + x^b F$$

with $x \in \hat{\mathcal{I}}_{C,q}$, $y, z \in \mathcal{O}_{X,q}$. There exists an étale neighborhood U of q such that (x, y, z) are uniformizing parameters on U, $x = 0$ is a local equation of E_X, $J = \Gamma(U, \mathcal{I}_C) = (x, f)$ is a complete intersection. 1. holds for all $q \neq q' \in U \cap E_X$ and

$$w = \frac{v - P_{b+r}(x)}{x^b} \in \Gamma(U, \mathcal{O}_U).$$

For $q' \in V(J) \subset U$ with $q' \neq q$,

$$u, v_1 = v - v(\Phi_X(q'))$$

are permissible parameters at $\Phi_X(q')$ and $(x, y - \alpha, z - \beta)$ are permissible parameters for (u, v_1) at q'.

$$F_{q'} = w - P_1(x) \in \hat{\mathcal{I}}^r_{C,q'}$$

where

$$P_1(x) = \sum_{i=0}^{\infty} \frac{1}{i!} \frac{\partial^i w}{\partial x^i}(0, \alpha, \beta) x^i = \sum_{i=0}^{\infty} a_i x^i.$$

Set

$$\Lambda = w - \sum_{i=0}^{r-1} a_i x^i \in \Gamma(U, \mathcal{O}_U).$$

$\Lambda \in \hat{\mathcal{I}}^r_{C,q'}$ implies $\Lambda \in J^r$, so that $\Lambda \in \hat{\mathcal{I}}^r_{C,p}$ implies

$$\frac{\partial^i \Lambda}{\partial x^i}(0, 0, 0) = 0$$

for $i < r$, and

$$F_p = \Lambda - \sum_{i=r}^{\infty} \frac{1}{r!} \frac{\partial^i \Lambda}{\partial x^i}(0, 0, 0) x^i \in \hat{\mathcal{I}}^r_{C,p}.$$

Now suppose that C is a 2 curve. Suppose that $q \in C$ is a 2 point. We have at q,

$$
\begin{aligned}
u &= (x^a y^b)^m \\
v &= P(x^a y^b) + x^c y^d F
\end{aligned}
$$

with $\hat{\mathcal{I}}_{C,q} = (x, y)$, $y, z \in \mathcal{O}_{X,q}$. There exists an étale neighborhood U of p such that (x, y, z) are uniformizing parameters on U, $xy = 0$ is a local equation of $E_X \cap U$, $J = \Gamma(U, \mathcal{I}_C) = (x, y)$. 2. holds for all 2 points $q \neq q' \in U \cap E_X$.

$$w = \frac{v - P_{c+d+r}(x^a y^b)}{x^c y^d} \in \Gamma(U, \mathcal{O}_U).$$

For $q' \in V(J) \subset U$ with $q' \neq q$, there exist permissible parameters $(x, y, z - \beta)$ at q' for $(u, v - v(\Phi_X(q)))$.

$$F_{q'} = w - \frac{P_1(x^a y^b)}{x^c y^d} \in \hat{\mathcal{I}}^r_{C,q'}$$

where

$$P_1(x^a y^b) = \sum_{i=0}^{\infty} a_i(x^a y^b)^i.$$

is a series. Set $\Lambda = w - \dfrac{\sum_{i=0}^{c+d+r-1} a_i(x^a y^b)^i}{x^c y^d} \in \Gamma(U, \mathcal{O}_U)$. $\Lambda \in \hat{\mathcal{I}}_{C,q}^r$ implies $\Lambda \in J^r$, so that $\Lambda \in \hat{\mathcal{I}}_{C,q}^r$ and

$$\frac{\partial^{(a+b)i-c-d}\Lambda}{\partial x^{ai-c}\partial y^{bi-d}}(0,0,0) = 0$$

for $(a+b)i - c - d < r$, so that

$$F_p = \Lambda - \sum_{(a+b)i-c-d \geq r} \frac{1}{(ai-c)!(bi-d)!} \frac{\partial^{(a+b)i-c-d}\Lambda}{\partial x^{ai-c}\partial y^{bi-d}}(0,0,0)x^{ai-c}y^{bi-d} \in \hat{\mathcal{I}}_{C,p}^r.$$

$\square$

Lemma 6.24. *Suppose that $r \geq 2$, $C \subset \bar{S}_r(X)$ is a nonsingular curve, and $p \in C$ is a 1 point, so that there exist permissible parameters x, y, z at p such that*

$$\begin{aligned} u &= x^a \\ v &= P(x) + x^c F \end{aligned}$$

where $\hat{\mathcal{I}}_{C,p} = (x, z)$. Then

$$F_p = a_r(x, y) + a_{r-1}(x, y)z + \cdots + a_1(x, y)z^{r-1} + g(x, y, z)z^r$$

where

$$x^i \mid a_i \text{ for } 1 \leq i \leq r - 1,$$

and $x^{r-1} \mid a_r$.

Proof. There exist permissible parameters $(x, \bar{y}, \bar{z})$ at p such that $\bar{y}, \bar{z} \in \mathcal{O}_{X,p}$ and $(x, \bar{z}) = \hat{\mathcal{I}}_{C,p}$. Then there exists $a, b \in \hat{\mathcal{O}}_{X,p}$ such that $\bar{z} = ax + bz$ where b is a unit. Assume that the conclusions of the Lemma are true for the variables $(x, \bar{y}, \bar{z})$. Then substituting for x, y, z we get the conclusions of the Lemma for (x, y, z), so we may suppose that $y, z \in \mathcal{O}_{X,p}$.

There exists an étale neighborhood U of p such that x, y, z are uniformizing parameters in U, $x = 0$ is a local equation of E_X in U, $x = z = 0$ are equations of $C \cap U$. If $p' \in U \cap C$, and $\alpha = y(p')$, then $(x, y_1 = y - \alpha, z)$ are permissible parameters at p'.

$$F_{p'} = \sum_{i \geq 0, j+k > 0} \frac{1}{(c+i)!j!k!} \frac{\partial^{c+i+j+k}v}{\partial x^{c+i}\partial y^j \partial z^k}(0, \alpha, 0)x^i y_1^j z^k.$$

$\nu(p') \geq r$ for $p' \in C \cap U$ implies that, if $j + k > 0$ and $i + j + k < r$, then

$$\frac{\partial^{c+i+j+k}v}{\partial x^{c+i}\partial y^j \partial z^k}(0, \alpha, 0) = 0$$

for infinitely many α, so that

$$\frac{\partial^{c+i+j+k}v}{\partial x^{c+i}\partial y^j \partial z^k}(0, y, 0) = 0$$

in U, if $j + k > 0$ and $i + j + k < r$. Thus

$$\frac{\partial^{c+i+k} v}{\partial x^{c+i} \partial z^k}(0, y, 0) = 0$$

in U if $i + k < r$, $k > 0$, so that

$$\frac{\partial^{c+i+j+k} v}{\partial x^{c+i} \partial y^j \partial z^k}(0, y, 0) = \frac{\partial^j}{\partial y^j}\left[\frac{\partial^{c+i+k} v}{\partial x^{c+i} \partial z^k}(0, y, 0)\right] = 0$$

if $i + k < r$, $k > 0$ and $j \geq 0$. Thus

$$\frac{\partial^{c+i+j+k} v}{\partial x^{c+i} \partial y^j \partial z^k}(0, 0, 0) = 0$$

if $k > 0$, $i < r - k$, $j \geq 0$,

$$\frac{\partial^{c+i+1} v}{\partial x^{c+i} \partial y}(0, y, 0) = 0$$

if $i < r - 1$, so that

$$\frac{\partial^{c+i+j} v}{\partial x^{c+i} \partial y^j}(0, 0, 0) = 0$$

if $i < r - 1$, $j > 0$, and the conclusions of the Lemma follow.   □

Lemma 6.25. *Suppose that* $r \geq 1$, $C \subset X$ *is a 2 curve such that* $\nu(p) \geq r$ *if* $p \in C$ *is a 2 point* $(C \subset \overline{S}_r(X)$ *if* $r \geq 2)$, *and* $p \in C$ *is a 2 point, so that there exist permissible parameters* x, y, z *at* p *such that*

$$\begin{aligned} u &= (x^a y^b)^m \\ v &= P(x^a y^b) + x^c y^d F \end{aligned}$$

where $\hat{\mathcal{I}}_{C,p} = (x, y)$. *Then there exists a series* $\tau(z)$ *with mult* $\tau(z) \geq 1$ *such that*

$$F = \begin{cases} \tau(z)x^{i_0}y^{j_0} + \sum_{i+j \geq r} a_{ij}(z)x^i y^j & \begin{aligned}&\textit{if there exist nonnegative} \\ &\textit{integers } (i_0, j_0) \textit{ such that} \\ &i_0 + j_0 = r - 1 \\ &\textit{and } a(d + j_0) - b(c + i_0) = 0\end{aligned} \\[2ex] \sum_{i+j \geq r} a_{ij}(z)x^i y^j & \textit{otherwise} \end{cases}$$

Proof. There exist permissible parameters $(\overline{x}, \overline{y}, \overline{z})$ at p such that $\overline{y}, \overline{z} \in \mathcal{O}_{X,p}$. $\sigma x = \overline{x}$, $\omega y = \overline{y}$, $z = \overline{z} + h$, $\sigma^a \omega^b = 1$ with $\sigma, \omega, h \in \hat{\mathcal{O}}_{X,p}$, $\sigma, \omega \equiv 1 \mod m_p^2 \hat{\mathcal{O}}_{X,p}$, $h \equiv 0 \mod m_p^2 \hat{\mathcal{O}}_{X,p}$. Suppose that the conclusions of the Lemma hold for $(\overline{x}, \overline{y}, \overline{z})$. Substituting for (x, y, z) we get the conclusions of the Lemma for (x, y, z). We may thus assume that $y, z \in \mathcal{O}_{X,p}$.

There exists an étale neighborhood U of p such that x, y, z are uniformizing parameters in U, $xy = 0$ is a local equation of $U \cap E_X$. Set

$$w = \frac{v - P_t(x^a y^b)}{x^c y^d}$$

where $t > c + d + r$. We have $w \in \Gamma(U, \mathcal{O}_{U,p})$ and

$$\begin{aligned} u &= (x^a y^b)^m \\ v &= P_t(x^a y^b) + x^c y^d w. \end{aligned}$$

If $p' \in U \cap C$, and $\alpha = z(p')$, then $(x, y, z_1 = z - \alpha)$ are permissible parameters at p'.

$$F_{p'} = \sum_{k>0,\, i,j\geq 0} \frac{1}{i!j!k!} \frac{\partial^{i+j+k}w}{\partial x^i \partial y^j \partial z^k}(0,0,\alpha) x^i y^j z_1^k$$
$$+ \sum_{a(i+c)-b(j+d)\neq 0} \frac{1}{i!j!} \frac{\partial^{i+j}w}{\partial x^i \partial y^j}(0,0,\alpha) x^i y^j$$

$\nu(p') \geq r$ for $p' \in C \cap U$ implies that for infinitely many α, we have

$$\frac{\partial^{i+j+k}w}{\partial x^i \partial y^j \partial z^k}(0,0,\alpha) = 0 \text{ if } k > 0 \text{ and } i + j + k < r$$

and

$$\frac{\partial^{i+j}w}{\partial x^i \partial y^j}(0,0,\alpha) = 0 \text{ if } a(i+c) - b(j+d) \neq 0 \text{ and } i + j < r.$$

Thus

$$\frac{\partial^{i+j+k}w}{\partial x^i \partial y^j \partial z^k}(0,0,z) = 0 \text{ if } a(i+c) - b(j+d) \neq 0 \text{ and } i + j < r$$

and

$$\frac{\partial^{i+j+1}w}{\partial x^i \partial y^j \partial z}(0,0,z) = 0 \text{ if } i + j + 1 < r,$$

so that

$$\frac{\partial^{i+j+k}w}{\partial x^i \partial y^j \partial z^k}(0,0,z) = 0$$

if $i + j < r - 1$, $k > 0$. Setting $z = 0$ in the above equations, we get the statement of the Lemma. $\qquad\square$

Lemma 6.26. *Suppose that $C \subset X$ is a nonsingular curve containing a 1 point, $p \in C$ is a 2 point such that C makes SNCs with the 2 curve through p, and (x, y, z) are permissible parameters at p such that $x = z = 0$ are local equations of C at p.*

1. *Write*
$$u = (x^a y^b)^m$$
$$v = P(x^a y^b) + x^c y^d F$$
   Then if $r \geq 2$ and $C \subset \bar{S}_r(X)$,
$$F = x^{r-1}\tau(y) + \sum_{i+k\geq r} a_{ijk} x^i y^j z^k$$
   where τ is a series with $\mathrm{mult}(\tau(y)) \geq 0$.
2. *If there exists a 1 point $q \in C$ such that $F_q \in \hat{\mathcal{I}}^r_{C,q}$ with $r \geq 1$, then $F_p \in \hat{\mathcal{I}}^r_{C,p}$.*

Proof. There exist permissible parameters $(\bar{x}, \bar{y}, \bar{z})$ at p such that $\bar{y}, \bar{z} \in \mathcal{O}_{X,p}$,

$$\sigma x = \bar{x}, \omega y = \bar{y}, \sigma^a \omega^b = 1,$$

with $\sigma, \omega \in \hat{\mathcal{O}}_{X,p}$, $\sigma, \omega \equiv 1 \bmod m_p \hat{\mathcal{O}}_{X,p}$, $\hat{\mathcal{I}}_{C,p} = (\bar{x}, \bar{z})$. Then $\bar{z} = \bar{a}x + \bar{b}z$ for some $\bar{a}, \bar{b} \in \hat{\mathcal{O}}_{X,p}$.

Suppose that the conclusions of the Lemma hold for $(\bar{x}, \bar{y}, \bar{z})$. Substituting back for (x, y, z), we get the conclusions of the Lemma for (x, y, z). We may thus assume that $y, z \in \mathcal{O}_{X,p}$

There exists an étale neighborhood U of p such that x, y, z are uniformizing parameters in U, $xy = 0$ is a local equation of $E_X \cap U$, $C \cap U = V(x, z)$ in U. Set

$$w = \frac{v - P_t(x^a y^b)}{x^c y^d}.$$

where $t > r + c + d$. We have $w \in \Gamma(U, \mathcal{O}_U)$,

$$u = (x^a y^b)^m$$
$$v = P_t(x^a y^b) + x^c y^d w$$

If $p' \in U \cap C$, and $\alpha = y(p')$, then $(x, y - \alpha, z)$ are regular parameters in $\hat{\mathcal{O}}_{X,p'}$. We have permissible parameters $\overline{x}, \overline{y}, z$ at $p' \neq p$ defined by

$$x = \overline{x}(\overline{y} + \alpha)^{-\frac{b}{a}}, y = \overline{y} + \alpha.$$

At p', we have

$$u = \overline{x}^{am}$$
$$v = P_t(\overline{x}^a) + \overline{x}^c(\overline{y} + \alpha)^{d - \frac{cb}{a}} w$$
(34)

$$\begin{aligned}
x^c y^d w &= \overline{x}^c(\overline{y} + \alpha)^{d - \frac{cb}{a}} w \\
&= \overline{x}^c(\overline{y} + \alpha)^{d - \frac{cb}{a}} \left[\sum_{i,j,k \geq 0} \frac{1}{i!j!k!} \frac{\partial^{i+j+k} w}{\partial x^i \partial y^j \partial z^k}(0, \alpha, 0) \overline{x}^i (\overline{y} + \alpha)^{-i\frac{b}{a}} \overline{y}^j z^k \right] \\
&= \sum_{i,k \geq 0} \left[\sum_{j \geq 0} \frac{1}{i!j!k!} \frac{\partial^{i+j+k} w}{\partial x^i \partial y^j \partial z^k}(0, \alpha, 0) \overline{y}^j \right] (\overline{y} + \alpha)^{d - \frac{b(c+i)}{a}} \overline{x}^{c+i} z^k.
\end{aligned}$$

Thus

$$\begin{aligned}
F_{p'} &= \sum_{i \geq 0} \left[\sum_{j \geq 0} \frac{1}{i!j!} \frac{\partial^{i+j} w}{\partial x^i \partial y^j}(0, \alpha, 0) \overline{y}^j \right] (\overline{y} + \alpha)^{d - \frac{b(c+i)}{a}} \overline{x}^i \\
&\quad - \sum_{i \geq 0} \frac{1}{i!} \frac{\partial^i w}{\partial x^i}(0, \alpha, 0) \alpha^{d - \frac{b(c+i)}{a}} \overline{x}^i \\
&\quad + \sum_{i \geq 0, k > 0} \left[\sum_{j \geq 0} \frac{1}{i!j!k!} \frac{\partial^{i+j+k} w}{\partial x^i \partial y^j \partial z^k}(0, \alpha, 0) \overline{y}^j \right] (\overline{y} + \alpha)^{d - \frac{b(c+i)}{a}} \overline{x}^i z^k.
\end{aligned}$$

$\nu(p') = r$ implies that

$$\frac{\partial^{i+j+k} w}{\partial x^i \partial y^j \partial z^k}(0, \alpha, 0) = 0$$
(35)

if $i + j + k < r$, $k > 0$, and for fixed $i < r$

$$\sum_{j < r - i} \frac{1}{i!j!} \frac{\partial^{i+j} w}{\partial x^i \partial y^j}(0, \alpha, 0) \overline{y}^j \equiv c_\alpha^i (\overline{y} + \alpha)^{\lambda_i} \mod \overline{y}^{r-i}$$
(36)

where $\lambda_i = \frac{b(c+i)}{a} - d$, and the $c_\alpha^i \in k$ depend on α and i. Since (35) holds for infinitely many α,

$$\frac{\partial^{i+k} w}{\partial x^i \partial z^k}(0, y, 0) = 0$$

if $i + k < r$ and $k > 0$. Thus

$$\frac{\partial^{i+j+k} w}{\partial x^i \partial y^j \partial z^k}(0, 0, 0) = \frac{\partial^j}{\partial y^j} \left[\frac{\partial^{i+k} w}{\partial x^i \partial z^k}(0, y, 0) \right](0, 0, 0) = 0$$

if $i + k < r$, $k > 0$.

If $i < r - 1$ we have

$$\frac{1}{i!}\frac{\partial^i w}{\partial x^i}(0, \alpha, 0) = c_\alpha^i \alpha^{\lambda_i}$$

and

$$\frac{1}{i!}\frac{\partial^{i+1} w}{\partial x^i \partial y}(0, \alpha, 0) = c_\alpha^i \lambda_i \alpha^{\lambda_i - 1}$$

for infinitely many α. Thus

$$\lambda_i \frac{\partial^i w}{\partial x^i}(0, \alpha, 0) = \alpha\frac{\partial}{\partial y}\frac{\partial^i w}{\partial x^i}(0, \alpha, 0)$$

for infinitely many α, and thus for all α. Set $\gamma_i(y) = \frac{\partial^i w}{\partial x^i}(0, y, 0)$. We have

$$\lambda_i \gamma_i(y) = y\frac{d\gamma_i}{dy}.$$

There is an expansion $\gamma_i(y) = \sum_{j=0}^\infty b_j y^j$ with $b_j \in k$. $\frac{d\gamma_i}{dy} = \sum_{j=1}^\infty j b_j y^{j-1}$.

$$y\frac{d\gamma_i}{dy} = \sum_{j=0}^\infty j b_j y^j$$

$$\lambda_i \gamma_i - y\frac{d\gamma_i}{dy} = \sum_{j=0}^\infty (\lambda_i b_j - j b_j) y^j = 0$$

so that $b_j(\lambda_i - j) = 0$ for all j, which implies that $\gamma_i = 0$, or $\lambda_i \in \mathbf{N}$ and $\gamma_i = b_{\lambda_i} y^{\lambda_i}$. Suppose that $\lambda_i \in \mathbf{N}$ and $\gamma_i(y) = \frac{\partial^i w}{\partial x^i}(0, y, 0) \neq 0$. Then

$$\frac{\partial^{i+\lambda_i} w}{\partial x^i \partial y^{\lambda_i}}(0, 0, 0) = \lambda_i! b_{\lambda_i} \neq 0.$$

But

$$b(c + i) - a(d + \lambda_i) = b(c + i) - a(\frac{b}{a}(c + i)) = 0$$

implies $i > r$, by our choice of t in p_t and the assumption that F is normalized, a contradiction. Thus

$$\frac{\partial^{i+j} w(0, 0, 0)}{\partial x^i \partial y^j} = 0$$

if $i < r - 1$.

Now suppose that there exists a 1 point $q' \in C$ such that $F_{q'} \in \hat{\mathcal{I}}^r_{C, q'}$. By 1. of Lemma 6.23, $F_q \in \hat{\mathcal{I}}^r_{C, q}$ at every 1 point $q \in C$. With the above notation, (trivially if $r = 1$)

$$F_p = x^{r-1}\tau(y) + \sum_{i+k \geq r} a_{ijk} x^i y^j z^k.$$

For $p \neq q \in C \cap U$ there exist permissible parameters $(\overline{x}, \overline{y}, z)$ at q such that

$$x = \overline{x}(\overline{y} + \alpha)^{-\frac{b}{a}}, y = \overline{y} + \alpha$$

$$F_q = \overline{x}^{r-1}\Lambda + \Omega$$

with

$$\Lambda = (\overline{y} + \alpha)^{d - \frac{b}{a}(c+r-1)}\tau(\overline{y} + \alpha) - \alpha^{d-(c+r-1)\frac{b}{a}}\tau(\alpha),$$

$\Omega \in \hat{\mathcal{I}}^r_{C,q}$. $F_q \in \hat{\mathcal{I}}^r_{C,q}$ implies $\tau = 0$ or $d - \frac{b}{a}(c + r - 1) = 0$ and $\tau \in k$. But $ad - b(c + r - 1) = 0$ and $\tau \in k$ is not possible since F is normalized. Thus $\tau = 0$. $\qquad\square$

Lemma 6.27. *Suppose that $r \geq 1$ $C \subset X$ is a 2 curve, such that $\nu(q) \geq r$ if $q \in C$ is a 2 point ($C \subset \overline{S}_r(X)$ if $r \geq 2$) and $p \in C$ is a 3 point, so that there exist permissible parameters x, y, z at p such that*

$$u = (x^a y^b z^c)^m$$
$$v = P(x^a y^b z^c) + x^d y^e z^f F$$

where $\hat{\mathcal{I}}_{C,p} = (x, y)$. Then

$$
F = \begin{cases}
\tau(z) x^{i_0} y^{j_0} + \sum_{i+j \geq r} a_{ij}(z) x^i y^j & \text{if there exist } (i_0, j_0) \text{ such that} \\
& i_0 + j_0 = r - 1 \\
& \text{and } a(e + j_0) - b(d + i_0) = 0 \\[4pt]
\sum_{i+j \geq r} a_{ij}(z) x^i y^j & \text{otherwise}
\end{cases}
$$

If there exists a 2 point $q \in C$ such that $F_q \in \hat{\mathcal{I}}^r_{C,q}$ and $r \geq 1$, then $F_p \in \hat{\mathcal{I}}^r_{C,p}$

Proof. There exist permissible parameters $(\overline{x}, \overline{y}, \overline{z})$ at p such that $\overline{y}, \overline{z} \in \mathcal{O}_{X,p}$,

$$\sigma x = \overline{x}, \omega y = \overline{y}, \mu z = \overline{z}$$

for some unit series $\sigma, \omega, \mu \in \hat{\mathcal{O}}_{X,p}$. Suppose that the conclusions of the Lemma are true for the parameters $(\overline{x}, \overline{y}, \overline{z})$. Substituting back for (x, y, z) we get the conclusions of the Lemma for (x, y, z). We may thus assume that $y, z \in \mathcal{O}_{X,p}$.

There exists an étale neighborhood U of p such that (x, y, z) are uniformizing parameters in U, $xyz = 0$ is a local equation of $E_X \cap U$. Set

$$w = \frac{v - P_t(x^a y^b z^c)}{x^d y^e z^f}$$

where $t \geq d + e + f + r$. We have $w \in \Gamma(U, \mathcal{O}_U)$ and

$$u = (x^a y^b z^c)^m, v = P_t(x^a y^b z^c) + x^d y^e z^f w.$$

If $p' \in U \cap C$ and $\alpha = z(p')$, then $(x, y, z - \alpha)$ are regular parameters in $\hat{\mathcal{O}}_{X,p'}$. If $\alpha \neq 0$, we have permissible parameters $(\overline{x}, y, \overline{z})$ at p' where $x = \overline{x}(\overline{z} + \alpha)^{\frac{-c}{a}}$, $\overline{z} = z - \alpha$. At p' we have

$$u = (\overline{x}^a y^b)^m, v = P_t(\overline{x}^a y^b) + \overline{x}^d y^e (\overline{z} + \alpha)^{f - \frac{dc}{a}} w.$$

$$
\begin{aligned}
\overline{x}^d y^e (\overline{z} + \alpha)^{f - \frac{cd}{a}} w \\
= \overline{x}^d y^e (\overline{z} + \alpha)^{f - \frac{cd}{a}} \left[\sum_{i,j,k \geq 0} \frac{1}{i!j!k!} \frac{\partial^{i+j+k} w}{\partial x^i \partial y^j \partial z^k}(0,0,\alpha) \overline{x}^i (\overline{z} + \alpha)^{\frac{-ic}{a}} y^j \overline{z}^k \right] \\
= \overline{x}^d y^e \left[\sum_{i,j \geq 0} \left(\sum_{k \geq 0} \frac{1}{i!j!k!} \frac{\partial^{i+j+k} w}{\partial x^i \partial y^j \partial z^k}(0,0,\alpha) \overline{z}^k \right) (\overline{z} + \alpha)^{f - \frac{c(i+d)}{a}} \overline{x}^i y^j \right].
\end{aligned}
$$

Thus

$$
\begin{aligned}
F_{p'} &= \sum_{i,j \geq 0} \left(\sum_{k \geq 0} \frac{1}{i!j!k!} \frac{\partial^{i+j+k} w}{\partial x^i \partial y^j \partial z^k}(0,0,\alpha) \overline{z}^k \right) (\overline{z} + \alpha)^{f - \frac{c(i+d)}{a}} \overline{x}^i y^j \\
&\quad - \sum_{i,j \text{ such that } b(i+d)-a(j+e)=0} \left(\frac{1}{i!j!} \frac{\partial^{i+j} w}{\partial x^i \partial y^j}(0,0,\alpha) \alpha^{f - \frac{c(i+d)}{a}} \right) \overline{x}^i y^j
\end{aligned}
$$

$\nu(p') = r$ implies

$$\frac{\partial^{i+j+k}w}{\partial x^i \partial y^j \partial z^k}(0,0,\alpha) = 0 \text{ if } i+j+k < r \text{ and } b(i+d) - a(j+e) \neq 0$$
(37)

and if $b(i+d) - a(j+e) = 0$ for fixed i, j with $i+j < r$,

$$\sum_{k < r-i-j} \frac{1}{i!j!k!} \frac{\partial^{i+j+k}w}{\partial x^i \partial y^j \partial z^k}(0,0,\alpha)\bar{z}^k \equiv c_\alpha^{ij}(\bar{z}+\alpha)^{\lambda_i} \bmod (\bar{z}^{r-i-j})$$
(38)

where $\lambda_i = \frac{c(i+d)}{a} - f$, $c_\alpha^{i,j} \in k$ depend on α, i and j.

Since (37) holds for infinitely many α,

$$\frac{\partial^{i+j+k}w}{\partial x^i \partial y^j \partial z^k}(0,0,z) = 0$$

if $i+j+k < r$ and $b(i+d) - a(j+e) \neq 0$. Thus

$$\frac{\partial^{i+j+k}w}{\partial x^i \partial y^j \partial z^k}(0,0,0) = \frac{\partial^k}{\partial z^k}\left[\frac{\partial^{i+j}w}{\partial x^i \partial y^j}(0,0,z)\right](0,0,0) = 0$$

if $i+j < r$, $k \geq 0$ and $b(i+d) - a(j+e) \neq 0$.

If $b(i+d) - a(j+e) = 0$ and $i+j < r-1$, we have

$$\frac{1}{i!j!} \frac{\partial^{i+j}w}{\partial x^i \partial y^j}(0,0,\alpha) = c_\alpha^{ij}\alpha^{\lambda_i}$$

and

$$\frac{1}{i!j!} \frac{\partial^{i+j+1}w}{\partial x^i \partial y^j \partial z}(0,0,\alpha) = c_\alpha^{ij}\lambda_i\alpha^{\lambda_i-1}$$

for infinitely many α. Thus

$$\lambda_i \frac{\partial^{i+j}w}{\partial x^i \partial y^j}(0,0,\alpha) = \alpha\frac{\partial^{i+j+1}w}{\partial x^i \partial y^j \partial z}(0,0,\alpha)$$

for infinitely many α, and thus for all α. Set

$$\gamma_{ij}(z) = \frac{\partial^{i+j}w}{\partial x^i \partial y^j}(0,0,z).$$

We have an expression

$$\gamma_{ij}(z) = \sum_{k=0}^{\infty} b_k z^k$$

with $b_k \in k$.

$$\frac{d\gamma_{ij}}{dz} = \sum_{k=1}^{\infty} k b_k z^{k-1}.$$

$$\lambda_i \gamma_{ij} - z\frac{d\gamma_{ij}}{dz} = \sum_{k=0}^{\infty}(\lambda_i b_k - k b_k)z^k = 0$$

implies $b_k(\lambda_i - k) = 0$ for all k, so that either $\gamma_{ij} = 0$, or $\lambda_i \in \mathbf{N}$ and $\gamma_{ij} = b_{\lambda_i} z^{\lambda_i}$. Suppose that

$$\gamma_{ij}(z) = \frac{\partial^{i+j} w}{\partial x^i \partial y^j}(0, 0, z) \neq 0.$$

Then

$$\frac{\partial^{i+j+\lambda_i} w}{\partial x^i \partial y^j \partial z^{\lambda_i}}(0, 0, 0) = \lambda_i! b_{\lambda_i} \neq 0$$

so that we have a nontrivial $x^{d+i} y^{e+j} z^{f+\lambda_i}$ term in $x^d y^e z^f F$. Recall that $\lambda_i = \frac{c(i+d)}{a} - f$. By assumption, $b(i + d) - a(j + e) = 0$. We further have $a(f + \lambda_i) - c(d + i) = 0$,

$$
\begin{aligned}
b(f + \lambda_i) - c(e + j) &= b(\tfrac{c(i+d)}{a}) - c(e + j) \\
&= b\tfrac{c}{a}(d + i) - c\tfrac{b}{a}(i + d) = 0
\end{aligned}
$$

a contradiction to the assumption that F is normalized. Thus

$$\frac{\partial^{i+j+k} w}{\partial x^i \partial y^j \partial z^k}(0, 0, 0) = 0$$

if $b(i + d) - a(j + e) = 0$, $i + j < r - 1$, $k \geq 0$.

Now suppose there exists a 2 point $q' \in C$ such that $F_{q'} \in \hat{\mathcal{I}}^r_{C, q'}$. By 2. of Lemma 6.23, $F_q \in \hat{\mathcal{I}}^r_{C,q}$ for all 2 points $q \in C$. With the above notation, if $F_p \notin \hat{\mathcal{I}}^r_{C,p}$, we have

$$F_p = \tau(z) x^{i_0} y^{j_0} + \sum_{i+j \geq r} a_{ij}(z) x^i y^j$$

where $i_0 + j_0 = r - 1$ and $a(e + j_0) - b(d + i_0) = 0$, $\tau(z) \neq 0$.

For $p \neq q \in C \cap U$, there exist permissible parameters $(\overline{x}, y, \overline{z})$ at q such that

$$x = \overline{x}(\overline{z} + \alpha)^{-\frac{c}{a}}, z = \overline{z} + \alpha$$

$$u = (\overline{x}^a y^b)^m = (\overline{x}^{\overline{a}} y^{\overline{b}})^{\overline{m}}$$

$$v = P_q(\overline{x}^{\overline{a}} y^{\overline{b}}) + \overline{x}^d y^e F_q$$

with $(\overline{a}, \overline{b}) = 1$.

$$F_q = \overline{x}^{i_0} y^{j_0} \Lambda + \Omega$$

with $\Omega \in \hat{\mathcal{I}}^r_{C,q}$ and

$$\Lambda = (\overline{z} + \alpha)^{f - \frac{c(d+i_0)}{a}} \tau(\overline{z} + \alpha) - \alpha^{f - \frac{c(d+i_0)}{a}} \tau(\alpha).$$

$$F_q \in \hat{\mathcal{I}}^r_{C,q}$$

implies $\tau = 0$, or $f - \frac{c(d+i_0)}{a} = 0$ and $\tau \in k$. $f - \frac{c(d+i_0)}{a} = 0$ and $\tau \in k$ is not possible since F is normalized. $\square$

Lemma 6.28. *Suppose that $r \geq 2$, $f_1, \ldots, f_{n-1}$ is a regular sequence in a n dimensional regular local ring A. Let $I = (f_1, \ldots, f_{n-1})$. Then*

$$\operatorname{depth} A/(I^r + (f_1)^{r-1}) = 1$$

for all $r \geq 2$.

Proof. Let m be the maximal ideal of A. There is an exact sequence of A modules

$$0 \to I^{r-1}/(I^r + (f_1)^{r-1}) \to A/(I^r + (f_1)^{r-1}) \to A/I^{r-1} \to 0$$

$I^{r-1}/(I^r + (f_1)^{r-1})$ is a free A/I module, since $f_1, \ldots, f_{n-1}$ is quasi regular by Theorem 27, [27]. depth $A/I^t = 1$ for all $t \geq 1$ by Proposition 16.F, [27]. Thus $\mathrm{Hom}_A(A/m, A/I^{r-1}) = 0$ and $\mathrm{Hom}_A(A/m, I^{r-1}/(I^r + (f_1)^{r-1})) = 0$, so that

$$\mathrm{Hom}_A(A/m, A/(I^r + (f_1)^{r-1})) = 0$$

and depth$(A/(I^r + (f_1)^{r-1})) = 1$. $\qquad\square$

Lemma 6.29. *Suppose that $r \geq 2$, $p \in X$ is a 2 point and (x, y, z) are permissible parameters at p such that $y, z \in \mathcal{O}_{X,p}$,*

$$u = (x^a y^b)^m$$
$$v = P(x^a y^b) + x^c y^d F_p$$

$p \in C \subset \bar{S}_r(X)$ is a curve such that $x \in \hat{\mathcal{I}}_{C,p}$ and C contains a 1 point q. Then there exists a polynomial g such that one of the following cases hold

Case 1): $\bar{y}^{ad-bc} F_p - g(x\bar{y}^b) \in \left(\hat{\mathcal{I}}^r_{C,p} + (x)^{r-1} \right) k[[x, \bar{y}, z]]$ *if $ad - bc \geq 0$.*

Case 2): $F_p - g(x\bar{y}^b)\bar{y}^{bc-ad} \in \left(\hat{\mathcal{I}}^r_{C,p} + (x)^{r-1} \right) k[[x, \bar{y}, z]]$ *if $ad - bc \leq 0$*

where $y = \bar{y}^a$.

Proof. x, y, z are uniformizing parameters in an étale neighborhood $U = \mathrm{spec}(R)$ of p and $xy = 0$ is a local equation of $E_X \cap U$, C is a complete intersection in U. For $t > c + d + r$,

$$w = \frac{1}{x^c y^d}[v - P_t(x^a y^b)] \in R.$$

If $q \in C \cap U$ is a 1 point, such that C is nonsingular at q, then q has permissible parameters (x_1, y_1, z_1) where $x = x_1(y_1 + \alpha)^{-\frac{b}{a}}$, $y = y_1 + \alpha$, $z = z_1 + \beta$ for some $\alpha, \beta \in k$ with $\alpha \neq 0$.

$$u = x_1^{am}$$
$$v = P_q(x_1) + x_1^c F_q$$

where

$$F_q = (y_1 + \alpha)^{\frac{ad-bc}{a}} w - g(x_1)$$

for some series g. $F_q \in \hat{\mathcal{I}}^r_{C,q} + (x_1)^{r-1}$ (by Lemma 6.24) implies

$$w - (y_1 + \alpha)^{\frac{bc-ad}{a}} g(x_1) \in \hat{\mathcal{I}}^r_{C,q} + (x_1)^{r-1}.$$

Let $y = \bar{y}^a$, $S = R[\bar{y}]$, $\lambda : V = \mathrm{spec}(S) \to \mathrm{spec}(R)$. Suppose that $q' \in \lambda^{-1}(q)$. Let $h(x_1) = g_r(x_1)$.

Suppose that $ad - bc \geq 0$. Then

$$\bar{y}^{ad-bc} w - h(x\bar{y}^b) \in \hat{\mathcal{I}}^r_{C,q'} + (x)^{r-1}.$$

$I = \Gamma(V, \mathcal{I}_C)$ is a complete intersection in V, so that (by Lemma 6.28)

$$\bar{y}^{ad-bc} w - h(x\bar{y}^b) \in (I^r + (x)^{r-1})S,$$

and

$$\overline{y}^{ad-bc}w - h(x\overline{y}^b) \in \left(\hat{\mathcal{I}}^r_{C,p} + (x)^{r-1}\right)k[[x, \overline{y}, z]].$$

$$\overline{y}^{ad-bc}w - h(x\overline{y}^b) \equiv \overline{y}^{ad-bc}F_p - h(x\overline{y}^b) \bmod (x^r)$$

implies

$$\overline{y}^{ad-bc}F_p - h(x\overline{y}^b) \in \left(\hat{\mathcal{I}}^r_{C,p} + (x)^{r-1}\right)k[[x, \overline{y}, z]].$$

The case when $ad - bc \leq 0$ is similar. $\qquad\square$

Lemma 6.30. *Suppose that $p \in X$ is a 1 or 2 point, D is a generic curve through p on a component of E_X containing p. Then $F_q \notin \hat{\mathcal{I}}_{D,q}$ for $q \in D$ (if F_q is computed with respect to permissible parameters (x, y, z) at q such that $x = z = 0$ are local equations of D at q).*

Proof. By Lemma 6.23 and Lemma 6.26, we need only check this at p. When p is a 1 point this follows from Lemma 6.13.

Suppose that p is a 2 point, (x, y, z) are permissible parameters at p such that $x \in \hat{\mathcal{I}}_{D,p}$. Then

$$\begin{aligned} u &= (x^a y^b)^m \\ v &= P(x^a y^b) + x^c y^d F_p \end{aligned}$$

There exists a series

$$\overline{z} = z - \sum_{i=1}^{\infty} \alpha_i y^i$$

with $\alpha_i \in k$ such that $\hat{\mathcal{I}}_{D,p} = (x, \overline{z})$. Let

$$v = \overline{P}(x^a y^b) + x^c y^d \overline{F}_p$$

be the normalized form of v with respect to the permissible parameters $(x, y, \overline{z})$. Then

$$\overline{F}_p = F_p - \sum b_i x^{\overline{a}_i} y^{\overline{b}_i}$$

with $b_i \in k$ and

$$a(d + \overline{b}_i) - b(c + \overline{a}_i) = 0$$

for all i. Suppose that $\overline{F}_p \in \hat{\mathcal{I}}_{D,p}$.

$$F_p = h(y, z) + x\Omega$$

with $h \neq 0$. We either have

$$\overline{z} \mid h(y, \overline{z} + \sum \alpha_i y^i)$$

or there exists $\overline{c} \in k$, $\overline{d} \in \mathbf{N}$ such that $a(d + \overline{d}) - bc = 0$ and

$$\overline{z} \mid (h(y, \overline{z} + \sum \alpha_i y^i) - \overline{c} y^{\overline{d}}).$$

Thus either $h(y, \sum \alpha_k y^k) = 0$ or $h(y, \sum \alpha_k y^k) = \overline{c} y^{\overline{d}}$.

Since D is generic, we can suppose that α_1, α_2 are independent generic points of k.

Let $e = \nu(h)$. Write $h = \sum_{i+j\geq e} a_{ij} y^i z^j$.

$$h(y, \textstyle\sum \alpha_k y^k) = \sum_{i+j=e} a_{ij} y^i (\alpha_1^j y^j + j\alpha_1^{j-1} \alpha_2 y^{j+1}) + \sum_{i+j=e+1} a_{ij} y^i (\alpha_1 y)^j$$
$$+ y^{e+2} \Omega$$
$$= \left(\sum_{i+j=e} a_{ij} \alpha_1^j \right) y^e$$
$$+ \left(\sum_{i+j=e} j a_{ij} \alpha_1^{j-1} \alpha_2 + \sum_{i+j=e+1} a_{ij} \alpha_1^j \right) y^{e+1}$$
$$+ y^{e+2} \Omega$$

We must have $h(y, \sum \alpha_k y^k) = \overline{c} y^{\overline{d}}$ and $e = \overline{d}$ since $\sum_{i+j=e} a_{ij} \alpha_1^j \neq 0$ as α_1 is a generic point of k. Since F_p is normalized, we must have $a_{e0} = 0$, and $\nu(h) = e$ implies there exists $a_{i_0 j_0} \neq 0$ such that $i_0 + j_0 = e$ and $j_0 > 0$.

We must have

$$\left(\sum_{i+j=e} j a_{ij} \alpha_1^{j-1} \right) \alpha_2 + \left(\sum_{i+j=e+1} a_{ij} \alpha_1^j \right) = 0.$$

which is a contradiction to the assumption that α_1, α_2 are independent generic points of k. $\qquad\square$

Definition 6.31. *Suppose that $\Phi_X : X \to S$ is weakly prepared.*

A monoidal transform $\pi : Y \to X$ is called weakly permissible if π is the blow-up of a point p on E_X, or a nonsingular curve C on E_X such that C makes SNCs with $\overline{B}_2(X)$.

Suppose that $\Phi_X : X \to S$ is weakly prepared, and $\pi : Y \to X$ is a weakly permissible monoidal transform. Define $\Phi_Y = \Phi_X \circ \pi : Y \to S$, $E_Y = \pi^{-1}(E_X)_{red}$. Φ_Y is weakly prepared.

Remark 6.32. *Most of the invariants of this section can probably be described intrinsicly in terms of the map*

$$\Phi_X^* : \Omega_S^2(\log D_S) \to \Omega_X^2(\log E_X)$$

7. The Invariant ν Under Quadratic Transforms

Throughout this section we will suppose that $\Phi_X : X \to S$ is weakly prepared.

Theorem 7.1. *Suppose that* $\nu(p) = r$, $\pi : X_1 \to X$ *is the blow-up of* p, $q \in \pi^{-1}(p)$ *with* $\nu(q) = r_1$.

Suppose that p is a 1 point. Then

1. *If q is a 1 point then $r_1 \le r$.*
2. *If q is a 2 point then $r_1 \le r$. $r_1 = r$ implies $\tau(q) > 0$.*

Suppose that p is a 2 point. Then

1. *If q is a 1 point then $r_1 \le r + 1$. $r_1 = r + 1$ implies $\gamma(q) = r + 1$.*
2. *If q is a 2 point then $r_1 \le r$.*
3. *If q is a 3 point then $r_1 \le r$.*

Suppose that p is a 3 point. Then

1. *If q is a 1 point then $r_1 \le r + 1$. $r_1 = r + 1$ implies $\gamma(q) = r + 1$.*
2. *If q is a 2 point then $r_1 \le r + 1$. $r_1 = r + 1$ implies $\tau(q) > 0$. Furthermore there are permissible parameters (x_1, y_1, z_1) at q such that*

$$
\begin{aligned}
u &= (x_1^a y_1^b)^m \\
v &= P(x_1^a y_1^b) + x_1^c y_1^d F_1
\end{aligned}
$$

and the leading form of F_1 is

$$
L_1 = c y_1^t z_1^{r+1-t} + x_1 \Omega
$$

where $0 \le t \le r$, $cb - (d+t)a = 0$. In this case, the leading form of F is

$$
L = y^t \left(\sum_{i+k=r-t} b_{ik} x^i z^k \right)
$$

where all $b_{ik} \ne 0$, and there are regular parameters $(\bar{x}_1, \bar{y}_1, \bar{z}_1)$ in $\hat{O}_{X_1,q}$ such that

$$
x = \bar{x}_1, y = \bar{x}_1 \bar{y}_1, z = \bar{x}_1(\bar{z}_1 + \beta)
$$

for some $0 \ne \beta \in k$.

3. *If q is a 3 point then $r_1 \le r$. If $r_1 = r$, and (x_1, y_1, z_1) are permissible parameters at q with*

$$
x = x_1, y = x_1 y_1, z = x_1 z_1,
$$

then the leading form of F is

$$
L = L(y_1, z_1).
$$

Proof. **Suppose that p is a 1 point**

$$
\begin{aligned}
u &= x^k \\
v &= P(x) + x^c F
\end{aligned}
$$

Write $F = \sum_{i+j+k \ge r} a_{ijk} x^i y^j z^k$.

Suppose that $q \in \pi^{-1}(p)$ is a 1 point. Then there are permissible parameters (x_1, y_1, z_1) at q such that

$$
\begin{aligned}
x &= x_1 \\
y &= x_1(y_1 + \alpha) \\
z &= x_1(z_1 + \beta)
\end{aligned}
$$

$$
\begin{aligned}
u &= x_1^k \\
v &= P(x_1) + x_1^{c+r} \frac{F}{x_1^r}
\end{aligned}
$$

$$
\frac{F}{x_1^r} = \sum_{j+k \le r} a_{jk}(y_1 + \alpha)^j (z_1 + \beta)^k + x_1 \Omega \tag{39}
$$

where $a_{jk} = a_{r-i-j,j,k}$. Suppose that $\nu(q) > \nu(p)$. Then

$$
\sum_{j+k \le r} a_{jk}(y_1 + \alpha)^j (z_1 + \beta)^k = \gamma \in k.
$$

and the leading form of F is

$$
L = x_1^r \Big(\sum_{j+k \le r} a_{jk}(y_1 + \alpha)^j (z_1 + \beta)^k \Big) = \gamma x_1^r = \gamma x^r
$$

a contradiction to the assumption that F is normalized. Thus $\nu(q) \le \nu(p)$.

Now suppose that $q \in \pi^{-1}(p)$ is a 2 point. Then, after possibly interchanging y and z, there are permissible parameters (x_1, y_1, z_1) at q such that

$$
\begin{aligned}
x &= x_1 y_1 \\
y &= y_1 \\
z &= y_1(z_1 + \alpha)
\end{aligned}
$$

$$
\begin{aligned}
u &= x_1^k y_1^k \\
v &= P(x_1 y_1) + x_1^c y_1^{c+r} \frac{F}{y_1^r}
\end{aligned}
$$

$$
\frac{F}{y_1^r} = \sum_{i+k \le r} a_{ik} x_1^i (z_1 + \alpha)^k + y_1 \Omega
$$

where $a_{ik} = a_{i,r-i-k,k}$. Suppose that $\nu(q) > \nu(p)$. Then

$$
\sum_{i+k \le r} a_{ik} x_1^i (z_1 + \alpha)^k = \sum \gamma_i x_1^{a_i}
$$

with $a_i + c = c + r$ for all i. $a_i = r$ is the only solution to this equation, so that if L is the leading form of F,

$$
x^c L = \gamma x_1^{r+c} y_1^{r+c} = \gamma x^{r+c}
$$

a contradiction to the assumption that F is normalized. Thus $\nu(q) \le \nu(p)$.

Suppose that $\nu(q) = r$. After making a permissible change of parameters, we may assume that $\alpha = 0$. We have

$$
F_q = \sum_{i+k \le r} a_{ik} x_1^i z_1^k + y_1 \Sigma \tag{40}
$$

Thus $a_{ik} = 0$ if $i + k < r$, and since L is normalized, we must have $a_{ik} \ne 0$ for some $k > 0$. Thus $\tau(q) > 0$.

Suppose that p is a 2 point

$$\begin{aligned} u &= (x^a y^b)^k \\ v &= P(x^a y^b) + x^c y^d F \end{aligned}$$

Write $F = \sum_{i+j+k \geq r} a_{ijk} x^i y^j z^k$.

Suppose that $q \in \pi^{-1}(p)$ is a 1 point. Then there are regular parameters (x_1, y_1, z_1) in $\hat{\mathcal{O}}_{X_1, q}$ such that

$$\begin{aligned} x &= x_1 \\ y &= x_1(y_1 + \alpha) \\ z &= x_1(z_1 + \beta) \end{aligned}$$

with $\alpha \neq 0$.

$$u = x_1^{(a+b)k}(y_1 + \alpha)^{bk} = \overline{x}_1^{(a+b)k}$$

where $x_1 = \overline{x}_1(y_1 + \alpha)^{-\frac{b}{a+b}}$.

$$v = P(\overline{x}_1^{a+b}) + \overline{x}_1^{c+d+r}(y_1 + \alpha)^\lambda \frac{F}{x_1^r}$$

where $\lambda = d - \frac{b(c+d+r)}{a+b}$.

$$\frac{F}{x_1^r} = \sum_{j+k \leq r} a_{jk}(y_1 + \alpha)^j (z_1 + \beta)^k + x_1 \Omega \tag{41}$$

where $a_{jk} = a_{r-i-j,j,k}$. Suppose that

$$(y_1 + \alpha)^\lambda \left(\sum_{j+k \leq r} a_{jk}(y_1 + \alpha)^j (z_1 + \beta)^k \right) \equiv \gamma \bmod (y_1, z_1)^{r+2}$$

for some $\gamma \in k$. Then

$$\sum_{j+k \leq r} a_{jk}(y_1 + \alpha)^j (z_1 + \beta)^k \equiv \gamma(y_1 + \alpha)^{-\lambda} \bmod (y_1, z_1)^{r+2} \tag{42}$$

Set

$$f(y_1) = (y_1 + \alpha)^{-\lambda} = \sum_{i=0}^\infty \alpha_i y_1^i$$

where $\alpha_0 = \alpha^{-\lambda}$ and

$$\alpha_i = \frac{-\lambda(-\lambda - 1) \cdots (-\lambda - i + 1)}{i!} \alpha^{-\lambda - i}$$

for $i \geq 1$. (42) implies $\alpha_{r+1} = 0$, so that $-\lambda \in \{0, 1, \ldots, r\}$, and

$$\sum_{j+k \leq r} a_{jk}(y_1 + \alpha)^j (z_1 + \beta)^k = \gamma(y_1 + \alpha)^{-\lambda}$$

Thus

$$\sum_{i+j+k=r} a_{ijk} x_1^r (y_1 + \alpha)^j (z_1 + \beta)^k = \gamma x_1^{r+\lambda} x_1^{-\lambda}(y_1 + \alpha)^{-\lambda}$$

which implies that the leading form of F is

$$L = \sum_{i+j+k=r} a_{ijk}x^i y^j z^k = \gamma x^{r+\lambda} y^{-\lambda}$$

$$x^c y^d L = \gamma x^{r+c+\lambda} y^{d-\lambda}$$

$$a(d-\lambda) - b(r+c+\lambda) = a\left[\frac{b(c+d+r)}{a+b}\right] - b\left[r+c+d-\frac{b(c+d+r)}{a+b}\right] = 0$$

Thus $x^{r+c+\lambda} y^{d-\lambda}$ is a power of $x^a y^b$, a contradiction to the assumption that F is normalized. We conclude that $\nu(q) \leq \nu(p) + 1$.

If $\nu(q) = \nu(p) + 1$, we must then have that

$$F_1 = (y_1 + \alpha)^\lambda \left(\sum_{j+k \leq r} a_{jk}(y_1 + \alpha)^j (z_1 + \beta)^k\right) - \gamma + \bar{x}_1 \Sigma$$

with $\gamma = \alpha^\lambda \sum_{j+k \leq r} a_{jk}\alpha^j \beta^k$. There is a nonzero degree $r+1$ term in $F_1(0, y_1, z_1)$, so that $\gamma(q) = r + 1$.

Now suppose that $q \in \pi^{-1}(p)$ is a 2 point. Then after possibly interchanging x and y, there are permissible parameters (x_1, y_1, z_1) at q such that

$$\begin{aligned} x &= x_1 \\ y &= x_1 y_1 \\ z &= x_1(z_1 + \beta) \end{aligned}$$

with $\beta \neq 0$.

$$\begin{aligned} u &= (x_1^{a+b} y_1^b)^k \\ v &= P(x_1^{a+b} y_1^b) + x_1^{c+d+r} y_1^d \frac{F}{x_1^r} \end{aligned}$$

$$\frac{F}{x_1^r} = \sum_{j+k \leq r} a_{jk} y_1^j (z_1 + \beta)^k + x_1 \Omega \tag{43}$$

with $a_{jk} = a_{r-j-k,j,k}$. Suppose that

$$\sum_{j+k \leq r} a_{jk} y_1^j (z_1 + \beta)^k = \sum \gamma_i y_1^{t_i}$$

with $\gamma_i \in k$, $(c+d+r)b - (a+b)(d+t_i) = 0$ for all i. There is at most one natural number $t = t_i$ which is a solution to this equation, which simplifies to

$$a(d+t) - b(c+r-t) = 0. \tag{44}$$

We have

$$\sum_{j+k \leq r} a_{jk} y_1^j (z_1 + \beta)^k = \gamma y_1^t$$

so that

$$\sum_{i+j+k=r} a_{ijk} x_1^r y_1^j (z_1 + \beta)^k = \gamma x_1^r y_1^t$$

Thus $L = \gamma x^{r-t} y^t$. But by (44) $x^{c+r-t} y^{d+t}$ is a power of $x^a y^b$, a contradiction to the assumption that F is normalized. Thus $\nu(q) \leq r$.

Now suppose that $q \in \pi^{-1}(p)$ is a 3 point. Then there are regular parameters (x_1, y_1, z_1) at Q such that

$$
\begin{aligned}
x &= x_1 z_1 \\
y &= y_1 z_1 \\
z &= z_1
\end{aligned}
$$

We have

$$
\begin{aligned}
u &= (x_1^a y_1^b z_1^{a+b})^k \\
v &= P(x_1^a y_1^b z_1^{a+b}) + x_1^c y_1^d z_1^{c+d+r} \frac{F}{z_1^r}
\end{aligned} \tag{45}
$$

$F_q = \frac{F}{z_1^r}$, so that $\nu(q) = \nu(\frac{F}{z_1^r}) \leq r$.

Suppose that p is a 3 point

$$
\begin{aligned}
u &= (x^a y^b z^c)^k \\
v &= P(x^a y^b z^c) + x^d y^e z^f F
\end{aligned}
$$

Write $F = \sum_{i+j+k \geq r} a_{ijk} x^i y^j z^k$. Suppose that $q \in \pi^{-1}(p)$ is a 1 point. Then there are regular parameters (x_1, y_1, z_1) in $\hat{\mathcal{O}}_{X_1, q}$ such that

$$
\begin{aligned}
x &= x_1 \\
y &= x_1(y_1 + \alpha) \\
z &= x_1(z_1 + \beta)
\end{aligned}
$$

with $\alpha, \beta \neq 0$.

$$
u = x_1^{(a+b+c)k}(y_1 + \alpha)^{bk}(z_1 + \beta)^{ck} = \overline{x}_1^{(a+b+c)k}.
$$

where $\overline{x}_1$ is defined by

$$
x_1 = \overline{x}_1(y_1 + \alpha)^{-\frac{b}{a+b+c}}(z_1 + \beta)^{-\frac{c}{a+b+c}}.
$$

$$
\begin{aligned}
v &= P(\overline{x}_1^{a+b+c}) + x_1^{d+e+f+r}(y_1 + \alpha)^e (z_1 + \beta)^f \frac{F}{x_1^r} \\
&= P(\overline{x}_1^{a+b+c}) + \overline{x}_1^{d+e+f+r}(y_1 + \alpha)^{\lambda_1}(z_1 + \beta)^{\lambda_2} \frac{F}{x_1^r}
\end{aligned}
$$

where

$$
\begin{aligned}
\lambda_1 &= e - \frac{b(d+e+f+r)}{a+b+c} \\
\lambda_2 &= f - \frac{c(d+e+f+r)}{a+b+c}
\end{aligned}
$$

$$
\frac{F}{x_1^r} = \sum_{j+k \leq r} a_{jk}(y_1 + \alpha)^j (z_1 + \beta)^k + \overline{x}_1 \Omega \tag{46}
$$

where $a_{jk} = a_{r-j-k,j,k}$. Suppose that

$$
(y_1 + \alpha)^{\lambda_1}(z_1 + \beta)^{\lambda_2}\left(\sum_{j+k \leq r} a_{jk}(y_1 + \alpha)^j (z_1 + \beta)^k\right) \equiv \gamma \mod (y_1, z_1)^{r+2}
$$

for some $\gamma \in k$. We first observe that we cannot have $\gamma = 0$, for $\gamma = 0$ implies

$$
\sum_{j+k \leq r} a_{jk}(y_1 + \alpha)^j (z_1 + \beta)^k \equiv 0 \mod (y_1, z_1)^{r+2}
$$

which implies

$$\sum_{j+k\leq r} a_{jk}(y_1+\alpha)^j(z_1+\beta)^k = 0,$$

a contradiction. Thus $\gamma \neq 0$.

$$\sum_{j+k\leq r} a_{jk}(y_1+\alpha)^j(z_1+\beta)^k \equiv \gamma(y_1+\alpha)^{-\lambda_1}(z_1+\beta)^{-\lambda_2} \mod (y_1,z_1)^{r+2}$$
(47)

Set $f(y_1,z_1) = (y_1+\alpha)^{-\lambda_1}(z_1+\beta)^{-\lambda_2}$,

$$\alpha_{ij} = \frac{1}{i!\,j!}\frac{\partial^{i+j}f}{\partial y_1^i \partial z_1^j}(0,0)$$

Then

$$(y_1+\alpha)^{-\lambda_1}(z_1+\beta)^{-\lambda_2} = \sum \alpha_{ij}y_1^i z_1^j.$$

$$\alpha_{ij} = \begin{cases} \left(\frac{-\lambda_1(-\lambda_1-1)\cdots(-\lambda_1-i+1)}{i!}\alpha^{-\lambda_1-i}\right)\left(\frac{-\lambda_2(-\lambda_2-1)\cdots(-\lambda_2-j+1)}{j!}\beta^{-\lambda_2-j}\right) \\ \quad \text{if } i,j > 0 \\ \alpha^{-\lambda_1}\left(\frac{-\lambda_2(-\lambda_2-1)\cdots(-\lambda_2-j+1)}{j!}\beta^{-\lambda_2-j}\right) \\ \quad \text{if } i = 0, j > 0 \\ \left(\frac{-\lambda_1(-\lambda_1-1)\cdots(-\lambda_1-i+1)}{i!}\alpha^{-\lambda_1-i}\right)\beta^{-\lambda_2} \\ \quad \text{if } j = 0, i > 0 \\ \alpha^{-\lambda_1}\beta^{-\lambda_2} \\ \quad \text{if } i = j = 0 \end{cases}$$

Thus $\alpha_{ij} = 0$ for $i+j = r+1$ by (47), and $-\lambda_1 \in \{0,1,\ldots,r\}$, $-\lambda_2 \in \{0,1,\ldots,r\}$ and $-\lambda_1 - \lambda_2 \leq r$. Thus

$$\sum_{j+k\leq r} a_{jk}(y_1+\alpha)^j(z_1+\beta)^k = \gamma(y_1+\alpha)^{-\lambda_1}(z_1+\beta)^{-\lambda_2}$$

so that

$$\begin{aligned} \sum_{i+j+k=r} a_{ijk}x_1^r(y_1+\alpha)^j(z_1+\beta)^k &= \gamma x_1^r(y_1+\alpha)^{-\lambda_1}(z_1+\beta)^{-\lambda_2} \\ &= \gamma x_1^{r+\lambda_1+\lambda_2}\left[x_1^{-\lambda_1}(y_1+\alpha)^{-\lambda_1}\right]\left[x_1^{-\lambda_2}(z_1+\beta)^{-\lambda_2}\right]\end{aligned}$$

and the leading form of F is

$$L = \sum_{i+j+k=r} a_{ijk}x^iy^jz^k = \gamma x^{r+\lambda_1+\lambda_2}y^{-\lambda_1}z^{-\lambda_2}$$

$$x^dy^ez^fL = \gamma x^{d+r+\lambda_1+\lambda_2}y^{e-\lambda_1}z^{f-\lambda_2}$$

Set

$$\begin{aligned} \underline{a} &= d+r+e-\frac{b(d+e+f+r)}{a+b+c}+f-\frac{c(d+e+f+r)}{a+b+c} \\ \underline{b} &= e-\left(e-\frac{b(d+e+f+r)}{a+b+c}\right) \\ \underline{c} &= f-\left(f-\frac{c(d+e+f+r)}{a+b+c}\right) \end{aligned}$$

Set $\tau = \frac{d+e+f+r}{a+b+c}$.

$$\underline{a} = \frac{(d+e+f+r)(a+b+c)-b(d+e+f+r)-c(d+e+f+r)}{a+b+c} = a\tau$$
$$\underline{b} = b\tau$$
$$\underline{c} = c\tau$$

$$b\underline{a} - a\underline{b} = (ba - ab)\tau = 0$$
$$a\underline{c} - c\underline{a} = (ac - ca)\tau = 0$$
$$c\underline{b} - b\underline{c} = (cb - bc)\tau = 0$$

thus $\gamma = 0$ since F is normalized. This contradiction shows that $\nu(q) \le r+1$.

We have shown that

$$F_1 = (y_1 + \alpha)^{\lambda_1}(z_1 + \beta)^{\lambda_2}\left(\sum_{j+k\le r} a_{jk}(y_1 + \alpha)^j(z_1 + \beta)^k\right) - \gamma + \overline{x}_1\Sigma$$

with

$$\gamma = \alpha^{\lambda_1}\beta^{\lambda_2}\sum_{j+k\le r} a_{jk}\alpha^j\beta^k.$$

Thus $r_1 = r+1$ implies there is a nonzero degree $r+1$ term in $F_1(0, y_1, z_1)$ so that $\gamma(q) = r+1$.

Now suppose that $q \in \pi^{-1}(p)$ is a 2 point. Then after possibly interchanging x, y, z, there are regular parameters (x_1, y_1, z_1) at q such that

$$x = x_1$$
$$y = x_1 y_1$$
$$z = x_1(z_1 + \beta)$$

with $\beta \ne 0$.

$$u = x_1^{(a+b+c)k} y_1^{bk}(z_1 + \beta)^{ck}$$

Set

$$x_1 = (z_1 + \beta)^{\frac{-c}{a+b+c}}\overline{x}_1$$

$$u = (\overline{x}_1^{a+b+c}y_1^b)^k$$
$$v = P(\overline{x}_1^{a+b+c}y_1^b) + x_1^{d+e+f+r}y_1^e(z_1 + \beta)^f\frac{F}{x_1^r}$$
$$= P(\overline{x}_1^{a+b+c}y_1^b) + \overline{x}_1^{d+e+f+r}y_1^e(z_1 + \beta)^{\lambda_1}\frac{F}{x_1^r}$$

where

$$\lambda_1 = f - \frac{c(d+e+f+r)}{a+b+c}. \tag{48}$$

$$\frac{F}{x_1^r} = \sum_{j+k\le r} a_{jk}y_1^j(z_1 + \beta)^k + \overline{x}_1\Omega \tag{49}$$

where $a_{jk} = a_{r-i-j,j,k}$. There is at most one natural number t such that

$$(d+e+f+r)b - (e+t)(a+b+c) = 0. \tag{50}$$

If $\nu(q) > r+1$ there exists a t satisfying (50), and $0 \neq \gamma \in k$ such that

$$(z_1 + \beta)^{\lambda_1} \left(\sum_{j+k \leq r} a_{jk} y_1^j (z_1 + \beta)^k \right) \equiv \gamma y_1^t \mod (y_1, z_1)^{r+2}.$$

Thus

$$\sum_{j+k \leq r} a_{jk} y_1^j (z_1 + \beta)^k \equiv \gamma (z_1 + \beta)^{-\lambda_1} y_1^t \mod (y_1, z_1)^{r+2}.$$

Set

$$\tau_j = \frac{-\lambda_1(-\lambda_1 - 1) \cdots (-\lambda_1 - j + 1)}{j!} \beta^{-\lambda_1 - j}.$$

$$\sum_{j+k \leq r} a_{jk} y_1^j (z_1 + \beta)^k \equiv \gamma y_1^t \left(\sum_{j=0}^{\infty} \tau_j z_1^j \right) \mod (y_1, z_1)^{r+2}.$$

implies

$$0 = \tau_{r+1-t} = \frac{-\lambda_1(-\lambda_1 - 1) \cdots (-\lambda_1 - r + t)}{(r + 1 - t)!} \beta^{-\lambda_1 - (r+1-t)}$$

so that $-\lambda_1 \in \{0, 1, \ldots, r - t\}$ and $t \leq r$. Thus

$$\sum_{j+k \leq r} a_{jk} y_1^j (z_1 + \beta)^k = \gamma (z_1 + \beta)^{-\lambda_1} y_1^t$$

$$\sum_{i+j+k=r} a_{ijk} x_1^r y_1^j (z_1 + \beta)^k = \gamma x_1^r (z_1 + \beta)^{-\lambda_1} y_1^t$$
$$= \gamma x_1^{r-t+\lambda_1} [x_1^t y_1^t] \left[x_1^{-\lambda_1} (z_1 + \beta)^{-\lambda_1} \right]$$

$$x^d y^e z^f L = \gamma x^{r-t+\lambda_1+d} y^{t+e} z^{f-\lambda_1}$$

where L is the leading form of F. Set

$$\begin{aligned} \underline{a} &= r - t + \lambda_1 + d \\ \underline{b} &= t + e \\ \underline{c} &= f - \lambda_1 \end{aligned}$$

We have the relations (50) and (48). (50) implies

$$t = \frac{(d + e + f + r)b - e(a + b + c)}{a + b + c}.$$

$$\begin{aligned} \underline{a} &= d + r - t + \lambda_1 = \frac{a(d+e+f+r)}{a+b+c} \\ \underline{b} &= t + e = \frac{(d+e+f+r)b}{a+b+c} \\ \underline{c} &= \frac{c(d+e+f+r)}{a+b+c} \\ 0 &= b\bar{a} - a\bar{b} = c\bar{a} - a\bar{c} = c\bar{b} - b\bar{c} \end{aligned}$$

Thus

$$x^{r-t+\lambda_1+d} y^{t+e} z^{f-\lambda_1} = (x^a y^b z^c)^m$$

for some $m \in \mathbb{N}$, a contradiction, since F is normalized. Thus $\nu(q) \leq r + 1$.
Suppose that $\nu(q) = r + 1$.

$$F_q = \sum_{j \leq r} \left[\sum_{k \leq r-j} a_{jk} (z_1 + \alpha)^{k+\lambda_1} \right] y_1^j - \gamma y_1^t + \bar{x}_1 \Sigma$$

with $t \leq r$, $\gamma \in k$ implies $a_{jk} = 0$ if $j \neq t$. Thus

$$F_q = y_1^t (\sum_{k \leq r-t} a_{tk}(z_1 + \alpha)^{k+\lambda_1} - \gamma) + \overline{x}_1 \Sigma$$

implies

$$L_q = c y_1^t z_1^{r+1-t} + \overline{x}_1 \Omega$$

where $0 \neq c \in k$.

Now suppose that $q \in \pi^{-1}(p)$ is a 3 point. After possibly permuting x, y, z, there are permissible parameters (x_1, y_1, z_1) at q such that

$$
\begin{aligned}
x &= x_1 \\
y &= x_1 y_1 \\
z &= x_1 z_1
\end{aligned}
$$

$$
\begin{aligned}
u &= (x_1^{a+b+c} y_1^b z_1^c)^k \\
v &= P(x_1^{a+b+c} y_1^b z_1^c) + x_1^{d+e+f+r} y_1^e z_1^f \frac{F}{x_1^r}
\end{aligned}
$$

$\nu(q) = \nu(\frac{F}{x_1^r}) \leq r$. $\square$

Example 7.2. $\nu(p)$ *can go up by 1 after a quadratic transform. We can construct the example as follows.*

$$u = xy, v = x^2 y$$

has $F = 1$. Blow up p and consider the point p_1 above p with regular parameters (x_1, y_1, z_1) defined by $x = x_1, y = x_1(y_1 + \alpha)$, $\alpha \neq 0$, $z = x_1 z_1$. Set $\overline{x}_1 = x_1(y_1 + \alpha)^{\frac{1}{2}}$, $\overline{y}_1 = (y_1 + \alpha)^{-\frac{1}{2}} - \alpha^{-\frac{1}{2}}$. Then

$$u = \overline{x}_1^2, v = \alpha^{-\frac{1}{2}} \overline{x}_1^3 + \overline{x}_1^3 \overline{y}_1,$$

so that $F_1 = \overline{y}_1$.

Theorem 7.3. *Suppose that $\nu(p) = r$, $\pi : X_1 \to X$ is the blow-up of p, $q \in \pi^{-1}(p)$ with $r_1 = \nu(q)$.*

If p is a 1 point then

1. *If q is a 1 point, then $r_1 < r$ if $\tau(p) < r$, and if $r_1 = r$ then $\tau(q) = r$.*
2. *If q is a 1 point, then $\gamma(q) \leq r$.*
3. *If q is a 2 point, and $r_1 = r$ then $\tau(p) \leq \tau(q)$.*
4. *If q is a 2 point and $\gamma(p) = r$, then $\gamma(q) \leq r$.*

If p is a 2 point and $1 \leq \tau(p)$ then

1. *If q is a 1 point then $r_1 \leq r$ and $\gamma(q) \leq r$.*
2. *If q is a 2 point and $r_1 = r$, then $\tau(p) \leq \tau(q)$.*
3. *If q is a 2 point and $\gamma(p) = r$, then $\gamma(q) \leq r$.*
4. *If q is a 3 point then $r_1 \leq r - \tau(p)$.*

Proof. **Suppose that p is a 1 point with $\gamma(p) = r$.** Suppose that $q \in \pi^{-1}(p)$ is a 1 point, and $r_1 = r$. After making a permissible change of parameters we can assume that $x = x_1, y = x_1 y_1, z = x_1 z_1$. We than have, with the notation of (39).

$$F_1 = \sum_{j+k \leq r} a_{r-j-k,j,k} y_1^j z_1^k + x_1 \Omega.$$

Now suppose that $q \in \pi^{-1}(p)$ is a 2 point, and $r_1 = r$. After making a permissible change of parameters we can assume that $x = x_1 y_1, y = y_1, z = y_1 z_1$. We than have, with the notation of (40)

$$F_1 = \sum_{i+k \leq r} a_{i,r-i-k,k} x_1^i z_1^k + y_1 \Sigma$$

Suppose that p is a 2 point and $1 \leq \tau(p)$. Suppose that $q \in \pi^{-1}(p)$ is a 1 point. After making a permissible change of parameters, we have $x = x_1, y = x_1(y_1 + \alpha), z = x_1 z_1$ with $\alpha \neq 0$. We then have, with the notation of (41),

$$F_1 = \sum_{j+k \leq r} a_{r-j-k,j,k} (y_1 + \alpha)^j (y_1 + \alpha)^\lambda z_1^k - \gamma + \overline{x}_1 \Omega.$$

There exists a_{ijk} with $i + j + k = r$ and $k = \tau(p) \geq 1$ such that $a_{ijk} \neq 0$. Thus $r_1 \leq r$ and $\gamma(q) \leq r$.

Now suppose that $q \in \pi^{-1}(p)$ is a 2 point. After making a permissible change of parameters, we have $x = x_1, y = x_1 y_1, z = x_1 z_1$ We then have, with the notation of (43),

$$F_1 = \sum_{j+k \leq r} a_{r-j-k,j,k} y_1^j z_1^k + x_1 \Omega.$$

and there exist i, j, k such that $i + j + k = r$ and $a_{ijk} \neq 0$ with $k = \tau(p)$. Thus if $r_1 = r$, we have $\tau(p) \leq \tau(q)$. If $\gamma(p) = r$, we have $\gamma(q) \leq r$.

Now suppose that $q \in \pi^{-1}(p)$ is a 3 point. Then $x = x_1 z_1, y = y_1 z_1, z = z_1$ We then have, with the notation of (45),

$$F_1 = \sum_{i+j \leq r} a_{i,j,r-i-j} x_1^i y_1^j + z_1 \Omega.$$

There exists a_{ijk} with $i + j + k = r$ and $k = \tau(p) \geq 1$ such that $a_{ijk} \neq 0$. Thus $r_1 \leq r - \tau(p)$. $\qquad \square$

Lemma 7.4. *Suppose that $r \geq 2$ and $p \in X$ is a 1 point. Suppose that (x, y, z) are permissible parameters at p and $C \subset \overline{S}_r(X)$ is a curve such that $p \in C$. Then $F_p \in \hat{\mathcal{I}}_{C,p}^r + (x^{r-1})$.*

Proof. $x \in \hat{\mathcal{I}}_{C,p}$ by Lemma 6.11. There exist permissible parameters $(x, \overline{y}, \overline{z})$ at p such that $\overline{y}, \overline{z} \in \mathcal{O}_{X,p}, \overline{y} = y + h_1, \overline{z} = z + h_2$ with $h_1, h_2 \in m^r$.

Suppose that the conclusions of the Lemma are true for the parameters $(x, \overline{y}, \overline{z})$.

$$u = x^a$$
$$v = \overline{P}(x) + x^b \overline{F}(x, \overline{y}, \overline{z})$$

and $\overline{F}(x, \overline{y}, \overline{z})$ is normalized with respect to the permissible parameters $(x, \overline{y}, \overline{z})$. We have an expression

$$
\begin{aligned}
u &= x^a \\
v &= P(x) + x^b F(x, y, z)
\end{aligned}
$$

where $F(x, y, z)$ is normalized with respect to the permissible parameters (x, y, z).

$$\overline{F}(x, \overline{y}, \overline{z}) = F(x, y, z) + \Omega$$

with Ω a series in x. Since $F(x, y, z)$ is normalized, $\nu(\overline{F}) = \nu(F) = r$ and only powers of x of order $\geq r$ can be removed from $\overline{F}(x, \overline{y}, \overline{z})$ to normalize to obtain $F(x, y, z)$. Thus the conclusions of the Lemma hold for (x, y, z)

We may thus assume that $y, z \in \mathcal{O}_{X,p}$.

There exists an étale neighborhood U of p such that (x, y, z) are uniformizing parameters in U, $x = 0$ is a local equation of $E_X \cap U$, $C \cap U$ is a complete intersection. Let $R = \Gamma(U, \mathcal{O}_U)$, $I_C = \Gamma(U, \mathcal{I}_C)$. Set

$$w = \frac{v - P_t(x)}{x^b}$$

where $t > b + r$. Thus

$$w \in (y, z, x^r)\mathcal{O}_{U,p} \tag{51}$$

and

$$w - F \in (x^r)\hat{\mathcal{O}}_{X,p}. \tag{52}$$

Let q be a smooth point of $C \cap U$. Then there exists $\alpha, \beta \in k$ such that $(x, y - \alpha, z - \beta)$ are permissible parameters at q. Lemma 6.24 implies

$$F_q \in \hat{\mathcal{I}}_{C,q}^r + (x^{r-1}).$$

$$F_q = w - \sum_{i=0}^{\infty} \frac{\partial^i w}{\partial x^i}(0, \alpha, \beta)x^i$$

Set

$$\Lambda = w - \sum_{i < r-1} \frac{\partial^i w}{\partial x^i}(0, \alpha, \beta)x^i.$$

$\Lambda \in (\mathcal{I}_{C,q}^r + (x^{r-1}))\hat{\mathcal{O}}_{X,q}$ implies by Theorem 7.1, Chapter VIII, section 4 [38] and Lemma 6.28,

$$\Lambda \in (\mathcal{I}_{C,q}^r + (x^{r-1})) \cap R = (I_C^r + (x^{r-1})R.$$

$\nu(p) = r$ and $t > b + r$ implies $\nu(w) \geq r$ and $\frac{\partial^i w}{\partial x^i}(0, \alpha, \beta) = 0$ if $i < r - 1$, so that $w \in \hat{\mathcal{I}}_{C,p}^r + (x^{r-1})$ which implies that $F_p \in \hat{\mathcal{I}}_{C,p}^r + (x^{r-1})$. $\quad\square$

Lemma 7.5. *Suppose that $p \in X$ is a 1 point and $\nu(p) = \gamma(p) = r \geq 2$. Then there exists at most one curve C in $\overline{S}_r(X)$ containing p. If C exists then it is nonsingular at p.*

Proof. Suppose that (x, y, z) are permissible parameters at p. Write $\hat{\mathcal{I}}_{C,p} = (x, f(y, z))$. By Lemma 7.4, $F_p \in \hat{\mathcal{I}}_{C,p}^r + (x^{r-1})$. $f^r \mid F_p(0, y, z)$ and $\gamma(p) = r$ implies $\nu(f) = 1$ and C is nonsingular at p.

Suppose that $D \subset \overline{S}_r(X)$ is another curve containing p. Then D is nonsingular at p. Lemma 6.24 implies there exist permissible parameters (x, y, z) at p such that $\hat{\mathcal{I}}_{C,p} = (x, z)$, there exist series a, b_{ij} such that

$$F_p = x^{r-1}a + \sum_{i+j=r} b_{ij}x^i z^j.$$

b_{0r} is a unit implies

$$\frac{\partial^{r-1}F_p}{\partial z^{r-1}} = x\phi + z\psi$$

where ψ is a unit. Since $F_p \in \hat{\mathcal{I}}_{D,p}^r + (x^{r-1})$, we have

$$\frac{\partial^{r-1}F_p}{\partial z^{r-1}} \in \hat{\mathcal{I}}_{D,p}$$

which implies $z \in \hat{\mathcal{I}}_{D,p}$, so that $C = D$. $\qquad\square$

Lemma 7.6. *Suppose that $r \geq 2$, p is a 2 point and $C \subset \overline{S}_r(X)$ is an irreducible curve containing a 1 point, such that $p \in C$. Then $\nu(p) \geq r - 1$. If $\tau(p) > 0$, then $\nu(p) \geq r$.*

Proof. First suppose that C is nonsingular at p and is transversal at p to the 2 curve through p. Then the result follows from Lemma 6.26. Now suppose that C does not make SNCs with the 2 curve through p. Let $s = \nu(p)$. There exists a sequence of quadratic transforms $\pi : X_1 \to X$ centered at 2 and 3 points such that the strict transform C' of C makes SNCs with $\overline{B}_2(X_1)$ at a 2 point $p_1 = C' \cap \pi^{-1}(p)$. We have $s_1 = \nu(p_1) \leq s + 1$, by Theorems 7.1 and 7.3. $s_1 = s + 1$ implies $\tau(p_1) > 0$ and $\tau(p) = 0$. $\tau(p) > 0$ implies $s_1 \leq s$ and if we further have $s_1 = s$, then $\tau(p_1) > 0$.

First suppose that $\tau(p) > 0$. If $s_1 = s$ then $\tau(p_1) > 0$ so that $r \leq s_1 = s$. If $s_1 < s$ then $s > s_1 \geq r - 1$, which implies $s \geq r$.

Now suppose that $\tau(p) = 0$. If $s_1 \leq s$ then $s \geq s_1 \geq r - 1$. If $s_1 = s + 1$ then $s + 1 = s_1 \geq r$ since $\tau(p_1) > 0$, so that $s \geq r - 1$. $\qquad\square$

Lemma 7.7. *Suppose that $r \geq 2$, p is a 3 point and $C \subset \overline{S}_r(X)$ is an irreducible curve containing a 1 point such that $p \in C$. Then $\nu(p) \geq r - 1$.*

Proof. Let $s = \nu(p)$. There exists a sequence of quadratic transforms $\pi : X_1 \to X$ centered at 2 and 3 points such that the strict transform C' of C makes SNCs with $\overline{B}_2(X_1)$ at the 2 point $p_1 = C' \cap \pi^{-1}(p)$. We have $s_1 = \nu(p_1) \leq s + 1$ by Theorems 7.1 and 7.3.

If $s_1 = s + 1$ then $\tau(p_1) > 0$, so that $s_1 \geq r$ by Lemma 7.6, so that $s \geq r - 1$. If $s_1 \leq s$, then $s \geq s_1 \geq r - 1$ by Lemma 7.6. $\qquad\square$

Theorem 7.8. *Suppose that $p \in X$ has $\nu(p) = r \geq 1$, and (x, y, z) are permissible parameters at p, $\pi : X_1 \to X$ is the blow-up of p.*

Suppose that p is a 3 point

1. Suppose that the leading form $L_p = L(x, y, z)$ depends on x, y and z. Then there are no curves C in $\pi^{-1}(p) \cap \overline{S}_{r+1}(X_1)$. No 2 curves C of $\pi^{-1}(p)$ satisfy $F_q \in \hat{\mathcal{I}}^r_{C,q}$ for $q \in C$.

2. Suppose that $L_p = L(x, y)$ depends on x and y. Then the curves in $\pi^{-1}(p) \cap \overline{S}_{r+1}(X_1)$ are a finite union of lines passing through a single 3 point of $\pi^{-1}(p)$. No 2 curves C of $\pi^{-1}(p)$ satisfy $F_q \in \hat{\mathcal{I}}^r_{C,q}$ for $q \in C$.

3. Suppose that $L_p = L(x)$ depends on x. Then there are no 1 points in $\pi^{-1}(p) \cap \overline{S}_{r+1}(X_1)$ and there is at most one curve C in $\pi^{-1}(p) \cap \overline{S}_{r+1}(X_1)$. It is the 2 curve D which is the intersection of the strict transform of $x = 0$ with $\pi^{-1}(p)$. D is the only 2 curve C in $\pi^{-1}(p)$ such that $F_q \in \hat{\mathcal{I}}^r_{C,q}$ for $q \in C$.

Suppose that p is a 2 point. Then the curves in $\pi^{-1}(p) \cap \overline{S}_{r+1}(X_1)$ are a finite union of lines passing through the 3 point. There are no 2 curves in $\pi^{-1}(p) \cap \overline{S}_{r+1}(X_1)$.

Proof. First suppose that p is a 3 point and $L_p = L(x, y, z)$ depends on x, y and z. There are no 3 points in $\pi^{-1}(p) \cap \overline{S}_{r+1}(X_1)$ by Lemma 7.7 and Lemma 6.27 since (by direct calculation) $\nu(q) \le r - 1$ at all 3 points in $\pi^{-1}(p)$. There are no 2 curves in $\pi^{-1}(p) \cap \overline{S}_{r+1}(X_1)$ and there are no 2 curves in $\pi^{-1}(p)$ such that $F_q \in \hat{\mathcal{I}}^r_{C,q}$ for $q \in C$ by Lemma 6.27. We will now show that there are no curves in $\overline{S}_{r+1}(X_1) \cap \pi^{-1}(p)$. Suppose that there is a curve C in $\overline{S}_{r+1}(X_1) \cap \pi^{-1}(p)$ containing a 1 point. C must contain a 2 point q. $\nu(q) = r$ or $r + 1$ by Lemma 7.6 and Theorem 7.1.

First suppose that $\nu(q) = r+1$. Then by Theorem 7.1, there exist permissible parameters (x, y, z) at p such that

$$L = y^t f(x, z) \tag{53}$$

for some t with $0 < t < r$, (since L depends on x, y, and z). Write

$$f = \sum_{i+k=r-t} b_{ik} x^i z^k$$

At a 1 point of C we have (with the notation of (46)) permissible parameters (x_1, y_1, z_1) such that $x = x_1, y = x_1(y_1 + \alpha), z = x_1(z_1 + \beta)$ with $\alpha, \beta \ne 0$

$$(y_1 + \alpha)^{\lambda_1} (z_1 + \beta)^{\lambda_2} \left[\sum_{i+j+k=r} a_{ijk}(y_1 + \alpha)^j (z_1 + \beta)^k \right] \equiv c_{\alpha,\beta} \bmod (y_1, z_1)^{r+1} \tag{54}$$

for some $0 \ne c_{\alpha,\beta} \in k$. Substituting (53), we have

$$(y_1 + \alpha)^{\lambda_1} (z_1 + \beta)^{\lambda_2} \left[(y_1 + \alpha)^t \sum_{i+k=r-t} b_{ik}(z_1 + \beta)^k \right] \equiv c_{\alpha,\beta} \bmod (y_1, z_1)^{r+1}$$
$$(y_1 + \alpha)^{t+\lambda_1} \left[\sum_{i+k=r-t} b_{ik}(z_1 + \beta)^k (z_1 + \beta)^{\lambda_2} \right] \equiv c_{\alpha,\beta} \bmod (y_1, z_1)^{r+1}$$

If $t \ne -\lambda_1$ this is a contradiction, since there is then a nonzero $y_1 z_1^s$ term for some $0 \le s \le r - t$. Thus $-\lambda_1 = t \in \{1, \dots, r-1\}$. But

$$\nu\left(\sum_{i+k=r-t} b_{ik}(z_1 + \beta)^k (z_1 + \beta)^{\lambda_2} - c_{\alpha,\beta} \right) \le r - t + 1 \le r$$

which is contradiction.

Now suppose that $\nu(q) = r$. We have from (49) (in the proof of Theorem 7.1) that there exist permissible parameters (x_1, y_1, z_1) at q such that

$$x = x_1, y = x_1 y_1, z = x_1(z_1 + \beta),$$

$$F_q = \left[\sum_{j+k \leq r} a_{r-j-k,j,k} y_1^j (z_1 + \beta)^k \right] (z_1 + \beta)^{\lambda_1} - \gamma y_1^t + \bar{x}_1 \Omega \qquad (55)$$

if there exists a natural number t such that $(d+e+f+r)b - (e+t)(a+b+c) = 0$, and

$$F_q = \left[\sum_{j+k \leq r} a_{r-j-k,j,k} y_1^j (z_1 + \beta)^k \right] (z_1 + \beta)^{\lambda_1} + \bar{x}_1 \Omega \qquad (56)$$

otherwise. Since $\nu(q) = r$, For fixed $j \neq t$, we have

$$\nu\left(\sum_{k \leq r-j} a_{r-j-k,j,k}(z_1 + \beta)^k \right) \geq r - j$$

Thus (for fixed $j \neq t$)

$$\sum_{k \leq r-j} a_{r-j-k,j,k}(z_1 + \beta)^k = \gamma_j z_1^{r-j}$$

for some $\gamma_j \in k$ and (for fixed $j \neq t$)

$$\begin{aligned}
\sum_{i+k=r-j} a_{ijk} x^i y^j z^k &= y^j \left[\sum_{k \leq r-j} a_{r-j-k,j,k} [x_1^k(z_1 + \beta)^k] x_1^{r-j-k} \right] \\
&= y^j x_1^{r-j} \left[\sum_{k \leq r-j} a_{r-j-k,j,k}(z_1 + \beta)^k \right] \\
&= \gamma_j x_1^{r-j} z_1^{r-j} y^j \\
&= \gamma_j x_1^{r-j} \left(\frac{z}{x} - \beta \right)^{r-j} y^j \\
&= \gamma_j (z - \beta x)^{r-j} y^j
\end{aligned}$$

For $j = t$, we have

$$\nu\left(\sum_{k \leq r-t} a_{r-t-k,t,k}(z_1 + \beta)^{k+\lambda_1} - \gamma \right) \geq r - t$$

Thus we either have

$$L_p = \sum_{j \neq t} \gamma_j y^j (z - \beta x)^{r-j} + y^t f(x, z)$$

where $(d + e + f + r)b - (e + t)(a + b + c) = 0$, and f is homogeneous of degree $r - t$, or

$$L_p = \sum_j \gamma_j y^j (z - \beta x)^{r-j}.$$

Thus

$$L_q = \sum_j \gamma_j y_1^j z_1^{r-j} + x_1 \Omega_1.$$

or

$$L_q = \sum_{j \neq t} \gamma_j y_1^j z_1^{r-j} + \gamma_t y_1^t z_1^{r-t} + x_1 \Omega_1.$$

for some $\gamma_t \in k$. If some $\gamma_j \neq 0$ with $j \neq r$, $\tau(q) > 0$, so that $\nu(q) \geq r + 1$ by Lemma 7.6, a contradiction.

The remaining case is

$$L_p = \gamma_r y^r + y^t f(x, z), \tag{57}$$

with $f \neq 0$, $t < r$ and $\gamma_r \neq 0$ if $t = 0$, since L_p depends on x, y and z.

$$f = \sum_{i+k=r-t} b_{ik} x^i z^k$$

At a 1 point of C we have (with the notation of (46)) regular parameters (x_1, y_1, z_1) such that $x = x_1, y = x_1(y_1 + \alpha), z = x_1(z_1 + \beta)$ with $\alpha, \beta \neq 0$

$$(y_1 + \alpha)^{\lambda_1} (z_1 + \beta)^{\lambda_2} \left[\sum_{i+j+k=r} a_{ijk}(y_1 + \alpha)^j (z_1 + \beta)^k \right] \equiv c_{\alpha,\beta} \bmod (y_1, z_1)^{r+1} \tag{58}$$

for some $c_{\alpha,\beta} \in k$. Substituting (57), we have

$$(y_1 + \alpha)^{\lambda_1} (z_1 + \beta)^{\lambda_2} \left[\gamma_r (y_1 + \alpha)^r + (y_1 + \alpha)^t \sum_{i+k=r-t} b_{ik}(z_1 + \beta)^k \right]$$
$$\equiv c_{\alpha,\beta} \bmod (y_1, z_1)^{r+1}$$

$$\gamma_r(y_1+\alpha)^{r-t} + \sum_{i+k=r-t} b_{ik}(z_1+\beta)^k \equiv (y_1+\alpha)^{-\lambda_1-t}(z_1+\beta)^{-\lambda_2} c_{\alpha,\beta} \bmod (y_1, z_1)^{r+1}$$

The LHS of the last equation has no $y_1 z_1$ term which implies $\lambda_1 = -t$ or $-\lambda_2 = 0$. $\lambda_1 = -t$ implies $\gamma_r = 0$ and $t > 0$,

$$\nu\left(\sum_{i+k=r-t} b_{ik}(z_1 + \beta)^k - c_{\alpha,\beta}(z_1 + \beta)^{-\lambda_2} \right) \leq r - t + 1 \leq r$$

which is contradiction. $\lambda_2 = 0$ implies $f = 0$, a contradiction.

Now suppose that p is a 3 point and $L_p = L(x, y)$. Suppose that $q \in \pi^{-1}(p)$ is a 1 point and $\nu(q) = r + 1$. $\hat{\mathcal{O}}_{X_1,q}$ has regular parameters x_1, y_1, z_1 such that

$$\begin{aligned} x &= x_1 \\ y &= x_1(y_1 + \alpha) \\ z &= x_1(z_1 + \beta) \end{aligned} \tag{59}$$

where $\alpha, \beta \neq 0$. Write

$$L(x, y) = \sum_{i+j=r} a_{ij} x^i y^j.$$

(with the notation of (46))

$$(y_1 + \alpha)^{\lambda_1} (z_1 + \beta)^{\lambda_2} \left[\sum a_{ij}(y_1 + \alpha)^j \right] \equiv c_{\alpha,\beta} \bmod (y_1, z_1)^{r+1} \tag{60}$$

which implies

$$\sum_{i+j=r} a_{ij}(y_1 + \alpha)^j \equiv (y_1 + \alpha)^{-\lambda_1}(z_1 + \beta)^{-\lambda_2} c_{\alpha,\beta} \bmod (y_1, z_1)^{r+1} \tag{61}$$

so that $\lambda_2 = 0$, and

$$\sum_{i+j=r} a_{ij}(y_1 + \alpha)^j \equiv (y_1 + \alpha)^{-\lambda_1} c_\alpha \bmod (y_1)^{r+1} \tag{62}$$

We will now show that there exist at most finitely many values of α such that an equation (62) holds. Set

$$g(t) = \sum_{i \le r} a_i t^i$$

where $a_i = a_{r-i,i}$.

Suppose there are infinitely many values of α such that (62) holds for some $q \in \pi^{-1}(p)$ with value β and regular parameters x_1, y_1, z_1 in $\hat{O}_{X_1,q}$ as in (59). Define g_α by

$$g_\alpha(y_1) = g(y_1 + \alpha) = g(\frac{y}{x}).$$

Set $\lambda = \lambda_1$. By assumption,

$$g_\alpha(y_1) = \sum_{i \le r} a_i(y_1 + \alpha)^i \equiv c_\alpha(y_1 + \alpha)^{-\lambda} \bmod y_1^{r+1} \qquad (63)$$

We can expand the RHS of (63) as

$$c_\alpha(y_1 + \alpha)^{-\lambda} = c_\alpha \alpha^{-\lambda} + c_\alpha(-\lambda)\alpha^{-\lambda-1}y_1$$
$$+ c_\alpha \frac{-\lambda(-\lambda-1)}{2}\alpha^{-\lambda-2}y_1^2 + \cdots$$
$$+ c_\alpha \frac{-\lambda(-\lambda-1)\cdots(-\lambda-r+1)}{r!}\alpha^{-\lambda-r}y_1^r + \cdots$$

We can expand the LHS of (63) as

$$g_\alpha(y_1) = g_\alpha(0) + \frac{dg_\alpha}{dy_1}(0)y_1 + \frac{1}{2}\frac{d^2 g_\alpha}{dy_1^2}(0)y_1^2 + \cdots + \frac{1}{r!}\frac{d^r g_\alpha}{dy_1^r}(0)y_1^r$$
$$= g(\alpha) + \frac{dg}{dt}(\alpha)y_1 + \frac{1}{2}\frac{d^2 g}{dt^2}(\alpha)y_1^2 + \cdots + \frac{1}{r!}\frac{d^r g}{dt^r}(\alpha)y_1^r$$

We get that

$$r!a_r = \frac{d^r g}{dt^r}(\alpha) = c_\alpha(-\lambda)(-\lambda-1)\cdots(-\lambda-r+1)\alpha^{-\lambda-r}$$

which implies that

$$c_\alpha = \frac{r!a_r}{(-\lambda)(-\lambda-1)\cdots(-\lambda-r+1)\alpha^{-\lambda-r}}$$

$$g(\alpha) = c_\alpha \alpha^{-\lambda} = \frac{r!a_r \alpha^{-\lambda}}{(-\lambda)(-\lambda-1)\cdots(-\lambda-r+1)\alpha^{-\lambda-r}}$$
$$= \frac{r!a_r \alpha^r}{(-\lambda)(-\lambda-1)\cdots(-\lambda-r+1)}$$

Since this holds for infinitely many α, and $g(t)$ is a polynomial,

$$g(t) = \frac{r!a_r t^r}{-\lambda(-\lambda-1)\cdots(-\lambda-r+1)}.$$

Thus

$$\sum_{i \le r} a_{r-i,i} t^i = \frac{r!a_r t^r}{-\lambda(-\lambda-1)\cdots(-\lambda-r+1)}$$

so that $a_{r-i,i} = 0$ if $i < r$. Thus $L_p = a_{0r} y^r$, a contradiction to the assumption that L depends on two variables.

Thus the only curves in $\overline{S_{r+1}(X_1)} \cap \pi^{-1}(p)$ which contain a 1 point are on the strict transforms of $y - \alpha x = 0$ for a finite number of nonzero α. These lines contain the 3 point of X which has permissible parameters (x_1, y_1, z_1) defined by $x = x_1 z_1, y = y_1 z_1, z = z_1$.

Since $L_p = L(x, y)$, there is at most one 3 point q in $\pi^{-1}(p)$ with $\nu(q) = r$. Thus there are no 2 curves in $\overline{S}_{r+1}(X_1) \cap \pi^{-1}(p)$ by Lemma 6.27.

Now suppose that p is a 3 point and $L_p = L(x)$. Suppose that $q \in \pi^{-1}(p)$ is a 1 point and $\nu(q) = r + 1$. $\hat{\mathcal{O}}_{X_1,q}$ has regular parameters x_1, y_1, z_1 such that

$$
\begin{aligned}
x &= x_1 \\
y &= x_1(y_1 + \alpha) \\
z &= x_1(z_1 + \beta)
\end{aligned}
$$

where $\alpha, \beta \neq 0$.

$$L(x) = \bar{a}x^r$$

With the notation of (46), we have

$$(y_1 + \alpha)^{\lambda_1}(z_1 + \beta)^{\lambda_2}\bar{a} \equiv c_{\alpha,\beta} \bmod (y_1, z_1)^{r+1}$$

for some $c_{\alpha,\beta} \in k$, which implies $\lambda_1 = \lambda_2 = 0$.

From equation (46) we have

$$
\begin{aligned}
e &= \frac{b(d+e+f+r)}{a+b+c} \\
f &= \frac{c(d+e+f+r)}{a+b+c}
\end{aligned}
$$

where $u = x^a y^b z^c$ and $x^d y^e z^f L_p = \bar{a}x^{d+r}y^e z^f$. Thus $ec - fb = 0$, $ae - b(d+r) = 0$ and $af - c(d+r) = 0$. It follows that F_p is not normalized, a contradiction.

The fact that there is at most one curve C in $\pi^{-1}(p) \cap \overline{S}_{r+1}(X_1)$, which is the 2 curve which is the intersection of the strict transform of $x = 0$ with $\pi^{-1}(p)$, follows from Lemma 6.27, since at the 3 point q with permissible parameters (x_1, y_1, z_1) defined by $x = x_1, y = x_1 y_1, z = x_1 z_1$, $\nu(q) = 0$.

Suppose that p is a 2 point. By Theorem 7.1, there are no 2 curves in $\overline{S}_{r+1}(X_1) \cap \pi^{-1}(p)$. Suppose that $\overline{S}_{r+1}(X_1) \cap \pi^{-1}(p)$ contains a 1 point. Then $\tau(p) = 0$ by Theorem 7.3. The leading form of F_p has an expression

$$L_p = \sum_{i+j=r} a_{ij}x^i y^j.$$

After possibly interchanging x and y, we may assume that $L \neq a_{0r}y^r$.

Suppose that there exist infinitely many distinct values of $\alpha \in k$ such that there exists a 1 point $q \in \overline{S}_{r+1}(X_1) \cap \pi^{-1}(p)$ with regular parameters (x_1, y_1, z_1) in $\hat{\mathcal{O}}_{X_1,q}$ defined by

$$x = x_1, y = x_1(y_1 + \alpha), z = x_1(z_1 + \beta)$$

for some $\alpha, \beta \in k$ with $\alpha \neq 0$, such that $\nu(q) = r + 1$.

With the notation of (41) of Theorem 7.1, there exist $c_\alpha \in k$ such that

$$\sum_{i+j=r} a_{ij}(y_1 + \alpha)^j = c_\alpha(y_1 + \alpha)^{-\lambda} \bmod y_1^{r+1}$$

Set $g(t) = \sum_{i+j=r} a_{ij}t^j$. $g(\alpha) = c_\alpha \alpha^{-\lambda}$.

$$r!a_{0r} = \frac{d^r g}{dt^r}(\alpha) = c_\alpha(-\lambda)(-\lambda - 1)\cdots(-\lambda - r + 1)\alpha^{-\lambda - r}$$

implies

$$g(\alpha) = \frac{r!a_{0r}\alpha^r}{(-\lambda)(-\lambda-1)\cdots(-\lambda-r+1)}$$

for infinitely many α, so that

$$L_p = \frac{r!a_{0r}}{(-\lambda)(-\lambda-1)\cdots(-\lambda-r+1)}y^r$$

a contradiction. Thus 1 curves in $\pi^{-1}(p)\cap\overline{S}_{r+1}(X_1)$ must be the intersection of the strict transform of $y - \alpha x = 0$ and $\pi^{-1}(p)$ for a finite number of $0 \neq \alpha \in k$. These lines intersect in the 3 point of $\pi^{-1}(p)$.

$\square$

Lemma 7.9. *Suppose that* $r \geq 2$ *and* $p \in X$ *is such that*

1. $\nu(p) \leq r$ *if* p *is a 1 point or a 2 point.*
2. *If* p *is a 2 point and* $\nu(p) = r$, *then* $\tau(p) > 0$.
3. $\nu(p) \leq r - 1$ *if* p *is a 3 point*

and $\pi : Y \to X$ *is the blow-up of a point* $p \in X$. *Then* C *is a line for every curve* C *in* $\overline{S}_r(Y)\cap\pi^{-1}(p)$ *containing a 1 point. Thus* C *intersects a 2 curve in at most one point, and this intersection must be transversal.*

If p *is a 1 or 2 point with* $\nu(p) = r$ *then there is at most one curve* C *in* $\overline{S}_r(Y)\cap\pi^{-1}(p)$ *containing a 1 point.*

Proof. Suppose that p is a 1 point. Suppose that $q \in \pi^{-1}(p)$ is a 1 point with $\nu(q) = r$. After a permissible change of parameters at p, we have permissible parameters x_1, y_1, z_1 at q defined by

$$x = x_1, y = x_1y_1, z = x_1z_1.$$

Write

$$F_p = \sum_{i+j+k\geq r} a_{ijk}x^iy^jz^k.$$

F_p has leading form

$$L_p = \sum_{i+j+k=r} a_{ijk}x^iy^jk^k.$$

Thus $L_p = L(y, z)$ depends only on y and z.

Suppose that $q' \in \pi^{-1}(p)$ is another 1 point with $\nu(q') = r$, with permissible parameters (x_1, y_1, z_1) defined by

$$x = x_1, y = x_1(y_1 + \alpha), z = x_1(z_1 + \beta)$$

for some $\alpha, \beta \in k$. Then there exists a form L_1 such that

$$L_p = L_1(y - \alpha x, z - \beta x) + cx^r$$

for some $c \in k$. There exist $\alpha_i, \beta_i, \gamma_i, \delta_i \in k$ such that

$$L_p(y, z) = \prod_{i=1}^{r}(\alpha_i y - \beta_i z)$$

$$L_1(y, z) = \prod_{i=1}^{r} (\gamma_i y - \delta_i z)$$

We can also assume that $\alpha_i \beta_j - \alpha_j \beta_i = 0$ implies $\alpha_i = \alpha_j$ and $\beta_i = \beta_j$.

$$\prod_{i=1}^{r} (\alpha_i y - \beta_i z) = \prod_{i=1}^{r} (\gamma_i (y - \alpha x) - \delta_i (z - \beta x)) + cx^r.$$

Set $x = 0$ to get that, after reindexing the (γ_i, δ_i), there exist $0 \neq \epsilon_i \in k$ such that

$$(\alpha_i, \beta_i) = \epsilon_i (\gamma_i, \delta_i)$$

for all i, and $\prod_{i=1}^{r} \epsilon_i = 1$. Thus

$$\prod_{i=1}^{r} (\alpha_i y - \beta_i z) = \prod_{i=1}^{r} (\alpha_i (y - \alpha x) - \beta_i (z - \beta x)) + cx^r.$$

First suppose that there exists (α_i, β_i), (α_j, β_j) such that $\alpha_i \beta_j - \alpha_j \beta_i \neq 0$. Suppose that $\alpha_i \neq 0$. There exist $t < r$ distinct values of (α_k, β_k) such that $\alpha_i \beta_k - \alpha_j \beta_k = 0$. Set $y = \frac{\beta_i z}{\alpha_i}$ to get

$$0 = (\beta_i \beta - \alpha_i \alpha)^t x^t \prod_{j \mid \alpha_i \beta_j - \alpha_j \beta_i \neq 0} \left(\left(\frac{\alpha_j}{\alpha_i} \beta_i - \beta_j \right) z + (\beta \beta_j - \alpha \alpha_j) x \right) + cx^r$$

We conclude that $\beta \beta_i - \alpha \alpha_i = 0$. If $\beta_i \neq 0$, we can set $z = \frac{\alpha_i}{\beta_i} y$ to again conclude that $\beta \beta_i - \alpha \alpha_i = 0$. Thus $\beta \beta_i - \alpha \alpha_i = 0$ for all i, and the 1 points $q' \in \pi^{-1}(p) \cap \overline{S}_r(X)$ must thus lie on the lines γ_i which are the intersection of the strict transform of $\beta_i z - \alpha_i y = 0$ and $\pi^{-1}(p)$.

Thus q' must be in the intersection $\cap \gamma_i \subset \pi^{-1}(p) \cong \mathbf{P}^2$, and there is at most one point $q \in \pi^{-1}(p)$ such that $\nu(q) = r$.

Now suppose that $\alpha_i \beta_j - \alpha_j \beta_i = 0$ for all i, j. Then after a permissible change of parameters at p, we have $L_p = z^r$.

$L_p = (z - \beta x)^r + cx^r$ and $r \geq 2$ implies $\beta = c = 0$, so q' is on the line $\gamma \subset \pi^{-1}(q) \subset \mathbf{P}^2$ which is the intersection of the strict transform of $z = 0$ and $\pi^{-1}(q)$.

Suppose that p is a 2 point such that $\nu(p) = r$ and $\tau(p) > 0$. Write

$$F_p = \sum_{i+j+k \geq r} a_{ijk} x^i y^j z^k.$$

Suppose there exists a 1 point $q \in \pi^{-1}(p)$ such that $\nu(q) = r$. After a permissible change of parameters at p, q has permissible parameters $(\overline{x}_1, y_1, z_1)$ at q such that, with the notation of (41) of Theorem 7.1,

$$x = x_1, y = x_1(y_1 + \alpha), z = x_1 z_1$$

$$x_1 = \overline{x}_1(y_1 + \alpha)^{-\frac{b}{a+b}}$$

$$F_q = \sum_{i+j+k=r} a_{ijk}(y_1 + \alpha)^{j+\lambda} z_1^k - \sum_{i+j=r} a_{ij0}\alpha^{j+\lambda} + \overline{x}_1 \Omega.$$

Let

$$L_p = \sum_{i+j+k=r} a_{ijk} x^i y^j z^k$$

be the leading form of F_p.

$$F_q = \sum_{k>0}(\sum_{i+j=r-k} a_{ijk}(y_1 + \alpha)^j)(y_1 + \alpha)^\lambda z_1^k$$
$$+(\sum_{i+j=r} a_{ij0}(y_1 + \alpha)^{j+\lambda} - \sum_{i+j=r} a_{ij0}\alpha^{j+\lambda}) + \bar{x}_1 \Omega.$$

$\nu(q) = r$ implies, for fixed $k > 0$,

$$\sum_{i+j=r-k} a_{ijk} x^i y^j = c_k(y - \alpha x)^{r-k}$$

for some $c_k \in k$, thus

$$L_p = \sum_{k>0} c_k(y - \alpha x)^{r-k} z^k + G(x, y).$$

$\tau(p) > 0$ implies some $c_k \neq 0$.

Suppose that there exists another 1 point $q' \in \pi^{-1}(p)$ with $\nu(q') = r$. $\hat{O}_{Y,q'}$ has regular parameters (x_1, y_1, z_1) such that

$$x = x_1, y = x_1(y_1 + \bar\alpha), z = x_1(z_1 + \bar\beta)$$

with $\bar\alpha \neq 0$. Then

$$L_p = \sum_{k>0} \bar c_k(y - \bar\alpha x)^{r-k}(z - \bar\beta x)^k + \bar G(x, y).$$

Thus

$$\sum_{k>0} c_k(y - \alpha x)^{r-k} z^k = \sum_{k>0} \bar c_k(y - \bar\alpha x)^{r-k}(z - \bar\beta x)^k + H(x, y). \qquad (64)$$

Set $x = 0$ in (64) to get $c_k = \bar c_k$ for all k. Let

$$k_0 = \max\{k \mid c_k \neq 0\} = \tau(p).$$

By assumption, $k_0 > 0$.

$$c_{k_0}(y - \alpha x)^{r-k_0} z^{k_0} = c_{k_0}(y - \bar\alpha x)^{r-k_0} z^{k_0},$$

and if $k_0 > 1$,

$$c_{k_0-1}(y - \alpha x)^{r-k_0+1} z^{k_0-1} = c_{k_0}(y - \bar\alpha x)^{r-k_0}(-\bar\beta k_0 x) z^{k_0-1}$$
$$+ c_{k_0-1}(y - \bar\alpha x)^{r-k_0+1} z^{k_0-1}.$$

If $k_0 < r$, then $\alpha = \bar\alpha$ implies all 1 points in $\bar S_r(X_1) \cap \pi^{-1}(p)$ are contained in the line which is the intersection of the strict transform of $y - \alpha x = 0$ and $\pi^{-1}(p)$. This line contains the 3 point of $\pi^{-1}(p)$.

If $k_0 = r \ (\geq 2)$,

$$c_{r-1}(y - \alpha x) = -c_r \bar\beta r x + c_{r-1}(y - \bar\alpha x)$$

so that

$$-\alpha c_{r-1} = -c_r \bar\beta r - \bar\alpha c_{r-1}$$

which implies that all 1 points in $\overline{S}_r(X_1) \cap \pi^{-1}(p)$ are contained in the line which is the intersection of the strict transform of

$$c_r rz + c_{r-1}y - \alpha c_{r-1}x = 0$$

and $\pi^{-1}(p)$.

Suppose that p is a 2 point or a 3 point, with $\nu(p) = r - 1$. Then by Theorem 7.8, the conclusions of the Theorem hold. □

8. PERMISSIBLE MONOIDAL TRANSFORMS CENTERED AT CURVES

Throughout this section we will assume that $\Phi_X : X \to S$ is weakly prepared.

Lemma 8.1. *Suppose that $C \subset X$ is a 2 curve. Then either $F_p \in \hat{\mathcal{I}}_{C,p}$ for all $p \in C$ or $F_p \notin \hat{\mathcal{I}}_{C,p}$ for all $p \in C$.*

Suppose that $r \geq 2$, $C \subset \overline{S}_r(X)$ is a 2 curve. Then either $F_p \in \hat{\mathcal{I}}_{C,p}^r$ for all $p \in C$ or $F_p \in \hat{\mathcal{I}}_{C,p}^{r-1}$, $F_p \notin \hat{\mathcal{I}}_{C,p}^r$ for all $p \in C$

Proof. This follows from Lemmas 6.23, 6.25, 6.27. □

Lemma 8.2. *Suppose that $r \geq 2$ and $C \subset \overline{S}_r(X)$ is a nonsingular curve such that C contains a 1 point and C makes SNCs with $\overline{B}_2(X)$. Then either $F_p \in \hat{\mathcal{I}}_{C,p}^r$ with respect to permissible parameters for C at p for all $p \in C$, or $F_p \in \hat{\mathcal{I}}_{C,p}^{r-1}$, $F_p \notin \hat{\mathcal{I}}_{C,p}^r$ with respect to permissible parameters for C at p for all $p \in C$.*

Proof. This follows from Lemmas 6.23, 6.24, 6.26. □

Definition 8.3. *Suppose that $r \geq 2$, $p \in X$, $C \subset \overline{S}_r(X)$ is a curve which contains p and makes SNCs with $\overline{B}_2(X)$ at p (or C is a component of $\overline{B}_2(X)$) and $C \not\subset \overline{S}_{r+1}(X)$. C is r big at p if $F_p \in \hat{\mathcal{I}}_{C,p}^r$ with respect to permissible parameters for C at p. C is r small at p if C is not r big at p.*

Suppose that C is a 2 curve, $\nu(q) \geq 1$ if $q \in C$ is a 2 point, $C \not\subset \overline{S}_2(X)$ and $p \in C$. Then C is 1 big at p if $F_p \in \hat{\mathcal{I}}_{C,p}$. F_p is 1 small at p if C is not 1 big at p.

Suppose that $r \geq 2$, $C \subset \overline{S}_r(X)$ is a curve which makes SNCs with $\overline{B}_2(X)$. We will say that C is r big if C is r big at p for all $p \in C$. We will say that C is r small if C is r small at p for all $p \in C$.

Suppose that C is a 2 curve, $\nu(q) \geq 1$ if $q \in C$ is a 2 point, $C \not\subset \overline{S}_2(X)$. We will say that C is 1 big if C is 1 big for all $p \in C$. We will say that C is 1 small if C is 1 small at p for all $p \in C$.

Lemma 8.4. *Suppose that C is a 2 curve on X, $p \in C$ is a 2 point, D_1 and D_2 are curves in E_X containing p such that $D_1 \cup D_2$ makes SNCs with C at p. Then there are regular parameters (x, y, z) in $\mathcal{O}_{X,p}$ such that*

$$\mathcal{I}_{C,p} = (x, y), \mathcal{I}_{D_1,p} = (x, z), \mathcal{I}_{D_2,p} = (y, z)$$

Proof. There exist regular parameters $(\tilde{x}, \tilde{y}, \tilde{z})$ in $\mathcal{O}_{X,p}$, and $\phi \in \mathcal{O}_{X,p}$ such that

$$\mathcal{I}_{C,p} = (\tilde{x}, \tilde{y}), \mathcal{I}_{D_1,p} = (\tilde{x}, \tilde{z}), \mathcal{I}_{D_2,p} = (\tilde{y}, \phi)$$

and $\phi \equiv a\tilde{x} + c\tilde{z} \mod m_p^2$, with $a, c \in k$, $c \neq 0$. In $\hat{\mathcal{O}}_{X,p}$, there exist series h, g such that

$$\phi = h(\tilde{x}, \tilde{y}, \tilde{z})\tilde{y} + g(\tilde{x}, \tilde{z})$$

$$g = u(\tilde{z} - \psi(\tilde{x}))$$

where u is a unit, ψ is a series.

$$\tilde{z} - \psi(\tilde{x}) \in \hat{\mathcal{I}}_{D_1,p} \cap \hat{\mathcal{I}}_{D_2,p} = (\mathcal{I}_{D_1,p} \cap \mathcal{I}_{D_2,p})\hat{\mathcal{O}}_{X,p},$$

where the last equality is by Corollary 2 to Theorem 11 of Chapter VIII, section 4 [38]). Suppose that

$$\mathcal{I}_{D_1,p} \cap \mathcal{I}_{D_2,p} = (f_1, \ldots, f_n).$$

$$\tilde{z} - \psi(\tilde{x}) = \sum \lambda_i f_i$$

implies there exists $f \in \mathcal{I}_{D_1,p} \cap \mathcal{I}_{D_2,p}$ such that

$$f \equiv \bar{a}\tilde{x} + \bar{c}\tilde{z} \bmod m_p^2$$

where $\bar{a}, \bar{c} \in k$, $\bar{c} \neq 0$. Since $f \in (\tilde{x}, \tilde{z})$, we have $f = \lambda\tilde{x} + \tau\tilde{z}$ where τ is a unit. Thus $(\tilde{x}, \tilde{z}) = (\tilde{x}, f)$. Since $f \in (\tilde{y}, \phi)$, we have $f = \alpha\tilde{y} + \beta\phi$ where β is a unit. Thus $(\tilde{y}, \phi) = (\tilde{y}, f)$. $(\tilde{x}, \tilde{y}, f)$ are the desired regular parameters. $\qquad\square$

Lemma 8.5. *Suppose that $p \in X$ is a 1 point or a 2 point with $\gamma(p) = r \geq 2$, and (u, v) are permissible parameters at $\Phi_X(p)$, such that $u = 0$ is a local equation of E_X at p. Then there exist regular parameters $(\tilde{x}, y, \tilde{z})$ in $R = \mathcal{O}_{X,p}$ and permissible parameters (x, y, z) at p with $x = \gamma\tilde{x}$, $z = \sigma\tilde{z}$ for some series $\gamma, \sigma \in \hat{\mathcal{O}}_{X,p}$ such that if p is a 1 point,*

$$\begin{aligned} u &= x^a \\ v &= P(x) + x^c F \end{aligned} \tag{65}$$

with $F = \tau z^r + \sum_{i=2}^r a_i(x, y)z^{r-i}$, τ a unit and some $a_i \neq 0$.
Further suppose that $\overline{S}_r(X) \cup \overline{B}_2(X)$ makes SNCs at p. Then there is at most one curve D in $\overline{S}_r(X)$ through p. If D exists, we can choose (x, y, z) so that $x = 0, z = 0$ are local equations of D at p.
If p is a 2 point,

$$\begin{aligned} u &= (x^a y^b)^m \\ v &= P(x^a y^b) + x^c y^d F \end{aligned} \tag{66}$$

with $F = \tau z^r + \sum_{i=2}^r a_i(x, y)z^{r-i}$, τ a unit, and some $a_i \neq 0$.
Further suppose that $\overline{S}_r(X) \cup \overline{B}_2(X)$ makes SNCs at p. Then there are at most 2 curves D_1 and D_2 in $\overline{S}_r(X)$ through p. If D_1 exists (or if D_1 and D_2 exist) then we can choose (x, y, z) so that $x = 0, z = 0$ are local equations of D_1 at p ($x = 0, z = 0$ are local equations of D_1 at p and $y = 0, z = 0$ are local equations of D_2 at p).

Proof. There exist regular parameters $\tilde{x}, y, \tilde{z}$ in R, and permissible parameters $(x = \gamma\tilde{x}, y, \tilde{z})$ at p such that $u = x^a$ or $u = (x^a y^b)^m$ in $\hat{R}$, and $\nu(F(0, 0, \tilde{z})) = r$.

If p is a 1 point, then there exists at most one curve D in $\overline{S}_r(X)$ containing p by Lemma 7.5. If D exists, we may assume that $x = 0, \tilde{z} = 0$ are local equations of D at p. If p is a 2 point, then there exist at most 2 curves D_1 and D_2 in $\overline{S}_r(X)$. If D_1 (or D_1 and D_2 exist) we may assume that $x = 0, \tilde{z} = 0$ are local equations of D_1 at p (or $\overset{,}{x} = 0, \tilde{z} = 0$ are local equations of D_1 at p and $y = 0, \tilde{z} = 0$ are local equations of D_2 at p by Lemma 8.4).

Set

$$\bar{z} = \frac{\partial^{r-1} F}{\partial \tilde{z}^{r-1}} = \omega(\tilde{z} - \phi(x, y))$$

where ω is a unit by the formal implicit function theorem. Set $z_1 = \tilde{z} - \phi(x, y)$, $G(x, y, z_1) = F(x, y, \tilde{z})$.

Suppose that p is a 1 point and there exists a curve $D \subset \overline{S}_r(X)$ containing p, so that D has local equations $x = 0, \tilde{z} = 0$. Then $F_p \in \hat{I}^r_{D,p} + (x^{r-1})$ by Lemma 6.24, so that

$$\frac{\partial^{r-1} F}{\partial \tilde{z}^{r-1}} \in \hat{I}_{D,p}$$

and $x \mid \phi(x, y)$. Thus $x = 0, z_1 = 0$ are local equations of D at p.

Suppose that p is a 2 point and there exist curves $D_1, D_2 \subset \overline{S}_r(X)$ containing p, so that D_1 has local equations $x = 0, \tilde{z} = 0$ and D_2 has local equations $y = 0, \tilde{z} = 0$. $F_p \in \hat{I}^r_{D_1,p} + (x^{r-1})$ and $F_p \in \hat{I}^r_{D_2,p} + (y^{r-1})$ by Lemma 6.26. Thus

$$\frac{\partial^{r-1} F}{\partial \tilde{z}^{r-1}} \in \hat{I}_{D_1,p}$$

and $\frac{\partial^{r-1} F}{\partial \tilde{z}^{r-1}} \in \hat{I}_{D_2,p}$, so that $xy \mid \phi(x, y)$, and $x = 0, z_1 = 0$ are local equations of D_1 at p and $y = 0, z_1 = 0$ are local equations of D_2 at p.

$$\begin{aligned} G &= G(x,y,0) + \tfrac{\partial G}{\partial z_1}(x,y,0)z_1 + \cdots + \tfrac{1}{(r-1)!}\tfrac{\partial^{r-1} G}{\partial z_1^{r-1}}(x,y,0)z_1^{r-1} \\ &\quad + \tfrac{1}{r!}\tfrac{\partial^r G}{\partial z_1^r}(x,y,0)z_1^r + \cdots \end{aligned}$$

$$\frac{\partial^{r-1} G}{\partial z_1^{r-1}}(x,y,0) = \frac{\partial^{r-1} F}{\partial \tilde{z}^{r-1}}(x,y,\phi(x,y)) = 0$$

$$\frac{\partial^r G}{\partial z_1^r}(x,y,0) = \frac{\partial^r F}{\partial \tilde{z}^r}(x,y,\phi(x,y))$$

is a unit. Thus with the regular parameters $(\tilde{x}, y, \tilde{z})$ in R and permissible parameters (x, y, z_1) at p, F has the desired form.

We cannot have $a_i = 0$ for all i, since $r \geq 2$ and x or $xy \in \sqrt{\hat{I}_{\text{sing}(\Phi_X),p}}$. $\square$

Lemma 8.6. *Suppose that $r \geq 2$, $C \subset X$ is a 2 curve such that C is $r - 1$ big or r small, $\pi : X_1 \to X$ is the blow-up of C.*

1. (a) *If $q \in C$ is a 2 point with $\nu(q) = r - 1$ and $q_1 \in \pi^{-1}(q)$, then*
   (i) *If q_1 is a 1 point then $\nu(q_1) \leq r$ and $\gamma(q_1) \leq r$.*
   (ii) *If q_1 is a 2 point then $\nu(q_1) \leq r - 1$.*
   (b) *If $q \in C$ is a 2 point with $\nu(q) = r$, $\tau(q) > 0$ and $q_1 \in \pi^{-1}(q)$, then*
   (i) *If q_1 is a 1 point then $\nu(q_1) \leq r$. $\nu(q_1) = r$ implies $\gamma(q_1) = r$.*
   (ii) *If q_1 is a 2 point then $\nu(q_1) \leq r$. $\nu(q_1) = r$ implies $\tau(q_1) > 0$.*
   (c) *If $q \in C$ is a 3 point with $\nu(q) = r - 1$ and $q_1 \in \pi^{-1}(q)$ then*
   (i) *q_1 a 2 point implies $\nu(q_1) \leq r$ and $\gamma(q_1) \leq r$.*
   (ii) *q_1 a 3 point implies $\nu(q_1) \leq r - 1$.*
2. *Suppose that $C \subset \overline{S}_r(X)$ (so that C is r small). If $q \in C$ is a 2 point with $\nu(q) = r$, $\tau(q) > 0$ and $q_1 \in \pi^{-1}(q)$, then*
   (a) *If q_1 is a 1 point then q_1 is resolved.*
   (b) *If q_1 is a 2 point then $\nu(q_1) \leq r$. $\nu(q_1) = r$ implies $\tau(q) > 0$.*

Proof. Suppose that $q \in C$ is a 2 point with $\nu(q) = r - 1$, and q has permissible parameters (x, y, z) with

$$
\begin{aligned}
u &= (x^a y^b)^m \\
v &= P(x^a y^b) + x^c y^d F_q \\
L_q &= \textstyle\sum_{i+j=r-1} a_{ij} x^i y^j
\end{aligned}
$$

Suppose that $q_1 \in \pi^{-1}(q)$ and $\hat{\mathcal{O}}_{Y_1, q_1}$ has regular parameters (x_1, y_1, z) such that

$$
x = x_1, y = x_1(y_1 + \alpha)
$$

with $\alpha \neq 0$. Set

$$
x_1 = \bar{x}_1 (y_1 + \alpha)^{-\frac{b}{a+b}}
$$

$$
\begin{aligned}
u &= \bar{x}_1^{(a+b)m} \\
v &= P_{q_1}(\bar{x}_1) + \bar{x}_1^{c+d+r-1} F_{q_1} \\
F_{q_1} &= \textstyle\sum_{i+j=r-1} a_{ij}(y_1 + \alpha)^{\lambda+j} - \textstyle\sum_{i+j=r-1} a_{ij}\alpha^{\lambda+j} + \bar{x}_1 \Omega + zG, \quad (67)
\end{aligned}
$$

$$
\lambda = d - \frac{b(c + d + r - 1)}{a + b}.
$$

Thus $\nu(q_1) \leq r$ and $\gamma(q_1) \leq r$.

Suppose that $q_1 \in \pi^{-1}(q)$ and q_1 has permissible parameters (x_1, y_1, z) such that

$$
x = x_1, y = x_1 y_1.
$$

Then

$$
\begin{aligned}
u &= (x_1^{a+b} y_1^b)^m \\
v &= P(x_1^{a+b} y_1^b) + x_1^{c+d+r-1} F_{q_1} \\
F_{q_1} &= \textstyle\sum_{i+j=r-1} a_{ij} y_1^j + x_1 \Omega + zG
\end{aligned}
$$

implies that $\nu(q_1) \leq r - 1$.

A similar argument holds at the point $q_1 \in \pi^{-1}(q)$ with permissible parameters (x_1, y_1, z) such that $x = x_1 y_1, y = y_1$.

Suppose that $q \in C$ is a 2 point with $\nu(q) = r$ and $\tau(q) > 0$. Then q has permissible parameters (x, y, z) with

$$
\begin{aligned}
u &= (x^a y^b)^m \\
v &= P(x^a y^b) + x^c y^d F_q \\
L_q &= z(\textstyle\sum_{i+j=r-1} a_{ij1} x^i y^j) + \textstyle\sum_{i+j=r} a_{ij0} x^i y^j
\end{aligned}
$$

with some $a_{ij1} \neq 0$.

Suppose that $q_1 \in \pi^{-1}(q)$ and $\hat{\mathcal{O}}_{Y_1, q_1}$ has regular parameters (x_1, y_1, z) such that $x = x_1, y = x_1(y_1 + \alpha)$ with $\alpha \neq 0$. Set

$$
x_1 = \bar{x}_1 (y_1 + \alpha)^{-\frac{b}{a+b}}.
$$

$$
\begin{aligned}
u &= \bar{x}_1^{(a+b)m} \\
v &= P_{q_1}(\bar{x}_1) + \bar{x}_1^{c+d+r-1} F_{q_1}
\end{aligned}
$$

with

$$
F_{q_1} = z(\textstyle\sum_{i+j=r-1} a_{ij1}(y_1 + \alpha)^j)(y_1 + \alpha)^\lambda + \bar{x}_1 \Omega + z^2 G, \quad (68)
$$

$$\lambda = d - \frac{b(c+d+r-1)}{a+b}.$$

Thus $\nu(q_1) \le r$ and $\nu(q_1) = r$ implies $\gamma(q_1) = r$.

Suppose that $q_1 \in \pi^{-1}(q)$ and q_1 has permissible parameters (x_1, y_1, z) such that

$$x = x_1, y = x_1 y_1.$$

Then

$$\begin{aligned}
u &= (x_1^{a+b} y_1^b)^m \\
v &= P(x_1^{a+b} y_1^b) + x_1^{c+d+r-1} y_1^d F_{q_1} \\
F_{q_1} &= \sum_{i+j=r-1} z a_{ij1} y_1^j + x_1 \Omega + z^2 G
\end{aligned}$$

implies $\nu(q_1) \le r$ and $\nu(q_1) = r$ implies $\tau(q_1) > 0$.

A similar analysis holds at the point $q_1 \in \pi^{-1}(q)$ with permissible parameters (x_1, y_1, z) such that $x = x_1 y_1, y = y_1$.

Suppose that $q \in C$ is a 3 point with $\nu(q) = r - 1$,

$$\begin{aligned}
u &= (x^a y^b z^c)^m \\
v &= P(x^a y^b z^c) + x^d y^e z^f F_q \\
F_q &= \sum_{i+j \ge r-1, k \ge 0} a_{ijk} x^i y^j z^k
\end{aligned}$$

some $a_{ij0} \ne 0$ with $i + j = r - 1$.

Suppose that $q_1 \in \pi^{-1}(q)$ is a 2 point.

$$x = x_1, y = x_1(y_1 + \alpha)$$

with $\alpha \ne 0$.

$$\begin{aligned}
x_1 &= \overline{x}_1 (y_1 + \alpha)^{-\frac{b}{a+b}} \\
u &= (\overline{x}_1^{a+b} z^c)^m = (\overline{x}_1^{\overline{a}} z^{\overline{c}})^{\overline{m}} \\
v &= P_{q_1}(\overline{x}_1^{\overline{a}} z^{\overline{c}}) + \overline{x}_1^{d+r-1+e} z^f F_{q_1}
\end{aligned}$$

with

$$\lambda = e - \frac{b(d+r-1+e)}{a+b}, (\overline{a}, \overline{c}) = 1$$

$$F_{q_1} = (y_1 + \alpha)^\lambda \frac{F_q}{x_1^{r-1}} - \frac{g(\overline{x}_1^{\overline{a}} z^{\overline{c}})}{\overline{x}_1^{d+r-1+e} z^f} \tag{69}$$

Thus

$$F_{q_1} = \sum_{i+j=r-1} a_{ij0}(y_1 + \alpha)^{j+\lambda} + zG + \overline{x}_1 \Omega,$$

or

$$F_{q_1} = \sum_{i+j=r-1} a_{ij0}(y_1 + \alpha)^{j+\lambda} - \sum a_{ij0} \alpha^{j+\lambda} + zG + \overline{x}_1 \Omega,$$

implies $\nu(q_1) \le r$, and $\gamma(q_1) \le r$.

Suppose that $q \in C$ is a 2 point with $\nu(q) = r$ and $\tau(q) > 0$ and $C \subset \overline{S}_r(X)$. By Lemma 6.25, q has permissible parameters (x, y, z) with

$$\begin{aligned}
u &= (x^a y^b)^m \\
v &= P(x^a y^b) + x^c y^d F_q \\
F_q &= \sum_{i+j \ge r, k \ge 0} c_{ijk} x^i y^j z^k + \overline{c} z x^{i_0} y^{j_0}
\end{aligned} \tag{70}$$

where $i_0 + j_0 = r - 1$, $(c + i_0)b - a(d + j_0) = 0$, $\overline{c} \ne 0$.

Suppose that $q_1 \in \pi^{-1}(q)$, and $\hat{\mathcal{O}}_{Y_1,q_1}$ has regular parameters (x_1, y_1, z) such that

$$x = x_1, y = x_1(y_1 + \alpha)$$

with $\alpha \neq 0$. Set $\overline{x}_1 = \overline{x}_1(y_1 + \alpha)^{-\frac{b}{a+b}}$. Then

$$
\begin{aligned}
u &= \overline{x}_1^{(a+b)m} \\
v &= P_{q_1}(\overline{x}_1) + \overline{x}_1^{c+d+r-1} F_{q_1} \\
F_{q_1} &= \overline{c}z(y_1 + \alpha)^{\lambda+j_0} + \overline{x}_1 \Omega
\end{aligned}
$$

where $\lambda = d - \frac{b(c+d+r-1)}{a+b}$. Thus q_1 is resolved.

Suppose that $q_1 \in \pi^{-1}(q)$, and $\hat{\mathcal{O}}_{Y_1,q_1}$ has regular parameters (x_1, y_1, z) such that

$$x = x_1, y = x_1 y_1.$$

Then

$$
\begin{aligned}
u &= (x_1^{a+b} y_1^b)^m \\
v &= P(x_1^{a+b} y_1^b) + x_1^{c+d+r-1} y_1^d F_{q_1} \\
F_{q_1} &= \frac{F_q}{x_1^{r-1}} = \overline{c}z y_1^{j_0} + x_1 \Omega
\end{aligned}
$$

q_1 satisfies the conclusions of the Theorem since $j_0 \leq r - 1$.

Suppose that $q_1 \in \pi^{-1}(q)$, and $\hat{\mathcal{O}}_{Y_1,q_1}$ has regular parameters (x_1, y_1, z) such that

$$x = x_1 y_1, y = y_1.$$

Then an argument similar to the above case shows that q_1 satisfies the conclusions of the Theorem (since $i_0 \leq r - 1$).

$\square$

Lemma 8.7. *Suppose that $r \geq 2$, $C \subset X$ is a 2 curve such that C is r-1 big and*

1. *$p \in C$ a 2 point implies $\nu(p) \leq r$, and if $\nu(p) = r$ then $\tau(p) > 0$.*
2. *$p \in C$ a 3 point implies $\nu(p) \leq r - 1$.*

Suppose that $\pi : X_1 \to X$ is the blow-up of C. Then

$$\pi^{-1}(C) \cap \overline{S}_r(X_1)$$

contains at most one curve. If $D \subset \pi^{-1}(C) \cap \overline{S}_r(X_1)$ is a curve, then D is a section over C, and D contains a 1 point.

Suppose that $D \subset \pi^{-1}(C) \cap \overline{S}_r(X_1)$ is a curve (which is necessarily a section over C). Supppose that $q \in C$ is a 2 point such that $\nu(q) = r - 1$. Then $\pi^{-1}(q) \cap D$ is a 1 point.

Proof. Suppose that $q \in C$ is a 2 point with $\nu(q) = r - 1$. Suppose, with the notation of (67) of Lemma 8.6, that there exists $q_1 \in \pi^{-1}(q)$ with $\nu(q_1) = r$. Then there exist regular parameters (x_1, y_1, z) in $\hat{\mathcal{O}}_{X_1,q_1}$ such that

$$x = x_1, y = x_1(y_1 + \alpha)$$

with $\alpha \neq 0$, and $\gamma_\alpha \in k$ such that

$$\sum_{i+j=r-1} a_{ij}(y_1 + \alpha)^j \equiv \gamma_\alpha(y_1 + \alpha)^{-\lambda} \mod y_1^r.$$

$-\lambda \notin \{0, \ldots, r-1\}$ since F_q is normalized.

Set $g(t) = \sum_{i+j=r-1} a_{ij} t^j$. We have

$$\frac{1}{i!} \frac{d^i g}{dt^i}(\alpha) = \gamma_\alpha \left(\frac{-\lambda(-\lambda-1) \cdots (-\lambda-i+1)}{i!} \right) \alpha^{-\lambda-i}$$

for $i \leq r-1$. Thus

$$a_{0,r-1} = \frac{1}{(r-1)!} \frac{d^{r-1} g}{dt^{r-1}}(\alpha) = \gamma_\alpha \left(\frac{-\lambda(-\lambda-1) \cdots (-\lambda-r+2)}{(r-1)!} \right) \alpha^{-\lambda-r+1}$$

$$a_{1r-2} + (r-1)a_{0r-1}\alpha = \frac{1}{(r-2)!} \frac{d^{r-2} g}{dt^{r-2}}(\alpha)$$
$$= \gamma_\alpha \left(\frac{-\lambda(-\lambda-1) \cdots (-\lambda-r+3)}{(r-2)!} \right) \alpha^{-\lambda-r+2}$$

$$a_{1r-2} = \gamma_\alpha \left[\frac{-\lambda(-\lambda-1) \cdots (-\lambda-r+3) - (-\lambda)(-\lambda-1) \cdots (-\lambda-r+2)}{(r-2)!} \right] \alpha^{-\lambda-r+2}$$
$$= \gamma_\alpha [\frac{\lambda(-\lambda-1) \cdots (-\lambda-r+3)(-\lambda-r+1)}{(r-2)!}] \alpha^{-\lambda-r+2}$$

$$\frac{-\lambda(-\lambda-1) \cdots (-\lambda-r+2)}{(r-1)!} \neq 0$$

and

$$\frac{\lambda(-\lambda-1) \cdots (-\lambda-r+3)(-\lambda-r+1)}{(r-2)!} \neq 0$$

since $-\lambda \notin \{0, \ldots, r-1\}$.

If $q_2 \in \pi^{-1}(q)$ has $\nu(q_2) = r$, and $q_2 \neq q_1$, then there exist $\alpha \neq \beta \in k$ such that

$$a_{0,r-1} = \gamma_\beta \left(\frac{-\lambda(-\lambda-1) \cdots (-\lambda-r+2)}{(r-1)!} \right) \beta^{-\lambda-r+1}$$

$$a_{1r-2} = \gamma_\beta [\frac{-\lambda(-\lambda-1) \cdots (-\lambda-r+3) - (-\lambda)(-\lambda-1) \cdots (-\lambda-r+2)}{(r-2)!}] \beta^{-\lambda-r+2}$$

which implies that

$$\gamma_\beta = \gamma_\alpha (\frac{\alpha}{\beta})^{-\lambda-r+1}$$

and

$$\gamma_\alpha \alpha^{-\lambda-r+2} = \gamma_\beta \beta^{-\lambda-r+2} = \gamma_\alpha \alpha^{-\lambda-r+1} \beta.$$

so that $\alpha = \beta$.

Thus there is at most one point $q_1 \in \pi^{-1}(q)$ with $\nu(q_1) = r$. q_1, if it exists, is a 1 point.

Suppose that $q \in C$ is 2 point with $\nu(q) = r$ and $\tau(q) > 0$. Suppose, with the notation of (68) of Lemma 8.6, that there exists a 1 point $q_1 \in \pi^{-1}(q)$ with $\nu(q_1) = r$. Then there exist regular parameters (x_1, y_1, z) in $\hat{\mathcal{O}}_{X_1,q_1}$ such that $x = x_1, y = x_1(y_1 + \alpha)$.

Set $g(t) = \sum_{i+j=r-1} a_{ij1} t^j$. By (68),

$$\nu(\sum_{i+j=r-1} a_{ij1}(y_1 + \alpha)^j) = r-1.$$

which implies $g(t+\alpha) = a_{0,r-1,1} t^{r-1}$ which implies $g(t) = a_{0,r-1,1}(t-\alpha)^{r-1}$.

Thus there is at most one 1 point $q_1 \in \pi^{-1}(q)$ with $\nu(q_1) = r$.

Suppose that $q \in C$ is a 3 point. Suppose that, with the notation of (69), of Lemma 8.6, that there exists a 2 point $q_1 \in \pi^{-1}(q)$ with $\nu(q_1) = r$. Then there exist regular parameters (x_1, y_1, z) in $\hat{\mathcal{O}}_{X_1, q_1}$ such that

$$x = x_1, y = x_1(y_1 + \alpha)$$

with $\alpha \neq 0$, and $\gamma_\alpha \in k$ such that

$$\sum_{i+j=r-1} a_{ij0}(y_1 + \alpha)^j \equiv \gamma_\alpha(y_1 + \alpha)^{-\lambda} \bmod y_1^r.$$

As in the argument for the case when q is a 2 point with $\nu(q) = r - 1$, we can conclude that there is at most one point $q_1 \in \pi^{-1}(q)$ with $\nu(q_1) = r$. q_1, if it exists, is a 2 point.

Suppose that $D \subset \pi^{-1}(C) \cap \overline{S}_r(X_1)$ is a curve, which is necessarily a section over C, and $q \in C$ is a 2 point such that $\nu(q) = r - 1$. Suppose there exists a 2 point $q' \in \pi^{-1}(q)$ such that $q' \in D$. Then $\nu(q') = r - 1$ by Lemma 7.6, so that (by the proof of Lemma 8.6), there exist permissible parameters (x, y, z) at q' such that

$$\begin{aligned} u &= (x^a y^b)^m \\ v &= P(x^a y^b) + x^c y^d F_{q'} \\ F_{q'} &= y^{r-1} + x\Omega + zG \end{aligned}$$

and there exists an irreducible series $f(y, z)$ such that $\hat{I}_{D,q'} = (x, f(y, z))$.

Case 1 or Case 2 of Lemma 6.29 must hold. Suppose that Case 1 holds. Set $x = 0$ in the formula of Case 1 to get that there exists a series $h(\overline{y}, z)$ such that

$$\overline{y}^{ad-bc}(\overline{y}^{(r-1)a} + zG(0, \overline{y}^a, z)) = hf(\overline{y}^a, z)^r$$

$\overline{y} \nmid f(\overline{y}^a, z)$ implies

$$a(r - 1) \geq \nu(f(\overline{y}^a, 0)^r) \geq ar$$

a contradiction.

Now suppose that Case 2 of Lemma 6.29 holds. $a(r-1) \neq bc - ad$ since $F_{q'}$ is normalized. Set $x = 0$ in the formula of Case 2 to get that there exists a series $h(\overline{y}, z)$ such that

$$\overline{y}^{a(r-1)} + zG(0, \overline{y}^a, z) - g(0)\overline{y}^{bc-ad} = hf(\overline{y}^a, z)^r$$

$$0 \neq \overline{y}^{a(r-1)} - g(0)\overline{y}^{bc-ad} = h(\overline{y}, 0)f(\overline{y}^a, 0)^r$$

Thus

$$a(r - 1) \geq \nu(\overline{y}^{a(r-1)} - g(0)\overline{y}^{bc-ad}) \geq r\nu(f(\overline{y}^a, 0)) \geq ra$$

which is a contradiction.

$\square$

Lemma 8.8. *Suppose that $r \geq 2$ and $C \subset \overline{S}_r(X)$ is a curve containing a 1 point such that C is r big. let $\pi : X_1 \to X$ be the blow-up of C.*

1. *Suppose that $p \in C$ is a 1 point with $\nu(p) = \gamma(p) = r$, and $q \in \pi^{-1}(p)$. Then*
   (a) *If q is a 1 point then $\gamma(q) \leq r$. There is at most one 1 point $q \in \pi^{-1}(p)$ such that $\gamma(q) > r - 1$.*
   (b) *If q is a 2 point then $\nu(q) = 0$.*

2. *Suppose that $p \in C$ is a 1 point with $\nu(p) = r$, $\gamma(p) \neq r$, and $q \in \pi^{-1}(p)$. Then*
   (a) *If q is a 1 point then $\gamma(q) < r$,*
   (b) *If $q \in \pi^{-1}(p)$ is a 2 point then $\nu(q) \leq r - 1$.*
3. *Suppose that $p \in C$ is a 2 point such that $\gamma(p) = \nu(p) = r$, and $q \in \pi^{-1}(p)$. Then*
   (a) *If q is a 2 point then $\nu(q) \leq r$ and $\gamma(q) \leq r$.*
   (b) *There is at most one 2 point $q \in \pi^{-1}(p)$ such that $\gamma(q) > r - 1$.*
   (c) *If q is a 3 point then $\nu(q) = 0$.*
4. *Suppose that $p \in C$ is a 2 point with $\nu(p) = r$ and $\tau(p) > 0$, and $q \in \pi^{-1}(p)$. Then*
   (a) *If q is a 2 point then $\gamma(q) \leq r$.*
   (b) *If q is the 3 point then $\nu(q) \leq r - \tau(p)$.*

Proof. First suppose that $p \in C$ is a 1 point such that $\nu(p) = \gamma(p) = r$. We have permissible parameters (x, y, z) at p such that $\hat{\mathcal{I}}_{C,p} = (x, z)$,

$$
\begin{aligned}
u &= x^a \\
v &= P(x) + x^b F_p \\
F_p &= \tau z^r + \sum_{i=2}^r \bar{a}_i(x, y) x^i z^{r-i}
\end{aligned}
\tag{71}
$$

where τ is a unit by Lemma 8.5.

Suppose that $q \in \pi^{-1}(p)$ and q is a 1 point. Then q has permissible parameters (x_1, y, z_1) such that $x = x_1$, $z = x_1(z_1 + \alpha)$. Then $\nu(F_q(0, 0, z_1)) \leq r$ and $\nu(F_q(0, 0, z_1)) < r$ if $\alpha \neq 0$.

If $q \in \pi^{-1}(p)$ is the 2 point then q has permissible parameters (x_1, y, z_1) such that $x = x_1 z_1$, $z = z_1$. Then $F_q = \frac{F_p}{z_1^r}$ is a unit.

Now suppose that $p \in C$ is a 1 point with $\nu(p) = r$ and $\gamma(p) \neq r$. Suppose that $q \in \pi^{-1}(p)$ is a 1 point. Then there exist permissible parameters (x, y, z) at p such that $x = z = 0$ are local equations of C at p, and permissible parameters (x_1, y, z_1) at q such that $x = x_1$, $z = x_1 z_1$.

$$
F_p = \sum_{i+j \geq r} a_{ij}(y) x^i z^j
$$

where $a_{r0}(0) = 0$, $a_{0r}(0) = 0$, and $a_{ij}(0) \neq 0$ for some i, j with $i + j = r$.

$$
F_q = \frac{F_p}{x_1^r} = \left(\sum_{i+j=r} a_{ij}(0) z_1^j \right) + x_1 \Omega + yG
$$

implies $\nu(F_q(0, 0, z_1)) \leq r - 1$.

At the 2 point $q \in \pi^{-1}(p)$, there exist permissible parameters (x, y, z) as above, and permissible parameters (x_1, y, z_1) at q such that $x = x_1 z_1$, $z = z_1$,

$$
F_q = \frac{F_p}{z_1^r} = \sum_{i+j=r} a_{ij}(0) x_1^i + z_1 \Omega + yG
$$

where $a_{ij}(0) \neq 0$ for some $i \leq r - 1$.

Now suppose that $p \in C$ is a 2 point such that $\nu(p) = r$ and $\gamma(p) = r$.

$$\begin{aligned} u &= (x^a y^b)^m \\ v &= P(x^a y^b) + x^c y^d F \end{aligned}$$

$\hat{\mathcal{I}}_{C,p} = (x, z)$. After a permissible change of parameters, we have by Lemma 8.5

$$F = \tau z^r + a_2(x, y) z^{r-2} + \cdots + a_r(x, y) \tag{72}$$

where τ is a unit and $x^i \mid a_i$ for all i.

If $p_1 \in \pi^{-1}(p)$ has permissible parameters (x_1, y, z_1) with

$$x = x_1 z_1, z = z_1$$

then

$$F_1 = \tau + x_1 \Omega$$

so that p_1 is resolved. Suppose that $p_1 \in \pi^{-1}(p)$ has regular parameters

$$x = x_1, z = x_1(z_1 + \alpha)$$

$$\frac{F}{x_1^r} = \tau(z_1 + \alpha)^r + \frac{a_2(x, y)}{x^2}(z_1 + \alpha)^{r-2} + \cdots + \frac{a_r(x, y)}{x^r}$$

Thus $\nu(F_1(0, 0, z_1)) \leq r$ and $\nu(F_1(0, 0, z_1)) \leq r - 1$ if $\alpha \neq 0$.

Suppose that $p \in C$ is a 2 point such that $\nu(p) = r$ and $\tau(p) > 0$.

$$\begin{aligned} u &= (x^a y^b)^m \\ v &= P(x^a y^b) + x^c y^d F \end{aligned}$$

where $F \in \hat{\mathcal{I}}_{C,p}^r = (x, z)^r$.

$$F = \sum_{i+k \geq r} a_{ijk} x^i y^j z^k.$$

Suppose that $q \in \pi^{-1}(p)$ is a 2 point. After a permissible change of parameters, replacing z with $z - \alpha x$, q_1 has permissible parameters (x_1, y_1, z_1) such that

$$x = x_1, z = x_1 z_1$$

$$\begin{aligned} u &= (x_1^a y^b)^m \\ v &= P(x_1^a y_1^b) + x_1^{c+r} y^d F_q \end{aligned}$$

$$F_q = \frac{F}{x_1^r} = \sum_{i+k \geq r} a_{ijk} x_1^{i+k-r} y^j z_1^k$$

$$F_q = \sum_{i+k=r} a_{i0k} z_1^k + x_1 \Omega_1 + yG$$

Thus $\gamma(q) \leq r$.

Now suppose that $q \in \pi^{-1}(p)$ has permissible parameters

$$x = x_1 z_1, z = z_1$$

so that q is a 3 point.

$$\begin{aligned} u &= (x_1^a y^b z_1^a)^m \\ v &= P(x_1^a y^b z_1^a) + x_1^c y^d z_1^{c+r} F_q \end{aligned}$$

$$F_q = \frac{F}{z_1^r} = \sum_{i+k \geq r} a_{ijk} x_1^i y^j z_1^{i+k-r}$$

$$F_q = \sum_{i+k=r} a_{i0k} x_1^i + yG + z_1 \Omega$$

$a_{r-k,0,k} \neq 0$ if $k = \tau(p)$ which implies that $\nu(q) \leq r - \tau(p)$. $\hfill\square$

Lemma 8.9. *Suppose that $r \geq 2$, $C \subset \bar{S}_r(X)$ is a curve containing a 1 point such that C is r small.*

1. *Let $\pi : Y \to X$ be the monoidal transform centered at C.*
   (a) *Suppose that $p \in C$ is a generic point. If $q \in \pi^{-1}(p)$ is a 1 point then $\nu(q) = 1$. If $q \in \pi^{-1}(p)$ is the 2 point then $\nu(q) \leq r$ and $\nu(q) = r$ implies $\tau(q) > 0$.*
   (b) *Suppose that $p \in C$ is a 2 point such that $\nu(q) = r - 1$. If $q \in \pi^{-1}(p)$ is a 2 point then $\nu(q) = 0$. If $q \in \pi^{-1}(p)$ is a 3 point then $\nu(q) \leq r - 1$.*
2. *Suppose that $p \in C$ is a 1 point such that $\nu(p) = r$ or a 2 point such that $\nu(p) = r$ and $\tau(p) > 0$. Then there exists a finite sequence of quadratic transforms $\sigma : Z \to X$ centered at points over p such that if $q \in \sigma^{-1}(p)$ is a 1 point then $\nu(q) \leq r$. $\nu(q) = r$ implies $\gamma(q) = r$. If $q \in \sigma^{-1}(p)$ is a 2 point then $\nu(q) \leq r$. $\nu(q) = r$ implies $\tau(q) > 0$. If $q \in \sigma^{-1}(p)$ is a 3 point then $\nu(q) \leq r - 1$. The strict transform of C intersects $\sigma^{-1}(p)$ in a 2 point p' such that $\nu(p') = r - 1$*

Proof. Suppose that $p \in C$ is a 2 point. By Lemma 6.26, there are permissible parameters (x, y, z) at p with $\hat{\mathcal{I}}_{C,p} = (x, z)$ such that

$$
\begin{aligned}
u &= (x^a y^b)^m \\
v &= P(x^a y^b) + x^c y^d F_p \\
F_p &= x^{r-1} y^n + \sum_{i+k \geq r} a_{ijk} x^i y^j z^k
\end{aligned}
\tag{73}
$$

with $n \geq 0$. Suppose that $\nu(p) = r$ and $\tau(p) > 0$. Then $n > 0$ and $a_{i0k} \neq 0$ for some i, k with $i + k = r$ and $k > 0$. Let $\pi' : X' \to X$ be the blow-up of p. Perform n quadratic transforms, $\pi_1 : X_1 \to X$, centered at the 2 point which is the intersection of the strict transform of C and the exceptional divisor. Then by Theorem 7.3

1. All 1 points q in $\pi_1^{-1}(p)$ with $\nu(q) = r$ have $\gamma(q) = r$.
2. All 2 points $q \in \pi^{-1}(p)$ with $\nu(q) = r$ have $\tau(q) > 0$.
3. All 3 points $q \in \pi^{-1}(p)$ have $\nu(q) \leq r - 1$.

If C_1 is the strict transform of C, and q is the exceptional point on C_1, then there are permissible parameters (x_1, y_1, z_1) at q such that

$$x = x_1 y_1^n, y = y_1, z = z_1 y_1^n$$

$$
\begin{aligned}
u &= (x_1^a y_1^{na+b})^m \\
F_q &= x_1^{r-1} + \sum_{i+k \geq r} a_{ijk} x_1^i y_1^{n(i+k-r)+j} z_1^k
\end{aligned}
\tag{74}
$$

where $\hat{\mathcal{I}}_{C_1, q} = (x_1, z_1)$. Thus $\nu(q) = r - 1$.

Suppose that $p \in C$ is a 1 point with $\nu(p) = r$. Then by Lemma 6.24 there are regular parameters (x, y, z) in $\hat{\mathcal{O}}_{X,p}$ such that $\hat{\mathcal{I}}_{C,p} = (x, z)$,

$$
\begin{aligned}
u &= x^a \\
F_p &= x^{r-1} y^n + \sum_{i+k \geq r} a_{ijk} x^i y^j z^k
\end{aligned}
\tag{75}
$$

with $n \geq 1$. There are only finitely many 1 points in C such that $n > 1$.

Suppose that $n > 1$. Then $a_{i0k} \neq 0$ for some a_{i0k} with $i + k = r$ and $k > 0$, so that $\tau(p) > 0$. Let $\lambda : Z \to X_2$ be the sequence of n quadratic transforms centered first at p, and then at the intersection of the strict transform of C and the exceptional fiber.

Let C' be the strict transform of C on Z. Let q' be the exceptional point of λ on C'. By Theorems 7.1 and 7.3, the conclusions of 2. of the Theorem hold at all points above p, except possibly at q'. q' has permissible parameters (x_1, y_1, z_1) such that

$$
x = x_1 y_1^n, y = y_1, z = z_1 y_1^n.
$$

$$
\begin{aligned}
u &= (x_1 y_1^n)^a \\
F_{q'} &= \frac{F_q}{y_1^{nr}} = x_1^{r-1} + \sum_{i+k \geq r} a_{ijk} x_1^i y_1^{(i+k-r)n+j} z_1^k
\end{aligned}
$$

Thus $\nu(q') = r - 1$ and C' has the form (73) with $n = 0$ at q'.

Let $\pi : Y \to X$ be the blow-up of C. Suppose that $p \in C$, and p is a 2 point such that $\nu(p) = r - 1$ so that (73) with $n = 0$ holds at p. Suppose that $q \in \pi^{-1}(p)$, and q has permissible parameters (x_1, y_1, z_1) such that

$$
x = x_1, z = x_1(z_1 + \alpha)
$$

After making a permissible change of variables, replacing z with $z - \alpha x$, we may assume that $\alpha = 0$. Then $F_q = \frac{F_p}{x_1^{r-1}}$, so that $\nu(q) = 0$.

Suppose that $q \in \pi^{-1}(p)$, and q has permissible parameters (x_1, y, z_1) such that

$$
x = x_1 z_1, z = z_1
$$

$$
\begin{aligned}
u &= (x_1^a y_1^b z_1^a)^m \\
F_q &= \frac{F_p}{z_1^{r-1}} = x_1^{r-1} + \sum_{i+k \geq r} a_{ijk} x_1^i y^j z_1^{i+k-(r-1)}
\end{aligned}
$$

so that $\nu(q) \leq r - 1$.

Now suppose that $p \in C$ is a generic point, so that (75) holds with $n = 1$ at p. Suppose that $q \in \pi^{-1}(p)$, and q has permissible parameters (x_1, y, z_1) such that

$$
x = x_1, z = x_1(z_1 + \alpha)
$$

After making a permissible change of variables, replacing z with $z - \alpha x$, we may assume that $\alpha = 0$. Then $F_q = \frac{F_p}{x_1^{r-1}}$, so that $\nu(q) = 1$.

Suppose that $q \in \pi^{-1}(p)$, and q has permissible parameters (x_1, y, z_1) such that

$$
x = x_1 z_1, z = z_1
$$

$$
\begin{aligned}
u &= (x_1 z_1)^a \\
F_q &= \frac{F_p}{z_1^{r-1}} = x_1^{r-1} y + \sum_{i+k \geq r} a_{ijk} x_1^i y^j z_1^{i+k-(r-1)}
\end{aligned}
$$

so that $\nu(q) \leq r$, $\nu(q) = r$ implies $\tau(q) > 0$. $\square$

Lemma 8.10. *Suppose that $r \geq 2$, $C \subset \bar{S}_r(X)$ is a curve containing a 1 point such that C is r small and $\gamma(q) = r$ for $q \in C$.*

1. *Let $\pi : Y \to X$ be the monoidal transform centered at C.*
   (a) *Suppose that $p \in C$ is a generic point. Then $\nu(q) \leq 1$ if $q \in \pi^{-1}(p)$.*
   (b) *Suppose that $p \in C$ is a 2 point such that $\nu(p) = r - 1$. Suppose that $q \in \pi^{-1}(p)$. Then $\nu(q) = 0$ if q is a 2 point, and $\nu(q) \leq 1$ if q is a 3 point.*
2. *Suppose that $p \in C$. Then there exists a finite sequence of quadratic transforms $\sigma : Z \to X$ centered at points over p such that $\nu(q) \leq r$, and $\gamma(q) \leq r$ if $q \in \sigma^{-1}(p)$ is a 1 or 2 point. $\nu(q) = 0$ if q is a 3 point, and the strict transform of C intersects $\sigma^{-1}(p)$ in a 2 point p' such that $\nu(p') = r - 1$ and $\gamma(p') = r$.*

Proof. Suppose that $p \in C$ is a 2 point. By Lemma 6.26, there exist permissible parameters (x, y, z) at p such that $x = z = 0$ are local equations of C at p.

$$\begin{aligned} u &= (x^a y^b)^m \\ v &= P(x^a y^b) + x^c y^d F_p \end{aligned} \tag{76}$$

$$F_p = \tau'(y) x^{r-1} y^n + \sum_{i+k \geq r} a_{ijk} x^i y^j z^k$$

with τ' a unit, $n \geq 0$, and $a_{00r} \neq 0$.

Suppose that $\nu(p) = r - 1$. Then $n = 0$ and $\tau(p) = 0$. Let $\pi : Y \to X$ be the monoidal transform centered at C.

If $q \in \pi^{-1}(p)$ is a 2 point, then after a permissible change of parameters at p, we have that q has permissible parameters (x_1, y, z_1) such that $x = x_1$, $z = x_1 z_1$.

$$F_q = \frac{F_p}{x_1^{r-1}} = \tau'(y) + x_1 \Omega$$

so that $\nu(q) = 0$.

If $q \in \pi^{-1}(p)$ is the 3 point, there exist permissible parameters (x_1, y, z_1) at q such that $x = x_1 z_1$, $z = z_1$.

$$F_q = \frac{F_p}{z_1^{r-1}} = \tau'(y) x_1^{r-1} + \sum_{i+k \geq r} a_{ijk} x_1^i y^j z_1^{i+k-(r-1)}$$

$a_{00r} \neq 0$ implies $\nu(q) \leq 1$.

Suppose that p is a 2 point and $\nu(p) = r$. Let $\sigma : Y \to X$ be the quadratic transform with center p. Suppose that $q \in \sigma^{-1}(p)$ is a 1 point or a 2 point. Then by Theorem 7.3, $\nu(q) \leq r$ and $\gamma(q) \leq r$. If q is a 3 point then $\nu(q) = 0$. At the 2 point q on the strict transform of C, we have permissible parameters (x_1, y_1, z_1) such that $x = x_1 y_1$, $y = y_1$, $z = x_1 y_1$. $x_1 = z_1 = 0$ are local equations of the strict transform of C at q.

$$F_q = \tau'(y) x_1^{r-1} y_1^{n-1} + \sum_{i+k \geq r} a_{ijk} x_1^i y_1^{i+j+k-r} z_1^k$$

of the form of (76) with n decreased by 1.

By induction on n, we achieve the conclusion of 2. after a finite sequence of quadratic transforms.

Suppose that $p \in C$ is a 1 point. By Lemma 6.24, there exist permissible parameters (x, y, z) at p such that $x = z = 0$ are local equations of C at p,

$$\begin{aligned}
u &= x^a \\
v &= P(x) + x^c F_p \\
F_p &= x^{r-1} \tau'(y) y^n + \textstyle\sum_{i+k \geq r} a_{ijk} x^i y^j z^k
\end{aligned} \tag{77}$$

with τ' a unit, $n \geq 1$, $a_{00r} \neq 0$.

Suppose that $p \in C$ is a generic point so that $n = 1$. Let $\pi : Y \to X$ be the monoidal transform centered at C. If $q \in \pi^{-1}(p)$ is a 1 point, then after making a permissible change of parameters at p, there are permissible parameters (x_1, y, z_1) at q such that $x = x_1, z = x_1 z_1$.

$$F_q = \frac{F_p}{x_1^{r-1}} = y\tau'(y) + x_1 \Omega$$

implies $\nu(q) = 1$. If $q \in \pi^{-1}(p)$ is the 2 point, then there are permissible parameters (x_1, y, z_1) at q such that $x = x_1 z_1, z = z_1$.

$$F_q = \frac{F_p}{z_1^{r-1}} = x_1^{r-1}\tau'(y)y + \sum_{i+k \geq r} a_{ijk} x_1^i y_1^j z_1^{i+k-(r-1)}$$

which implies that $\nu(q) \leq 1$ since $a_{00r} \neq 0$.

Suppose that $n \geq 2$ in (77). Let $\sigma : Y \to X$ be the quadratic transform with center p.

Suppose that $q \in \sigma^{-1}(p)$. Then q is a 1 or 2 point, and $\nu(q) \leq r$, $\gamma(q) \leq r$ by Theorem 7.3.

At the 2 point $q \in \pi^{-1}(p)$ which is contained in the strict transform of C, there are permissible parameters (x_1, y_1, z_1) at q such that $x = x_1 y_1, y = y_1, z = y_1 z_1$.

$$\begin{aligned}
u &= (x_1 y_1)^a \\
F_q &= \frac{F_p}{y_1^r} = x_1^{r-1} y_1^{n-1} \tau'(y_1) + \textstyle\sum_{i+k \geq r} a_{ijk} x_1^i y_1^{i+j+k-r} z_1^k
\end{aligned}$$

The strict transform of C has local equations $x_1 = z_1 = 0$ at q. We are thus at a point of the form of (76) with n decreased by 1.

We thus achieve the conclusions of 2. after a finite number of quadratic transforms. $\qquad\square$

Lemma 8.11. *Suppose that $r = 2$ in Lemma 8.10, $C \subset \overline{S}_2(X)$ is a curve containing a 1 point such that C is 2 small, $\gamma(p) = 2$ if $p \in C$, $\nu(p) = 1$ if $p \in C$ is a 2 point and p is a generic point of C ($n = 1$ in (77)) if $p \in C$ is a 1 point. Let $\pi : Y \to X$ be the monoidal transform centered at C. Suppose that there exists a 2 point $p \in C$ such that $\nu(p) = r - 1 = 1$, and $q \in \pi^{-1}(p)$ is a 3 point such that $\nu(q) = 1$, or $p \in C$ is a generic point of C ($n = 1$ in (77)) and $q \in \pi^{-1}(p)$ is a 2 point such that $\nu(q) = 1$. Let $\overline{C}$ be the 2 curve through q which is a section over C. Then $F_{q'} \in \hat{I}_{\overline{C}, q'}$ for all $q' \in \overline{C}$.*

Suppose that there does exist a 2 curve $\overline{C}$ which is a section over C such that $F_{q'} \in \hat{I}_{\overline{C}, q'}$ for $q' \in \overline{C}$. Let $\pi_1 : Z \to Y$ be the blow-up of $\overline{C}$. Then

1. *Suppose that $q \in \overline{C}$ is a 2 point such that $q \in \pi^{-1}(p)$ where p is a generic point of C ($n = 1$ in (77)), and $q' \in \pi_1^{-1}(q)$.*
   (a) *If q' is a 1 point then $\nu(q') = 1$.*
   (b) *If $q' \in \pi_1^{-1}(q)$ is a 2 point then $\gamma(q') \leq 1$.*
2. *Suppose that $q \in \overline{C}$ is a 3 point such that $q \in \pi^{-1}(p)$ where $p \in C$ is a 2 point such that $\nu(p) = 1$ and $q' \in \pi^{-1}(q)$.*
   (a) *If q' is a 2 point then $\gamma(q') \leq 1$.*
   (b) *If q' is a 3 point then $\nu(q') = 0$.*

Proof. If $p \in C$ is a 2 point with $\nu(p) = 1$ (and $\gamma(p) = 2$), then there exist permissible parameters (x, y, z) at p such that

$$
\begin{aligned}
u &= (x^a y^b)^m \\
v &= P(x^a y^b) + x^c y^d F_p \\
F_p &= x + z^2
\end{aligned}
$$

where $x = z = 0$ are local equations of C at p. There exist permissible parameters (x_1, y, z_1) at the 3 point $q \in \pi^{-1}(p)$ such that $x = x_1 z_1, z = z_1$.

$$
\begin{aligned}
u &= (x_1^a y^b z_1^a)^m \\
v &= P(x_1^a y^b z_1^a) + x_1^c y^d z_1^{c+1} F_q \\
F_q &= x_1 + z_1
\end{aligned}
\tag{78}
$$

and $x_1 = z_1 = 0$ are local equations of $\overline{C}$ at q. We have $F_q \in \hat{I}_{\overline{C},q}$ which implies $F_{q'} \in \hat{I}_{\overline{C},q'}$ if $q' \in \overline{C}$ by Lemma 8.1.

If $p \in C$ is a generic point, then p is a 1 point and there exist permissible parameters (x, y, z) at p such that

$$
F_p = xy + \sum_{i+k \geq 2} a_{ijk} x^i y^j z^k
$$

with $a_{002} \neq 0$ and $x = z = 0$ are local equations of C at p. There exist permissible parameters (x_1, z_1, y) at the 2 point $q \in \pi^{-1}(p)$ such that $x = x_1 z_1, z = z_1$.

$$
F_q = x_1 y_1 + z_1 \left(\sum_{i+k=2} a_{ijk} x_1^i y_1^j \right) + z_1^2 \Omega
$$

and $x_1 = z_1 = 0$ are local equations of $\overline{C}$ at q. Since $a_{002} \neq 0$, there exist permissible parameters $(x_1, \overline{z}_1, y_1)$ at q such that

$$
F_q = x_1 y_1 + \overline{z}_1
\tag{79}
$$

and $x_1 = \overline{z}_1 = 0$ are local equations of $\overline{C}$ at q.

We have $F_q \in \hat{I}_{\overline{C},q}$ implies $F_{q'} \in \hat{I}_{\overline{C},q'}$ for all $q' \in \overline{C}$ by Lemma 8.1.

1. follows from (79).

Suppose that $q \in \overline{C}$ is a 3 point, with permissible parameters (x_1, y, z_1) such that (78) holds at q. Suppose that $q' \in \pi_1^{-1}(q)$. If q' is a 3 point, then $\nu(q') = 0$. Suppose that q' is a 2 point. Then there exist regular parameters (x_2, y, z_2) in $\hat{O}_{Z,q'}$ such that

$$
x_1 = x_2, z_1 = x_2(z_2 + \alpha)
$$

with $\alpha \neq 0$.
$$u = (x_2^{2a} y^b (z_2 + \alpha)^a)^m = (\overline{x}_2^{2a} y^b)^m = (\overline{x}_2^{\overline{a}} y^{\overline{b}})^{\overline{m}}$$
where $x_2 = \overline{x}_2 (z_2 + \alpha)^{-\frac{1}{2}}$, $(\overline{a}, \overline{b}) = 1$.
$$v = P_{q'}(\overline{x}_2^{\overline{a}} y^{\overline{b}}) + \overline{x}_2^{2c+2} y^d (1 + \alpha + z_2)$$
Thus $\gamma(q') \leq 1$ and 2. follows.

□

9. Power Series in 2 Variables

Lemma 9.1. *Suppose that $R = k[[x, y]]$ is a power series ring in two variables and $u(x, y), v(x, y) \in R$ are series. Suppose that $R \to R'$ is a quadratic transform. Set $R_1 = \hat{R}'$. Then (u, v) are analytically independent in R_1 if and only if u and v are analytically independent in R.*

Proof. By Zariski's Subspace Theorem (Theorem 10.6 [3]), $R \to R_1$ is an inclusion, and the Lemma follows. $\qquad\square$

Lemma 9.2. *Suppose that $R = k[[x, y]]$ is a power series ring in two variables over an algebraically closed field k of characteristic 0 and $u(x, y), v(x, y) \in R$ are series such that either*

$$u = x^a$$

or

$$u = (x^a y^b)^m$$

with $(a, b) = 1$. Then u and v are analytically dependent if and only if there exists a series $p(t)$ such that $v = p(x)$ in the first case and $v = p(x^a y^b)$ in the second case.

Proof. First suppose that $u = x^a$ and $v = p(x)$ is a series. Let ω be a primitive a-th root of unity.

$$0 = \prod_{i=0}^{a-1} (v - p(\omega^i x)) \in k[[u, v]]$$

implies u and v are analytically dependent.

Now suppose that $u = x^a$ and u, v are analytically dependent. Suppose that v is not a series in x. Write

$$v = q(x) + x^b F$$

where $q(x)$ is a polynomial, $x \nmid F$ and $F(0, y)$ is a nonzero series with no constant term.

$$F(0, y) = y^r \mu(y)$$

for some $r > 0$ where $\mu(y)$ is a unit series. x and $x^b F$ are thus analytically dependent, and x and F are analytically dependent. There exists an irreducible series

$$p(s, t) = \sum a_{ij} s^i t^j$$

such that

$$0 = \sum a_{ij} x^i F^j$$

which implies that

$$0 = \sum a_{0j} F(0, y)^j = \sum a_{0j} y^{rj} \mu(y)^j,$$

a contradiction, since $p(s, t)$ irreducible implies some $a_{0j} \neq 0$. Thus v is a series in x.

Now suppose that $u = (x^a y^b)^m$ and $v = p(x^a y^b)$ is a series in $x^a y^b$. Let ω be a primitive m-th root of unity.

$$0 = \prod_{i=0}^{m-1} (v - p(\omega^i x^a y^b)) \in k[[u, v]]$$

implies that u, v are analytically dependent.

Suppose that

$$u = (x^a y^b)^m$$

and u, v are analytically dependent. Consider the quadratic transform

$$R \to R_1 = R[x_1, y_1]_{(x_1, y_1)}$$

where $x = x_1, y = x_1(y_1 + 1)$. $\hat{R}_1$ has regular parameters $(\overline{x}_1, y_1)$ where

$$x_1 = \overline{x}_1 (y_1 + 1)^{-\frac{b}{a+b}}.$$

Thus in $\hat{R}_1$, $u = \overline{x}_1^{(a+b)m}$. Since u, v must be analytically dependent in $\hat{R}_1$, there exists a series $q(\overline{x}_1)$ such that $v = q(\overline{x}_1)$, by the first part of the proof.

Suppose that v is not a series in $x^a y^b$. Write $v = \sum a_{ij} x^i y^j$. There exists a smallest r such that there exists i_0, j_0 such that $i_0 + j_0 = r$, $b i_0 - a j_0 \neq 0$ and $a_{i_0 j_0} \neq 0$.

$$v = \sum_{i+j<r, aj-bi=0} a_{ij} \overline{x}_1^{i+j} + \overline{x}_1^r \left(\sum_{i+j=r} a_{ij} (y_1 + \alpha)^{j - \frac{br}{a+b}} \right) + \overline{x}_1^{r+1} \Omega.$$

$$\sum_{i+j=r} a_{ij} (y_1 + \alpha)^{j - \frac{br}{a+b}} \in k$$

implies

$$(y_1 + \alpha)^{\frac{-br}{a+b}} \left(\sum_{i+j=r} a_{ij} (y_1 + \alpha)^j \right) = c \in k$$

so that

$$\sum_{i+j=r} a_{ij} (y_1 + \alpha)^j = c(y_1 + \alpha)^{\frac{br}{a+b}}.$$

Thus

$$\frac{br}{a+b} \in \{0, 1, \cdots, r\}$$

and $j_0 = \frac{br}{a+b}$. This implies that $a j_0 - b i_0 = 0$, a contradiction. Thus v is a series in $x^a y^b$. $\square$

Lemma 9.3. *Suppose that $R = k[[x, y]]$ is a power series in two variables over an algebraically closed field k of characteristic 0, $u = x^a$ or $u = (x^a y^b)^m$, and (u, v) are analytically independent. let $\pi : X \to \operatorname{spec}(R)$ be the blow-up of $m = (x, y)$. Then for all but finitely many points $q \in \pi^{-1}(m)$ there exist regular parameters $(\overline{x}, \overline{y})$ in $\hat{\mathcal{O}}_{X,q}$ such that there is an expansion*

$$\begin{aligned} u &= \overline{x}^a \\ v &= P(\overline{x}) + \overline{x}^b \overline{y} \end{aligned} \tag{80}$$

Proof. First suppose that $u = x^a$. Write $v = P(x) + x^b F$ where $x \nmid F$ and F has no terms which are powers of x. Write

$$F = \sum_{i+j \geq r} a_{ij} x^i y^j$$

where $r = \nu(F)$. There exists $j_0 > 0$ such that $i_0 + j_0 = r$ and $a_{i_0 j_0} \neq 0$. For all but one point $q \in \pi^{-1}(m)$ there are regular parameters (x_1, y_1) in $\hat{O}_{X,q}$ such that

$$x = x_1, y = x_1(y_1 + \alpha)$$

with $\alpha \in k$.

$$\begin{aligned} u &= x_1^a \\ v &= P(x_1) = x_1^{b+r}(\textstyle\sum_{i+j=r} a_{ij}(y_1 + \alpha)^j + x_1 \Omega) \end{aligned} \qquad (81)$$

v has an expansion (80) if and only if

$$\frac{d}{dy_1}(\textstyle\sum_{i+j=r} a_{ij}(y_1 + \alpha)^j)\,|_{y_1=0} = \textstyle\sum_{j \leq r} j a_{r-j,j}(-\alpha)^{j-1} \neq 0. \qquad (82)$$

Since

$$\sum_{i+j=r} j a_{r-j,j}(-\alpha)^{j-1}$$

has at most finitely many roots, all but finitely many $q \in \pi^{-1}(m)$ have an expansion (80).

Now suppose that $u = (x^a y^b)^m$. Write

$$v = P(x^a y^b) + x^c y^d F$$

where $x, y \nmid F$ and $x^c y^d F$ has no terms which are powers of $x^a y^b$. Write

$$F = \sum_{i+j \geq r} a_{ij} x^i y^j$$

where $r = \nu(F)$.

For all but two points $q \in \pi^{-1}(m)$ there are regular parameters (x_1, y_1) in $\hat{O}_{X,q}$ such that

$$x = x_1, y = x_1(y_1 + \alpha)$$

with $\alpha \neq 0$. There are regular parameters $(\overline{x}_1, y_1)$ in $\hat{O}_{X,q}$ such that

$$x_1 = \overline{x}_1(y_1 + \alpha)^{-\frac{b}{a+b}}.$$

$$\begin{aligned} u &= \overline{x}_1^{(a+b)m} \\ v &= P(\overline{x}_1^{(a+b)m}) + \overline{x}_1^{c+d+r}(y_1 + \alpha)^\lambda \frac{F}{x_1^r} \end{aligned}$$

where

$$\lambda = d - \frac{b(c+d+r)}{a+b}$$

$$(y_1 + \alpha)^\lambda \frac{F}{x_1^r} = \sum_{i+j=r} a_{ij}(y_1 + \alpha)^{j+\lambda} + \overline{x}_1 \Omega.$$

v does not have an expression (80) at q if and only if there exists $c_\alpha \in k$ such that

$$\sum_{j=0}^{r} a_{r-j,j}(y_1 + \alpha)^j \equiv c_\alpha(y_1 + \alpha)^{-\lambda} \mod (y_1)^2.$$

Set $a_j = a_{r-j,j}$. Suppose that q does not have a form (80). Then

$$\sum_{j=0}^{r} a_j \alpha^j = c_\alpha \alpha^{-\lambda}$$

and

$$\sum_{j=0}^{r} j a_j \alpha^{j-1} = -c_\alpha \lambda \alpha^{-\lambda-1}$$

implies

$$(-\lambda) \sum_{j=0}^{r} a_j \alpha^j = \sum_{j=0}^{r} j a_j \alpha^j. \qquad (83)$$

If there are infinitely many values of α satisfying (83), then $(-\lambda - j)a_j = 0$ for $0 \le j \le r$, which implies that $-\lambda \in \{0, \dots, r\}$ and the leading form of F is

$$L = \sum_{i+j=r} a_{ij} x^i y^j = a_{r+\lambda, -\lambda} x^{r+\lambda} y^{-\lambda}.$$

Thus $x^c y^d F$ has a nonzero $x^{c+r+\lambda} y^{d-\lambda}$ term.

$$\begin{aligned}
a(d - \lambda) - b(c + r + \lambda) &= ad - b(c + r) - (a + b)\lambda \\
&= ad - b(c + r) - (a + b)(d - \tfrac{b(c+d+r)}{a+b}) \\
&= ad - b(c + r) - (a + b)d + b(c + d + r) = 0
\end{aligned}$$

which is impossible since F is normalized (contains no terms which are powers of $x^a y^b$). Thus there are at most a finite number of points $q \in \pi^{-1}(m)$ where the form (80) does not hold. $\qquad \square$

Theorem 9.4. *Suppose that k is an algebraically closed field of characteristic zero, B is a powerseries ring in 2 variables over k. Suppose that $u, v \in B$ are analytically independent, and there exist regular parameters (x, y) in B such that $u = x^a$ or $u = x^a y^b$. Let $A = \text{spec}(B)$. Then there exists a sequence of quadratic transforms $\pi : X \to A$ such that for all points $q \in X$, there exist regular parameters $(\bar{x}, \bar{y})$ in $\hat{\mathcal{O}}_{X,q}$ such that either*

$$\begin{aligned}
u &= \bar{x}^a \\
v &= P(\bar{x}) + \bar{x}^b \bar{y}^c
\end{aligned} \qquad (84)$$

or

$$\begin{aligned}
u &= (\bar{x}^a \bar{y}^b)^m \\
v &= P(\bar{x}^a \bar{y}^b) + \bar{x}^c \bar{y}^d
\end{aligned} \qquad (85)$$

where $(a, b) = 1$ and $ad - bc \ne 0$.

Theorem 9.4 will follow from Theorem 9.15. Throughout this section, we will use the notations of the statement of Theorem 9.4.

If $A \to \mathrm{spec}(k[[u,v]])$ is weakly prepared, then a stronger result than the conclusions of Theorem 9.4 are true in B.

Remark 9.5. *With the assumptions of Theorem 9.4, further suppose that*

$$\sqrt{(\frac{\partial u}{\partial x}\frac{\partial v}{\partial y} - \frac{\partial u}{\partial y}\frac{\partial v}{\partial x})} = \sqrt{(u)}.$$

Then there exist regular parameters $(\overline{x}, \overline{y})$ in B, and a power series P in B such that one of the following forms holds.

$$\begin{aligned} u &= \overline{x}^a \\ v &= P(\overline{x}) + \overline{x}^c \overline{y} \end{aligned} \tag{86}$$

$$\begin{aligned} u &= (\overline{x}^a \overline{y}^b)^m \\ v &= P(\overline{x}) + \overline{x}^c \overline{y}^d \end{aligned} \tag{87}$$

where $(a,b) = 1$ and $ad - bc \neq 0$.

Proof. (7.4 [8]) With our assumptions, one of the following must hold.

$$\begin{aligned} u &= x^a \\ u_x v_y - u_y v_x &= \delta x^e \end{aligned} \tag{88}$$

where δ is a unit or

$$\begin{aligned} u &= (x^a y^b)^m \\ u_x v_y - u_y v_x &= \delta x^e y^f \end{aligned} \tag{89}$$

where $a, b, e, f > 0$, $(a,b) = 1$ and δ is a unit.

Write $v = \sum a_{ij} x^i y^j$. First suppose that (88) holds. Then $a x^{a-1} v_y = \delta x^e$ implies we have the form (86). Now suppose that (89) holds.

$$u_x v_y - u_y v_x = \sum m(aj - bi) a_{ij} x^{am+i-1} y^{bm+j-1} = \delta x^e y^f.$$

Thus

$$v = \sum_{aj-bi=0} a_{ij} x^i y^j + \epsilon x^c y^d$$

where ϵ is a unit. After making a change of variables, multiplying x by a unit, and multiplying y by a unit, we get the form (87). $\square$

Definition 9.6. *Suppose that $\Phi : X \to A$ is a product of quadratic transforms, p is a point of X. We will say that (u, v) are 1-resolved at p if there exist regular parameters (x, y) in $\hat{\mathcal{O}}_{X,p}$ such that one of the forms (84) or (85) hold at p.*

For the rest of this section, we will assume that

$$\Phi : X \to A$$

is a sequence of quadratic transforms.

Suppose that $p \in X$ is a point. Then there are regular parameters (x, y) of $\hat{\mathcal{O}}_{X,p}$ such that $u = x^{\overline{a}} y^{\overline{b}}$, and $\overline{a} > 0$, $\overline{b} \geq 0$.

Suppose that $\bar{b} > 0$. Let $m = (\bar{a}, \bar{b})$, let $a = \frac{\bar{a}}{m}$, $b = \frac{\bar{b}}{m}$. There are power series $P(t)$ and $F(x, y)$ such that x does not divide F, y does not divide F, $x^c y^d F$ has no nonzero terms which are powers of $x^a y^b$ and (in $\hat{\mathcal{O}}_{X,p}$)

$$\begin{aligned} u &= (x^a y^b)^m \\ v &= P(x^a y^b) + x^c y^d F(x, y) \end{aligned} \tag{90}$$

In this case, we will say that p is a 2 point.

If $\bar{b} = 0$, there are power series $P(t)$ and $F(x, y)$ such that x does not divide F, F has no nonzero terms which are powers of x and (in $\hat{\mathcal{O}}_{X,p}$)

$$\begin{aligned} u &= x^a \\ v &= P(x) + x^c F(x, y) \end{aligned} \tag{91}$$

In this case we will say that p is a 1 point.

Suppose that $p \in X$, and (x, y) are regular parameters in $\hat{\mathcal{O}}_{X,p}$ such that (u, v) have one of the forms (90) or (91). Set

$$\bar{\nu}(p) = \begin{cases} \text{mult}(F) - 1 & \text{if } p \text{ is a 1 point} \\ \text{mult}(F) & \text{if } p \text{ is a 2 point} \end{cases}$$

Lemma 9.7. $\bar{\nu}(p)$ is independent of the choice of regular parameters (x, y) in (90) or (91).

Proof. First suppose that p is a 2 point. To express u and v in the form (90) we can only make a permissible change of variables in x and y, where a permissible change of variables is one of the following two forms:

$$x = \omega_x \bar{x}, y = \omega_y \bar{y} \text{ where } \omega_x^{ma} \omega_y^{mb} = 1 \tag{92}$$

or

$$y = \omega_y \bar{x}, x = \omega_x \bar{y} \text{ where } \omega_x^{ma} \omega_y^{mb} = 1 \tag{93}$$

where ω_x, ω_y are unit series. $\bar{\nu}(p)$ does not change after a change of variables of one of these forms.

Now suppose p is a 1 point. To preserve the form (91) we can only make a permissible change of variables, where a permissible change of variables is of the form:

$$x = \omega_x \bar{x}, y = \phi(\bar{x}, \bar{y}) \text{ where } \text{mult}(\phi(0, \bar{y})) = 1. \tag{94}$$

and $\omega_x \in k$ is an a-th root of unity. Then

$$\phi(\bar{x}, \bar{y}) = \bar{\phi}(\bar{x}, \bar{y})(\bar{y} + \psi(\bar{x}))$$

where $\bar{\phi}$ is a unit. Write

$$\psi(\bar{x}) = \sum b_i \bar{x}^i.$$

$$\begin{aligned} u &= \bar{x}^a \\ v &= \bar{P}(\bar{x}) + \bar{x}^c \bar{F}(\bar{x}, \bar{y}) \end{aligned} \tag{95}$$

where

$$\begin{aligned} \bar{F} &= \omega_x^c (F(\omega_x \bar{x}, \phi(\bar{x}, \bar{y})) - F(\omega_x \bar{x}, \phi(\bar{x}, 0))) \\ \bar{P}(\bar{x}) &= P(\omega_x \bar{x}) + \bar{x}^c \omega_x^c F(\omega_x \bar{x}, \phi(\bar{x}, 0)) \end{aligned}$$

Suppose that the leading form of F is

$$L = \sum_{i+j=r} a_{ij} x^i y^j.$$

The leading form $\overline{L}$ of $\overline{F}$ is then

$$\omega_x^c \left(\sum_{i+j=r} a_{ij} \omega_x \overline{x}^i (e(\overline{y} - b_1 \overline{x}))^j - \sum_{i+j=r} a_{ij} \omega_x \overline{x}^i (-eb_1 \overline{x})^j \right)$$

where $e = \overline{\phi}(0,0)$. $\overline{L}$ is nonzero since $a_{ij} \neq 0$ for some $j > 0$. $\qquad \square$

(u,v) are 1-resolved at a 2 point p if and only if F is a unit. (u,v) are 1-resolved at a 1 point p if and only if $F(x,y) = g(x,y)^d + h(x)$ for some series $g(x,y)$ with $\text{mult}(g(0,y)) = 1$, and positive integer d.

Theorem 9.8. *Suppose that $g : X_1 \to X$ is a quadratic transform, centered at a point p of X, and $p_1 \in X_1$ is a point such that $g(p_1) = p$. Then*

$$\overline{\nu}(p_1) \leq \overline{\nu}(p).$$

If (u,v) are 1-resolved at p then (u,v) are 1-resolved at p_1.

Proof. First suppose that p is a 2 point. Write

$$F = \sum_{i+j \geq r} a_{ij} x^i y^j.$$

in $\hat{\mathcal{O}}_{X,p}$, where $r = \text{mult}(F) = \overline{\nu}(p)$. Suppose that $\hat{\mathcal{O}}_{X_1,p_1}$ has regular parameters (x_1, y_1) such that $x = x_1, y = x_1(y_1 + \alpha)$ with $\alpha \neq 0$. Define $\overline{x}_1$ by

$$x_1 = \overline{x}_1 (y_1 + \alpha)^{\frac{-b}{a+b}}.$$

Then $(\overline{x}_1, y_1)$ are regular parameters in $\hat{\mathcal{O}}_{X_1,p_1}$.

$$u = x_1^{m(a+b)} (y_1 + \alpha)^{mb} = \overline{x}_1^{m(a+b)}.$$

$$v = P(\overline{x}_1^{a+b}) + \overline{x}_1^{c+d+r} (y_1 + \alpha)^{\lambda} \left(\frac{F}{x_1^r} \right)$$

where $\lambda = d - \frac{b(c+d+r)}{a+b}$.

$$F = \sum_{i+j=r} a_{ij} x_1^r (y_1 + \alpha)^j + x_1^{r+1} \Omega.$$

$$\frac{F}{x_1^r} = \sum_{j=0}^{r} a_j (y_1 + \alpha)^j + x_1 \Omega.$$

where $a_j = a_{r-j,j}$. We have

$$\begin{aligned} u &= \overline{x}_1^{m(a+b)} \\ v &= \overline{P}(\overline{x}_1) + \overline{x}_1^{c+d+r} \overline{F}(\overline{x}_1, y_1) \end{aligned} \qquad (96)$$

where

$$\overline{P} = P(\overline{x}_1^{a+b}) + \overline{x}_1^{c+d+r} (\alpha)^{\lambda} \left(\frac{F(\alpha^{\frac{-b}{a+b}} \overline{x}_1, \alpha^{\frac{a}{a+b}} \overline{x}_1)}{\alpha^{\frac{-rb}{a+b}} \overline{x}_1^r} \right)$$

$$\overline{F} = (y_1 + \alpha)^\lambda \left(\frac{F((y_1 + \alpha)^{\frac{-b}{a+b}} \overline{x}_1, (y_1 + \alpha)^{\frac{-a}{a+b}} \overline{x}_1)}{(y_1 + \alpha)^{\frac{-rb}{a+b}} \overline{x}_1^r} \right) - (\alpha)^\lambda \left(\frac{F(\alpha^{\frac{-b}{a+b}} \overline{x}_1, \alpha^{\frac{-a}{a+b}} \overline{x}_1)}{\alpha^{\frac{-rb}{a+b}} \overline{x}_1^r} \right)$$

Set

$$\beta = \left(\sum_{j=0}^r a_j \alpha^j \right) \alpha^\lambda.$$

Suppose that $\overline{\nu}(p_1) > \overline{\nu}(p) = r$, so that

$$\text{mult}(\overline{F}) \geq \text{mult}(F) + 2 = r + 2.$$

Then

$$(y_1 + \alpha)^\lambda \left(\sum_{j=0}^r a_j (y_1 + \alpha)^j \right) - \beta \equiv 0 \mod (y_1)^{r+2}.$$

$\beta \neq 0$ since $\sum_{j=0}^r a_j (y_1 + \alpha)^j \neq 0$. We have

$$\sum_{j=0}^r a_j (y_1 + \alpha)^j \equiv \beta (y_1 + \alpha)^{-\lambda} \mod (y_1)^{r+2} \qquad (97)$$

First suppose that $-\lambda \in \{0, 1, \dots, r\}$. Then

$$\sum_{j=0}^r a_j (y_1 + \alpha)^j = \beta (y_1 + \alpha)^{-\lambda}$$

where $t = -\lambda \leq r$. Thus the leading form of F is

$$\begin{aligned} L &= \textstyle\sum_{i+j=r} a_{ij} x_1^r (y_1 + \alpha)^j \\ &= \beta x_1^r (y_1 + \alpha)^{-\lambda} \\ &= \beta x^{r+\lambda} y^{-\lambda} \\ &= \beta x^{r-t} y^t \end{aligned}$$

So the leading form of F is $\beta x^{r-t} y^t$. Thus $\beta x^{c+r-t} y^{d+t}$ is a nonzero term of $x^c y^d F$. Since

$$t = \frac{b(c + d + r)}{a + b} - d,$$

we have

$$b(c + r - t) - a(d + t) = 0$$

so that $x^{c+r-t} y^{d+t}$ is a power of $x^a y^b$, a contradiction.

We must then have $-\lambda \notin \{0, 1 \dots, r\}$. But then the y_1^{r+1} coefficient of $\beta (y_1 + \alpha)^{-\lambda}$ is non zero, a contradiction to (97).

Now suppose that p is a 2 point and $\hat{\mathcal{O}}_{X_1, p_1}$ has regular parameters (x_1, y_1) such that $x = x_1, y = x_1 y_1$. Write

$$F = \sum_{i+j \geq r} a_{ij} x^i y^j.$$

Then

$$\begin{aligned} u &= x_1^{m(a+b)} y_1^{mb} \\ v &= \overline{P}(x_1^{a+b} y_1^b) + x_1^{c+d+r} y_1^d \overline{F}(x_1, y_1) \end{aligned} \qquad (98)$$

where $\overline{P} = P$, $\overline{F} = \frac{F}{x_1^r}$. We need only check that $\frac{F}{x_1^r}$ has no nonzero $x_1^\alpha y_1^\beta$ terms with $b(c+d+r+\alpha) = (a+b)(d+\beta)$. We have that $a_{ij} = 0$ if $b(c+i) - a(d+j) = 0$.

$$\frac{F}{x_1^r} = \sum a_{ij} x_1^{i+j-r} y_1^j.$$

Suppose that $b(c + d + r + \alpha) = (a + b)(d + \beta)$. Set $i = \alpha - \beta + r$, $j = \beta$. Then $b(c + i) - a(d + j) = 0$, and $a_{ij} = 0$. But this is the coefficient of $x_1^\alpha y_1^\beta$ in $\frac{F}{x_1^r}$. We have

$$\mathrm{mult}(\overline{F}) \leq \mathrm{mult}(F).$$

The above argument also works, by interchanging the variables x and y, in the case where p is a 2 point and $\hat{\mathcal{O}}_{X_1,p_1}$ has regular parameters (x_1, y_1) such that $x = x_1 y_1, y = y_1$.

Now suppose that p is a 1 point and $\hat{\mathcal{O}}_{X_1,p_1}$ has regular parameters (x_1, y_1) such that $x = x_1 y_1, y = y_1$. Write

$$F = \sum_{i+j\geq r} a_{ij} x^i y^j.$$

Then

$$u = x_1^a y_1^a$$
$$v = \overline{P}(x_1 y_1) + x_1^c y_1^{c+r} \overline{F}(x_1, y_1) \qquad (99)$$

where $\overline{P} = P$, $\overline{F} = \frac{F}{y_1^r}$. We must show that $\overline{F}$ has no nonzero terms $x_1^\alpha y_1^\beta$ terms with $\alpha = r + \beta$. But this is impossible since F has no nonzero x^i terms, with $i \geq 0$.

The leading form of $\overline{F}$ is

$$\overline{F} = \sum_{i=0}^{r-1} a_{i,r-i} x_1^i + y_1 \Omega$$

since $a_{r0} = 0$, where some $a_{ij} \neq 0$ with $i + j = r$, $j > 0$. Thus $\mathrm{mult}(\overline{F}) \leq \mathrm{mult}(F) - 1$.

Now suppose that p is a 1 point and $\hat{\mathcal{O}}_{X_1,p_1}$ has regular parameters (x_1, y_1) such that $x = x_1, y = x_1(y_1 + \alpha)$. By making if necessary a permissible change of variables at p, replacing y with $y - \alpha x$, we may assume that $x = x_1, y = x_1 y_1$. Write

$$F = \sum_{i+j\geq r} a_{ij} x^i y^j.$$

where $a_{i0} = 0$ for all i.

$$u = x_1^a$$
$$v = \overline{P}(x_1) + x_1^{c+r} \overline{F}(x_1, y_1)$$

where $\overline{P} = P$, $\overline{F} = \frac{F}{x_1^r}$. $\overline{F}$ has no nonzero terms which are powers of x_1. Thus

$$\mathrm{mult}(\overline{F}) \leq \mathrm{mult}(F).$$

$\square$

Suppose that $p \in X$. Set

$$\sigma(p) = \begin{cases} 0 & \text{if } p \text{ is a 1 point and } \text{mult}(F) = \text{mult}(F(0,y)) \\ \frac{1}{2} & \text{if } p \text{ is a 2 point} \\ 1 & \text{if } p \text{ is a 1 point and } \text{mult}(F) < \text{mult}(F(0,y)) \end{cases}$$

Lemma 9.9. $\sigma(p)$ *is independent of the choice of permissible parameters* (x, y) *at* p.

Proof. The proof of Lemma 9.7 shows that $\text{mult}(F(0,y))$ is independent of the choice of permissible parameters at a 1 point. $\qquad\square$

Lemma 9.10. *Suppose that* $g : X_1 \to X$ *is a quadratic transform, centered at a point* p *of* X, *and* $p_1 \in X_1$ *is a point such that* $g(p_1) = p$. *Further suppose that* p *is a 2 point,* p_1 *is a 1 point and* $\overline{\nu}(p_1) = \overline{\nu}(p)$. *Then* $\sigma(p_1) = 0$.

Proof. $\hat{\mathcal{O}}_{X_1,p_1}$ has regular parameters (x_1, y_1) such that $x = x_1, y = x_1(y_1 + \alpha)$ with $\alpha \neq 0$. Let $r = \text{mult}(F)$. $\text{mult}(F_1) = \text{mult}(F) + 1 = r + 1$. Let

$$F = \sum_{i+j \geq r} a_{ij} x^i y^j$$

As in the analysis leading to (97),

$$F_1 \equiv (y_1 + \alpha)^\lambda \left(\sum_{j=0}^{r} a_{r-j,j}(y_1 + \alpha)^j \right) - \beta \bmod (x_1, y_1^{r+1}) \qquad (100)$$

for some $\beta \in k$. If $F_1 \equiv 0 \bmod (x_1, y_1^{r+1})$, then $\beta \neq 0$ and $-\lambda \notin \{0, 1, \ldots, r\}$, as in the proof of Theorem 9.8. Then

$$\begin{aligned} F_1 &\equiv (y_1 + \alpha)^\lambda \left(-\beta \frac{-\lambda(-\lambda-1)\cdots(-\lambda-r)}{(r+1)!} \alpha^{-\lambda-r-1} \right) y_1^{r+1} \bmod (x_1, y_1^{r+2}) \\ &\equiv -\beta \frac{-\lambda(-\lambda-1)\cdots(-\lambda-r)}{(r+1)!} \alpha^{-r-1} y_1^{r+1} \bmod (x_1, y_1^{r+2}) \end{aligned} \qquad (101)$$

Thus $\text{mult}(F_1(0, y_1, z_1)) = r + 1$. $\qquad\square$

Lemma 9.11. *Suppose that* $g : X_1 \to X$ *is a quadratic transform, centered at a 1 point* p *of* X *and* p_1 *is a point above* p *such that* $g(p_1) = p$.

If p_1 *is a 1 point and* $\overline{\nu}(p_1) = \overline{\nu}(p)$, *then* $\sigma(p_1) = 0$. *If* $\sigma(p) = 0$ *and* p_1 *is a 2 point then* $\overline{\nu}(p_1) = 0$.

Proof. First suppose that $\sigma(p) = 0$ and p_1 is a 2 point. Then $\hat{\mathcal{O}}_{X_1,p_1}$ has regular parameters (x_1, y_1) such that $x = x_1 y_1, y = y_1$.

$$F = \sum_{i+j \geq r} a_{ij} x^i y^j$$

with $a_{0r} \neq 0$.

$$F_1 = \sum_{i=0}^{r-1} a_{i,r-i} x_1^i + y_1 \Omega.$$

is then a unit.

Now suppose that p_1 is a 1 point and $\nu(p_1) = \nu(p)$. After appropriate choice of permissible variables (x, y) at p, $\hat{\mathcal{O}}_{X_1,p_1}$ has regular parameters (x_1, y_1) such that $x = x_1, y = x_1 y_1$. Set $r = \text{mult}(F) = \text{mult}(F_1)$. Then $F_1 = \frac{F(x_1, x_1 y_1)}{x_1^r}$ and $\text{mult}(F_1(0, y_1)) = r$. $\qquad \square$

Theorem 9.12. *Suppose that $g : X_1 \to X$ is a quadratic transform, centered at a point p of X, and p_1 is a closed point such that $g(p_1) = p$. If $\overline{\nu}(p_1) = \overline{\nu}(p)$, then $\sigma(p_1) \le \sigma(p)$.*

Proof. This is immediate from Lemmas 9.10 and 9.11. $\qquad \square$

Suppose that

$$F = \sum_{i+j \ge r} a_{ij} x^i y^j$$

has multiplicity r. Define

$$\delta(F; x, y) = \min(\frac{i}{r-j} \mid j < r, a_{ij} \ne 0).$$

$\delta(F; x, y) = \infty$ if and only if $F = y^r \omega$, where ω is a unit. If $\delta(F; x, y) < \infty$, then $\delta(F; x, y) \in \frac{1}{r!} \mathbf{N}$.

Suppose that $p \in X$. If (x, y) are permissible parameters at p with one of the forms (90) or (91), set

$$\delta(p; x, y) = \delta(F; x, y).$$

Then set

$$\delta(p) = \sup(\delta(p; x, y))$$

where the sup is over all permissible parameters at p. Note that if p is a 2 point, then

$$\delta(p) = \max(\delta(p; x, y), \delta(p; y, x))$$

if (x, y) are a particular choice of permissible parameters at p.

If p is a 2 point and $\nu(p) > 0$, then $\delta(p) < \infty$. If p is a 1 point and $\sigma(p) = 1$, then $\delta(p) = 1$, since $\delta(p; x, y) = 1$ for all permissible parameters (x, y).

Lemma 9.13. *Suppose that p is a 1 point, $\sigma(p) = 0$ and (x, y) are fixed permissible parameters at p. Then there exists a power series $t(x)$ such that*

$$\delta(p) = \delta(p; x, y - t(x)).$$

If $\delta(p) < \infty$, then $t(x)$ is a polynomial.
$\delta(p) > \delta = \delta(p; x, y)$ *if and only if $\delta \in \mathbf{N}$ and*

$$\sum_{i+\delta j = r\delta} a_{ij} x^i y^j = \tau(y - cx^\delta)^r + \lambda x^{r\delta}$$

for some $\tau, c, \lambda \in k$ with $c \ne 0$ (so that $\lambda = -\tau(-c)^r$).

Proof. Suppose that $(\overline{x}, \overline{y})$ are also permissible parameters at p. Then $\overline{x} = \lambda x$, with $\lambda^a = 1$ and $\overline{y} = \overline{\phi}(y - t(x))$ for some unit series $\overline{\phi}$ and series $t(x)$.

$$\delta(p; \overline{x}, \overline{y}) = \delta(p; x, y - t(x))$$

Thus

$$\delta(p) = \sup(\delta(p; x, y - t(x)) \mid t(x) \text{ is a polynomial of positive order}).$$

$$(102)$$

Let $\delta = \delta(p; x, y)$,

$$\overline{L} = \sum_{i+\delta j = r\delta} a_{ij} x^i y^j.$$

so that

$$F = \overline{L} + \sum_{i+\delta j > r\delta} a_{ij} x^i y^j.$$

Suppose that

$$\overline{L} = \tau (y - cx^\delta)^r + \lambda x^{r\delta}$$

for some $\tau, c, \lambda \in k$ with $0 \neq c$. Set $y_1 = y - cx^\delta$. Then $\delta \in \mathbf{N}$ and $\delta(p; x, y_1) > \delta(p; x, y)$ since

$$F_1 = \tau y_1^r + \sum_{i+\delta j > r\delta} \overline{a}_{ij} x^i y_1^j.$$

where

$$v = P_1(x) + x_1^c F_1$$

is the normalized form of v with respect to (x, y_1). We can repeat this process, with y replaced by $y - cx^\delta$. The process will either produce a polynomial $t(x)$ such that if $y_1 = y - t(x)$, and $\delta_1 = \delta(p; x, y_1)$, then $\delta_1 \notin \mathbf{N}$, or $\delta_1 \in \mathbf{N}$ and

$$\sum_{i+\delta_1 j = r\delta_1} \overline{a}_{ij} x^i y_1^j \neq \tau (y_1 - cx^{\delta_1})^r + \lambda x^{r\delta_1} \qquad (103)$$

for any $\tau, c, \lambda \in k$ with $0 \neq c$, or we will produce a series $t(x)$ such that if $y_1 = y - t(x)$, then $\delta(p; x, y_1) = \delta(p) = \infty$, so that $F_1 = y_1^r \phi$, where ϕ is a unit series.

Suppose that we have produced y_1 such that $\delta(p; x, y_1) \notin \mathbf{N}$ or $\delta(p; x, y_1) \in \mathbf{N}$ and (103) holds. We will show that $\delta(p) = \delta(p; x, y_1)$. Suppose that $\delta_1 = \delta(p; x, y_1) < \delta(p)$. By (102), there is a polynomial

$$t(x) = \sum e_i x^i$$

such that if $y_2 = y_1 - t(x)$, then $\delta(p; x, y_2) > \delta(p; x, y_1)$. Substitute $y_1 = y_2 + t(x)$ into

$$F_1 = \sum_{i+\delta_1 j = r\delta_1} \overline{a}_{ij} x^i y_1^j + \sum_{i+\delta_1 j > r\delta_1} \overline{a}_{ij} x^i y_1^j,$$

and normalize with respect to the permissible parameters to get

$$v = P_2(x) + x^c F_2(x, y_2).$$

Let $\overline{d} = \mathrm{ord}\,(t(x))$. $x^i y_1^j = x^i (y_2 + t(x))^j$ has nonzero $x^{i+m\overline{d}} y_2^{j-m}$ terms with $0 \leq m \leq j$, and may have other nonzero $x^{i+m\overline{d}+\gamma} y_2^{j-m}$ terms with $0 \leq m \leq j$, $\gamma \geq 0$.

Suppose that $\overline{d} < \delta_1 = \delta(p; x, y_1)$. The expansion of y_1^r has a nontrivial $x^{\overline{d}} y_1^{r-1}$ term. Suppose that $x^i y_1^j$ is such that its expansion has a nontrivial $x^{\overline{d}} y_1^{r-1}$ term. Then $\overline{d} = i + m\overline{d} + \gamma$, $r - 1 = j - m$ with $0 \leq m \leq j$, $i, \gamma \geq 0$. $\overline{d}(1 - m) = i + \gamma \geq 0$ implies $m = 0$ or 1. $m = 0$ implies $j = r - 1$, $i \leq \overline{d}$. $\overline{a}_{ij} = 0$ in this case since

$$i + \delta_1 j \leq \overline{d} + \delta_1 (r - 1) < \delta_1 r.$$

$m = 1$ implies $i = 0$, $j = r$. Thus there exists a nontrivial $x^{\bar{d}} y_1^{r-1}$ term in $F_2(x, y_1)$ which implies that $\delta_2 < \delta_1$, a contradiction. Thus $\bar{d} \geq \delta_1$.

We then see that if $i + \delta_1 j > r\delta_1$, then all terms $x^\alpha y^\beta$ in the expansion of $x^i y_1^j = x^i(y_2 + t(x))^j$ satisfy $\alpha + \delta_1 \beta > r\delta_1$. Since $\delta_2 < \delta_1$, we see that

$$\sum_{i+\delta_1 j = r\delta_1} \bar{a}_{ij} x^i (y_2 + t(x))^j = \begin{cases} c y_2^r + \text{terms with } i + \delta_1 j > r\delta_1 \text{ if } \delta_1 \notin \mathbf{N}, \\ c y_2^r + dx^{r\delta_1} + \text{terms with } i + \delta_1 j > r\delta_1 \text{ if } \delta_1 \in \mathbf{N}. \end{cases}$$

Thus $\text{mult}(t) = \delta_1$ and

$$\sum_{i+\delta_1 j = r\delta_1} \bar{a}_{ij} x^i y_1^j = c(y_1 - e_{\delta_1} x^{\delta_1})^r + dx^{r\delta_1}$$

a contradiction. □

Lemma 9.14. *Suppose that $g : X_1 \to X$ is a quadratic transform, centered at a point p of X, and $p_1 \in X_1$ is a closed point above p such that $g(p_1) = p$ and $\bar{\nu}(p_1) = \bar{\nu}(p)$.*

Suppose that p and p_1 are both 2 points. Then $\delta(p_1) = \delta(p) - 1$.

Suppose that p and p_1 are both 1 points, $\sigma(p) = 0$ and $\delta(p) < \infty$. Then $\delta(p_1) = \delta(p) - 1$.

Proof. Suppose that $r = \text{mult}(F)$.

First suppose that p and p_1 are both 2 points. Then p has permissible parameters (x, y) and $\hat{O}_{X_1, p_1}$ has permissible parameters (x_1, y_1) such that $x = x_1, y = x_1 y_1$. Since $F_1 = \frac{F}{x_1^r}$, $\delta(p_1; x_1, y_1) = \delta(p; x, y) - 1$. Since $\bar{\nu}(p_1) = \bar{\nu}(p)$, we have $F = \sum a_{ij} x^i y^j$ with $a_{ij} = 0$ if $i + j \leq r$ and $j < r$. Thus $a_{0r} \neq 0$, so that $\delta(p; y, x) = 1$ and $\delta(p; x, y) > 1$. Thus $\delta(p) = \delta(p; x, y)$. Since $\text{mult}(F_1) = r$ and $\text{mult}(F_1(0, y_1)) = r$, $\delta(p_1; y_1, x_1) = 1$ and $\delta(p_1; x_1, y_1) \geq 1$. Then $\delta(p_1) = \delta(p_1; x_1, y_1) = \delta(p) - 1$.

Now suppose that p and p_1 are both 1 points, $\sigma(p) = 0$ and $\delta(p) < \infty$. We can suppose that we have permissible coordinates (x, y) at p such that $\delta = \delta(p) = \delta(F; x, y)$ and $\text{mult}(F(0, y)) = \text{mult}(F)$. p_1 has permissible parameters (x_1, y_1) such that $x = x_1, y = x_1(y_1 + \gamma)$ for some $\gamma \in k$.

First suppose that $\gamma \neq 0$.

$$F_1 = \sum_{i+j=r} a_{ij}(y_1 + \gamma)^j - \bar{a} + x_1 \Omega$$

where

$$\bar{a} = \sum_{i+j=r} a_{ij} \gamma^j.$$

$\text{mult}(F_1) = \text{mult}(F)$ implies

$$\sum_{i+j=r} a_{ij}(y_1 + \gamma)^j - \bar{a} = a_{0r} y_1^r.$$

Thus

$$\sum_{i+j=r} a_{ij} x^i y^j = a_{0r}(y - \gamma x)^r + \bar{a} x^r.$$

This is a contradiction to the assumption that $\delta(p; x, y) = \delta(p)$ by Lemma 9.13.

Now suppose that $\gamma = 0$. Then $F_1 = \frac{F}{x_1^r}$ and $\delta(p_1; x_1, y_1) = \delta(p; x, y) - 1$. If $\delta(p_1; x_1, y_1) < \delta(p_1)$, then we must also have $\delta(p; x, y) < \delta(p)$ By Lemma 9.13. Thus $\delta(p_1) = \delta(p) - 1$.   □

If $p \in X$ is a 2 point, then (u, v) are 1-resolved at p precisely when $\overline{\nu}(p) = 0$. If $p \in X$ is a 1 point then (u, v) are 1-resolved at p precisely when $\delta(p) = \infty$. Thus (u, v) are not 1-resolved at $p \in X$ if and only if $\overline{\nu}(p) > 0$ and $\delta(p) < \infty$.

We can define an invariant

$$\text{Inv}(p) = (\overline{\nu}(p), \sigma(p), \delta(p))$$

for $p \in X$.

Theorem 9.15. *Suppose that $g : X_1 \to X$ is a quadratic transform, centered at a point p of X, and $p_1 \in X_1$ is such that $g(p_1) = p$. Suppose that $\overline{\nu}(p) > 0$ and $\delta(p) < \infty$. Then*

$$\text{Inv}(p_1) < \text{Inv}(p)$$

in the lexicographic ordering.

Proof. The Theorem follows from Theorem 9.8, Lemmas 9.10, 9.11, 9.14.   □

The proof of Theorem 9.4 is immediate from Theorem 9.15.

Lemma 9.16. *Suppose that $f(x, y) \in T_0 = k[[x, y]]$ is a series. Suppose that we have an infinite sequence of quadratic transforms*

$$T_0 \to T_1 \to \cdots \to T_n \to \cdots$$

Then there exists n_0 such that $n \geq n_0$ implies there exist regular parameters (x_n, y_n) in T_n, $\alpha_n, \beta_n \in \mathbf{N}$ and a unit $u_n \in T_n$ such that $f = x_n^{\alpha_n} y_n^{\beta_n} u_n$.

Proof. This follows directly from Zariski's proof of resolution of surface singularities along a valuation ([36]), or can be deduced easily after blowing up enough to make $f = 0$ a SNC divisor.   □

Lemma 9.17. *Suppose that $\alpha_j + \beta_j \geq j$, $x^{\alpha_j} y^{\beta_j} \in T_0 = k[x, y]_{(x,y)}$ for $2 \leq j \leq r$ (or $1 \leq j \leq r$). Suppose that we have a sequence of quadratic transforms*

$$T_0 \to T_1 \to \cdots \to T_n \to \cdots$$

where each T_n has regular parameters (x_n, y_n) such that either $x_{n-1} = x_n$, $y_{n-1} = x_n y_n$, or $x_{n-1} = x_n y_n$, $y_{n-1} = y_n$. There are natural numbers $\alpha_{n,i}$, $\beta_{n,i}$ such that

$$x^{\alpha_j} y^{\beta_j} = x_n^{\alpha_{n,i}} y_n^{\beta_{n,i}}.$$

Define

$$\delta_{n,i,j} = \left(\frac{\alpha_{n,i}}{i} - \frac{\alpha_{n,j}}{j} \right) \left(\frac{\beta_{n,i}}{i} - \frac{\beta_{n,j}}{j} \right)$$

Then

1. $\delta_{n+1,i,j} \geq \delta_{n,i,j}$
2. $\delta_{n,i,j} < 0$ *implies* $\delta_{n+1,i,j} - \delta_{n,i,j} \geq \frac{1}{r^4}$.

Proof. We will first verify 1. Suppose that $x_n = x_{n+1}y_{n+1}$, $y_n = y_{n+1}$. The proof when $x_n = x_{n+1}$, $y_n = x_{n+1}y_{n+1}$ is the same. 1. is immediate from

$$\delta_{n+1,i,j} = \delta_{n,i,j} + \left(\frac{\alpha_{n,i}}{i} - \frac{\alpha_{n,j}}{j}\right)^2$$

Now suppose that $\delta_{n,i,j} < 0$. Then $\left(\frac{\alpha_{n,i}}{i} - \frac{\alpha_{n,j}}{j}\right)$ and $\left(\frac{\beta_{n,i}}{i} - \frac{\beta_{n,j}}{j}\right)$ are nonzero. We can suppose that $x_n = x_{n+1}y_{n+1}$, $y_n = y_{n+1}$.

$$\delta_{n+1,i,j} - \delta_{n,i,j} = \left(\frac{\alpha_{n,i}}{i} - \frac{\alpha_{n,j}}{j}\right)^2$$

$$= \left(\frac{j\alpha_{n,i} - i\alpha_{n,j}}{ij}\right)^2 \geq \frac{1}{r^4}$$

since $i, j \leq r$ implies $(ij)^2 \leq r^4$. □

Corollary 9.18. *Suppose that $\alpha_j + \beta_j \geq j$ and $x^{\alpha_j}y^{\beta_j} \in T_0 = k[x,y]_{(x,y)}$ for $2 \leq j \leq r$ (or $1 \leq j \leq r$) and*

$$T_0 \to T_1 \to \cdots \to T_n \to \cdots$$

is a sequence of quadratic transformations as in the statement of Lemma 9.17. Then

1. *There exists n_0 and i such that $n \geq n_0$ implies*

$$\frac{\alpha_{n,i}}{i} \leq \frac{\alpha_{n,j}}{j} \text{ and } \frac{\beta_{n,i}}{i} \leq \frac{\beta_{n,j}}{j}$$

   for $2 \leq j \leq r$ (or $1 \leq j \leq r$).
2. *There exists an $n_1 \geq n_0$ such that*

$$\left\{\frac{\alpha_{n_1,i}}{i}\right\} + \left\{\frac{\beta_{n_1,i}}{i}\right\} < 1$$

Proof. By Lemma 9.17, there exists n_0 such that $n \geq n_0$ implies $\delta_{n,i,j} \geq 0$ for all i, j. Let $\lambda_1 = \min\left(\frac{\alpha_{n,j}}{j}\right)$. Let $\lambda_2 = \min\left(\frac{\beta_{n,j}}{j} \text{ such that } \frac{\alpha_{n,j}}{j} = \lambda_1\right)$. Choose i such that $\frac{\alpha_{n,i}}{i} = \lambda_1$, $\frac{\beta_{n,i}}{i} = \lambda_2$. Then

$$\frac{\alpha_{n,i}}{i} \leq \frac{\alpha_{n,j}}{j}, \frac{\beta_{n,i}}{i} \leq \frac{\beta_{n,j}}{j}$$

for $2 \leq j \leq r$ (or $1 \leq j \leq r$).

Now we will prove 2. Suppose that $n \geq n_0$. Then

$$\left\{\frac{\alpha_{n,i}}{i}\right\} + \left\{\frac{\beta_{n,i}}{i}\right\} \in \frac{1}{i}\mathbf{N}.$$

Suppose that

$$\left\{\frac{\alpha_{n,i}}{i}\right\} + \left\{\frac{\beta_{n,i}}{i}\right\} \geq 1.$$

Without loss of generality,

$$x_n = x_{n+1}y_{n+1}, y_n = y_{n+1}.$$

Then

$$\left\{\frac{\alpha_{n+1,i}}{i}\right\} + \left\{\frac{\beta_{n+1,i}}{i}\right\} = \left\{\frac{\alpha_{n,i}}{i}\right\} + \left\{\frac{\alpha_{n,i}}{i}\right\} + \left\{\frac{\beta_{n,i}}{i}\right\} - 1$$
$$< \left\{\frac{\alpha_{n,i}}{i}\right\} + \left\{\frac{\beta_{n,i}}{i}\right\}$$

Thus there exists $n_1 \geq n_0$ such that 2. holds. $\qquad\square$

Remark 9.19. *The conditions* $\alpha_i + \beta_i \geq i$ *and* $\left\{\frac{\alpha_i}{i}\right\} + \left\{\frac{\beta_i}{i}\right\} < 1$ *imply either* $\alpha_i \geq i$ *or* $\beta_i \geq i$.

Lemmas 9.20 and 9.21 are used in Abhyankar's Good Point proof of resolution of singularities [4], [26].

Lemma 9.20. *Suppose that* $\alpha_{j_0} + \beta_{j_0} \geq j$, $(\alpha_{j_0}, \beta_{j_0})$ *are nonnegative integers for* $1 \leq j \leq r$. *Suppose that we have pairs of nonnegative integers* $(\alpha_{n,j}, \beta_{n,j})$ *for all positive* n *and* $1 \leq j \leq r$ *such that either*

$$(\alpha_{n+1,j}, \beta_{n+1,j}) = (\alpha_{n,j} + \beta_{n,j} - j, \beta_{n,j})$$

or

$$(\alpha_{n+1,j}, \beta_{n+1,j}) = (\alpha_{n,j}, \alpha_{n,j} + \beta_{n,j} - j).$$

Define

$$\delta_{n,i,j} = \left(\frac{\alpha_{n,i}}{i} - \frac{\alpha_{n,j}}{j}\right)\left(\frac{\beta_{n,i}}{i} - \frac{\beta_{n,j}}{j}\right)$$

Then

1. $\delta_{n+1,i,j} \geq \delta_{n,i,j}$
2. $\delta_{n,i,j} < 0$ *implies* $\delta_{n+1,i,j} - \delta_{n,i,j} \geq \frac{1}{r^4}$.

Lemma 9.21. *Suppose that the assumptions are as in Lemma 9.20.*

Suppose that $\alpha_j + \beta_j \geq j$, $x^{\alpha_j} y^{\beta_j} \in T_0 = k[x,y]_{(x,y)}$ *for* $1 \leq j \leq r$. *Suppose that we have a possibly infinite sequence of quadratic transforms*

$$T_0 \to T_1 \to \cdots \to T_n \to \cdots$$

where each T_n *has regular parameters* (x_n, y_n) *such that either* $x_{n-1} = x_n, y_{n-1} = x_n y_n$ *or* $x_{n-1} = x_n y_n, y_{n-1} = y_n$ *and* (α_n, β_n) *are defined by the respective rules of Lemma 9.20. Then*

1. *There exists* n_0 *and* i *such that* $n \geq n_0$ *implies*

$$\frac{\alpha_{n,i}}{i} \leq \frac{\alpha_{n,j}}{j} \quad and \quad \frac{\beta_{n,i}}{i} \leq \frac{\beta_{n,j}}{j}$$

   for $1 \leq j \leq r$.
2. *There exists* $n_1 \geq n_0$ *such that*

$$\left\{\frac{\alpha_{n_1,i}}{i}\right\} + \left\{\frac{\beta_{n_1,i}}{i}\right\} < 1$$

10. $\mathbf{A_r(X)}$

Throughout this section we will assume that $\Phi_X : X \to S$ is weakly prepared.

Definition 10.1. *Suppose that* $r \geq 2$. *$\overline{A}_r(X)$ holds if*

1. $\nu(p) \leq r$ *if* $p \in X$ *is a 1 point or a 2 point.*
2. *If* $p \in X$ *is a 1 point and* $\nu(p) = r$, *then* $\gamma(p) = r$.
3. *If* $p \in X$ *is a 2 point and* $\nu(p) = r$, *then* $\tau(p) > 0$.
4. $\nu(p) \leq r - 1$ *if* $p \in X$ *is a 3 point*

Definition 10.2. *Suppose that* $r \geq 2$. *$A_r(X)$ holds if*

1. $\overline{A}_r(X)$ *holds.*
2. $\overline{S}_r(X)$ *is a union of nonsingular curves and isolated points.*
3. $\overline{S}_r(X) \cap (X - \overline{B}_2(X))$ *is smooth.*
4. $\overline{S}_r(X)$ *makes SNCs with* $\overline{B}_2(X)$ *on the open set* $X - B_3(X)$.
5. *The curves in* $\overline{S}_r(X)$ *passing through a 3 point* $q \in X$ *have distinct tangent directions at* q. *(They are however, allowed to be tangent to a 2 curve).*

Lemma 10.3. *Suppose that* X *satisfies* $\overline{A}_r(X)$ *with* $r \geq 2$. *Then there exists a sequence of quadratic transforms* $X_1 \to X$ *such that* $A_r(X_1)$ *holds.*

Proof. Let $\pi : X_1 \to X$ be a sequence of quadratic transforms so that the strict transform of $\overline{S}_r(X)$ makes SNCs with $\overline{B}_2(X)$. Then $\overline{A}_r(X_1)$ holds by Theorems 7.1 and 7.3, and $A_r(X_1)$ holds by Lemma 7.9 and Theorem 7.8. $\qquad \square$

Definition 10.4. *Suppose that* $A_r(X)$ *holds. A weakly permissible monoidal transform* $\pi : X_1 \to X$ *is called permissible if* π *is the blow-up of a point, a 2 curve or a curve* C *containing a 1 point such that* $C \cup \overline{S}_r(X)$ *makes SNCs with* $\overline{B}_2(X)$ *at all points of* C.

Remark 10.5.　1. *If* $A_r(X)$ *holds and* $\pi : X_1 \to X$ *is a permissible monoidal transform, then the strict transform of* $\overline{S}_r(X)$ *on* X_1 *makes SNCs with* $\overline{B}_2(X_1)$ *at 1 and 2 points, and has distinct tangent directions at 3 points.*

2. *If* $\pi : X_1 \to X$ *is a quadratic transform centered at a point* $p \in X$ *with* $\nu(p) = r$ *and* $A_r(X)$ *holds, then* $A_r(X_1)$ *holds.*

3. *If* $A_r(X)$ *holds and all 3 points* q *of* X *satisfy* $\nu(q) \leq r - 2$, *then* $\overline{S}_r(X)$ *makes SNCs with* $\overline{B}_2(X)$.

The Remark follows from Lemmas 7.9 and 7.7, and the observation that the strict transforms of nonsingular curves with distinct tangent directions at a point p intersect the exceptional fiber of the blow-up of p transversally in distinct points.

11. REDUCTION OF ν IN A SPECIAL CASE

Throughout this section we will assume that $\Phi_X : X \to S$ is weakly prepared.

Lemma 11.1. *Suppose that $r \geq 2$ and $A_r(X)$ holds, $p \in X$ is a 1 point or a 2 point with $\nu(p) = \gamma(p) = r$. Let $R = \mathcal{O}_{X,p}$. Suppose that (x, y, z) are permissible parameters at p as in Lemma 8.5. Then there exists a finite sequence of permissible monoidal transforms $\pi : Y \to \mathrm{Spec}(\hat{R})$ centered at sections over $C = V(x, y)$, such that for $q \in \pi^{-1}(p)$, there exist permissible parameters $(\overline{x}, \overline{y}, z)$ at q such that F_q has one of the following forms.*

$$
\begin{aligned}
u &= \overline{x}^a \\
v &= P(\overline{x}) + \overline{x}^c F_q
\end{aligned}
\tag{104}
$$

or

$$
\begin{aligned}
u &= (\overline{x}^a \overline{y}^b)^m \\
v &= P(\overline{x}^a \overline{y}^b) + \overline{x}^c \overline{y}^d F_q
\end{aligned}
\tag{105}
$$

with

$$
F_q = \tau z^r + \sum_{i=2}^{r-1} \overline{a}_i(\overline{x}, \overline{y}) \overline{x}^{\alpha_i} \overline{y}^{\beta_i} z^{r-i} + \epsilon \overline{x}^{\alpha_r} \overline{y}^{\beta_r}
$$

with τ a unit, $\overline{a}_i$ a unit (or zero), $\alpha_i + \beta_i \geq i$ for all i, and $\epsilon = 0$ or 1.

Proof. We have one of the forms (65) or (66) of Lemma 8.5 at p. By Lemma 9.2 and Theorem 9.4 applied to

$$
\overline{u} = x^a, \overline{v} = P(x) + x^c F_p(x, y, 0)
$$

or

$$
\overline{u} = (x^a y^b)^m, \overline{v} = P(x^a y^b) + x^c y^d F_p(x, y, 0)
$$

there exists a sequence of permissible blow-ups of sections over C such that for all q over p, there are permissible parameters $(\overline{x}, \overline{y}, z)$ at q such that

$$
\begin{aligned}
u &= \overline{x}^a \\
v &= P(\overline{x}) + \overline{x}^c(\tau z^r + \sum_{i=2}^{r-1} a_i(\overline{x}, \overline{y}) z^{r-i} + \epsilon \overline{x}^{e_0} \overline{y}^{f_0})
\end{aligned}
\tag{106}
$$

with τ a unit, $\epsilon = 0$ or 1 and $f_0 > 0$, or

$$
\begin{aligned}
u &= (\overline{x}^a \overline{y}^b)^m \\
v &= P(\overline{x}^a \overline{y}^b) + \overline{x}^c \overline{y}^d(\tau z^r + \sum_{i=2}^{r-1} a_i(\overline{x}, \overline{y}) z^{r-i} + \epsilon \overline{x}^{e_0} \overline{y}^{f_0})
\end{aligned}
\tag{107}
$$

with τ a unit, $\epsilon = 0$ or 1 and $a(d + f_0) - b(c + e_0) \neq 0$.

By further permissible blowing up (of sections over C) we can make

$$
u \prod_{2 \leq i \leq r-1, a_i \neq 0} a_i = 0
$$

a SNC divisor, while preserving the forms (106) and (107). At points q over p satisfying (107) we have then achieved the conclusions of the Lemma.

Suppose that q is a point over p satisfying (106) such that the conclusions of the Lemma do not hold. We then have $\epsilon = 1$, $f_0 > 0$ and

$$
u \prod_{2 \leq i \leq r, a_i \neq 0} a_i = 0
$$

is not a SNC divisor. Since

$$u \prod_{2 \leq i \leq r-1, a_i \neq 0} a_i = 0$$

is a SNC divisor, there exists a nonzero, nonunit series $g(\overline{x})$ such that

$$a_i = \overline{a}_i(\overline{x}, \overline{y})\overline{x}^{\alpha_i}(\overline{y} - g(\overline{x}))^{\beta_i} \tag{108}$$

for $2 \leq i \leq r-1$, where the $\overline{a}_i$ are units (or 0), and some $\beta_i > 0$ with $\overline{a}_i \neq 0$. If $f_0 = 1$, we can set $\tilde{y} = \overline{y} - g(\overline{x})$ and renormalize with respect to $(\overline{x}, \tilde{y}, z)$ to get in the form of the conclusions of the Lemma.

Otherwise $f_0 > 1$. Let $t = \nu(g(\overline{x}))$, so that

$$g(\overline{x}) = \alpha \overline{x}^t + \text{ higher order terms}$$

for some $0 \neq \alpha$. Now blow up $V(\overline{x}, \overline{y})$. under $\overline{x} = x_1 y_1$, $\overline{y} = y_1$, we have

$$u = x_1^a y_1^a$$
$$v = P(x_1 y_1) + x_1^c y_1^c (\tau z^r + \sum_{i=2}^{r-1} \tilde{a}_i(x_1, y_1) x_1^{\alpha_i} y_1^{\beta_i + \alpha_i} z^{r-i} + x_1^{e_0} y_1^{e_0 + f_0})$$

in the form of the conclusions of the Lemma. Under $\overline{x} = x_1$, $\overline{y} = x_1(y_1 + \beta)$, with $\beta \neq 0$, we have

$$u = x_1^a$$
$$v = P(x_1) + x_1^c (\tau z^r + \sum_{i=2}^{r-1} \overline{a}_i x_1^{\alpha_i + \beta_i} (y_1 + \beta - \tfrac{g(x_1)}{x_1})^{\beta_i} z^{r-i}$$
$$\quad + x_1^{e_0 + f_0} (y_1 + \beta)^{f_0})$$
$$= P(x_1) + \beta^{f_0} x_1^{c + e_0 + f_0}$$
$$\quad + x_1^c (\tau z^r + \sum_{i=2}^{r-1} \overline{a}_i x_1^{\alpha_i + \beta_i} ((\overline{y}_1 + \beta^{f_0})^{\frac{1}{f_0}} - \tfrac{g(x_1)}{x_1})^{\beta_i} z^{r-i} + x_1^{e_0 + f_0} \overline{y}_1)$$

where $\overline{y}_1 = (y_1 + \beta)^{f_0} - \beta^{f_0}$. If we are not in the form of the conclusions of the Lemma, then

$$(\overline{y}_1 + \beta^{f_0})^{\frac{1}{f_0}} - \frac{g(x_1)}{x_1} = a(x_1, \overline{y}_1)(\overline{y}_1 - \phi(x_1))$$

where $\nu(\phi) \geq 1$. We can make a change of variable in $\overline{y}_1$, replacing $\overline{y}_1$ with $\overline{y}_1 - \phi(x_1)$, and renormalize, to get in the form of the conclusions of the Lemma.

Under $\overline{x} = x_1$, $\overline{y} = x_1 y_1$, we have

$$u = x_1^a$$
$$v = P(x_1) + x_1^c (\tau z^r + \sum_{i=2}^{r-1} \overline{a}_i(x, y) x_1^{\alpha_i + \beta_i} (y_1 - \tfrac{g(x_1)}{x_1})^{\beta_i} z^{r-i} + x_1^{e_0 + f_0} y_1^{f_0})$$

the coefficients of z^i are in the form of (108), but we have a reduction $\nu\left(\frac{g(x_1)}{x_1}\right) = t - 1$. If $\nu\left(\frac{g(x_1)}{x_1}\right) = 0$ we are in the form of the conclusions of the Lemma. Thus after t blow-ups, centered at the intersection of the strict transform of the surface $y = 0$ with the exceptional divisor, we achieve the conclusions of the Lemma. $\square$

Theorem 11.2. *Suppose that $r \geq 2$ and $A_r(X)$ holds, $p \in X$ is a 1 point or a 2 point with $\nu(p) = \gamma(p) = r$. Let $R = \mathcal{O}_{X,p}$. Suppose that (x, y, z) are permissible parameters at p as in Lemma 8.5, where $z = \sigma \tilde{z}$ for some $\tilde{z} \in R$ and unit $\sigma \in \hat{R}$. Then there exists a finite sequence of permissible monoidal transforms $\pi : Y \to \mathrm{Spec}(\hat{R})$ centered at sections over $C = V(x, y)$, such that* *for $q \in \pi^{-1}(p)$, q has permissible parameters $(\bar{x}, \bar{y}, z)$ such that F_q has one of the following forms:*

1.

$$
\begin{aligned}
u &= \bar{x}^a \\
v &= P(\bar{x}) + \bar{x}^c F_q \text{ with} \\
F_q &= \tau z^r + \sum_{i=2}^{r-1} \bar{a}_i(\bar{x}, \bar{y}) \bar{x}^{\alpha_i} z^{r-i} + \epsilon \bar{x}^{\alpha_r} \bar{y}
\end{aligned}
\tag{109}
$$

where τ is a unit, $\alpha_i \geq i$ for $2 \leq i \leq r-1$, $\alpha_r \geq r-1$, $\epsilon = 0$ or 1, and $\bar{a}_i$ are units (or 0), or

2.

$$
\begin{aligned}
u &= (\bar{x}^a \bar{y}^b)^m \\
v &= P(\bar{x}^a \bar{y}^b) + \bar{x}^c \bar{y}^d F_q \text{ with} \\
F_q &= \tau z^r + \sum_{j=2}^{r-1} \bar{a}_j(\bar{x}, \bar{y}) \bar{x}^{\alpha_j} \bar{y}^{\beta_j} z^{r-j} + \epsilon \bar{x}^{\alpha_r} \bar{y}^{\beta_r}
\end{aligned}
\tag{110}
$$

where τ is a unit, $\alpha_j + \beta_j \geq j$ and $\bar{a}_j$ are units or 0 for all j, $\epsilon = 0$ or 1, there exists an i such that $\bar{a}_i \neq 0$, $2 \leq i \leq r$ and

$$
\frac{\alpha_i}{i} \leq \frac{\alpha_j}{j}, \quad \frac{\beta_i}{i} \leq \frac{\beta_j}{j}
$$

for $2 \leq j \leq r$. We further have

$$
\left\{ \frac{\alpha_i}{i} \right\} + \left\{ \frac{\beta_i}{i} \right\} < 1
$$

or

3.

$$
\begin{aligned}
u &= \bar{x}^a \\
v &= P(\bar{x}) + \bar{x}^b F_q \text{ with} \\
F_q &= \tau z^r + \sum_{j=2}^{r-1} \bar{a}_j(\bar{x}, \bar{y}) \bar{x}^{\alpha_j} \bar{y}^{\beta_j} z^{r-j} + \epsilon \bar{x}^{\alpha_r} \bar{y}^{\beta_r}
\end{aligned}
\tag{111}
$$

where τ is a unit, $\alpha_j + \beta_j \geq j$ and $\bar{a}_j$ are units or 0 for all j, $\epsilon = 0$ or 1, there exists an i such that $\bar{a}_i \neq 0$, $2 \leq i \leq r$ and

$$
\frac{\alpha_i}{i} \leq \frac{\alpha_j}{j}, \quad \frac{\beta_i}{i} \leq \frac{\beta_j}{j}
$$

for $2 \leq j \leq r$. We further have

$$
\left\{ \frac{\alpha_i}{i} \right\} + \left\{ \frac{\beta_i}{i} \right\} < 1.
$$

Proof. We can first construct a sequence of monoidal transforms $\pi : Y \to \text{Spec}(\hat{R})$ satisfying the conclusions of Lemma 11.1. (109) holds at all but finitely many points $q \in \pi^{-1}(p)$.

Suppose that $q \in \pi^{-1}(p)$ and (104) holds at q, with $\epsilon = 1$, but F_q is not in the form of (109) or (111). Perform a monoidal transform $\pi' : Y' \to Y$ centered at the section over C through q with local equations $\bar{x} = \bar{y} = 0$ in (104). Suppose that $q' \in (\pi')^{-1}(q)$. Suppose that there are permissible parameters (x_1, y_1, z) at q' such that $\bar{x} = x_1$, $\bar{y} = x_1(y_1 + \alpha)$ where $\alpha \neq 0$. Then

$$
F_q = \tau z^r + \sum_{i=2}^{r-1} \bar{a}_i x_1^{\alpha_i + \beta_i}(y_1 + \alpha)^{\beta_i} z^{r-i} + x_1^{\alpha_r + \beta_r}(y_1 + \alpha)^{\beta_r}
$$

Set $\tilde{y}_1 = (y_1 + \alpha)^{\beta_r} - \alpha^{\beta_r}$. Then

$$F_{q'} = \tau z^r + \sum_{i=2}^{r-1} \tilde{a}_i(x_1, \tilde{y}_1) x_1^{\tilde{\alpha}_i} z^{r-i} + x_1^{\tilde{\alpha}_r} \tilde{y}_1$$

in the form of (109). Thus the only points $q' \in (\pi')^{-1}(q)$ which might not satisfy the conclusions of Theorem 11.2 are the points q' which have regular parameters (x_1, y_1, z) such that $\overline{x} = x_1$, $\overline{y} = x_1 y_1$ or $\overline{x} = x_1 y_1$, $\overline{y} = y_1$.

The analysis of the case when (104) holds at q, with $\epsilon = 0$, is simpler. We again conclude that the only points in the blow-up of the curve with local equations $\overline{x} = \overline{y} = 0$ above q which may not satisfy the conclusions of Theorem 11.2 are the points which have regular parameters (x_1, y_1, z) such that $\overline{x} = x_1$, $\overline{y} = x_1 y_1$ or $\overline{x} = x_1 y_1$, $\overline{y} = y_1$.

Suppose that $q \in \pi^{-1}(p)$ and (105) holds at q, with $\epsilon = 1$, but F_q is not in the form of (110). Perform a monoidal transform $\pi' : Y' \to Y$ centered at the section over C through q with local equations $\overline{x} = \overline{y} = 0$. Suppose that $q' \in (\pi')^{-1}(q)$. Suppose that there are regular parameters (x_1, y_1, z) at q' such that $\overline{x} = x_1$, $\overline{y} = x_1(y_1 + \alpha)$ where $\alpha \neq 0$. Then

$$u = \overline{x}_1^{(a+b)m}$$
$$(y_1 + \alpha)^\lambda F_q = \tau(y_1 + \alpha)^\lambda z^r + \sum_{i=2}^{r-1} \tilde{a}_i(y_1 + \alpha)^{\beta_i + \lambda - (\alpha_i + \beta_i)\frac{b}{a+b}} \overline{x}_1^{\alpha_i + \beta_i} z^{r-i}$$
$$+ \overline{x}_1^{\alpha_r + \beta_r}(y_1 + \alpha)^{\lambda + \beta_r - (\alpha_r + \beta_r)\frac{b}{a+b}}.$$

Thus

$$F_{q'} = \tilde{\tau} z^r + \sum_{i=2}^{r-1} \tilde{a}_i(\overline{x}_1, \overline{y}_1) \overline{x}_1^{\tilde{\alpha}_i} z^{r-i} + \overline{x}_1^{\tilde{\alpha}_r} \overline{y}_1$$

is in the form of (109), where $x_1 = \overline{x}_1(y_1 + \alpha)^{-\frac{b}{a+b}}$, $\overline{y}_1 = (y_1 + \alpha)^{\lambda_1} - \alpha^{\lambda_1}$, where $\lambda = d - \frac{b(c+d)}{a+b}$, $\lambda_1 = \lambda + \beta_r - (\alpha_r + \beta_r)\frac{b}{a+b} \neq 0$ since F_q is normalized implies

$$a(d + \beta_r) - b(c + \alpha_r) \neq 0.$$

Thus the only points $q' \in (\pi')^{-1}(q)$ which might not satisfy the conclusions of Theorem 11.2 are the points which have permissible parameters (x_1, y_1, z) such that $\overline{x} = x_1$, $\overline{y} = x_1 y_1$ or $\overline{x} = x_1 y_1$, $\overline{y} = y_1$.

The analysis of the case when (105) holds at q, with $\epsilon = 0$, is simpler. We again conclude that the only points in the blow-up of the curve with local equations $\overline{x} = \overline{y} = 0$ above q which may not satisfy the conclusions of Theorem 11.2 are the points which have regular parameters (x_1, y_1, z) such that $\overline{x} = x_1$, $\overline{y} = x_1 y_1$ or $\overline{x} = x_1 y_1$, $\overline{y} = y_1$.

We can construct a sequence of monoidal transforms

$$Y_n \to \cdots \to Y_1 \to Y$$

with maps $\pi_i : Y_i \to Y$ such that $Y_i \to Y_{i-1}$ are centered at sections C_i over C, such that $\pi_i^{-1}(p) \cap C_i$ does not satisfy (109), (110) or (111). By the above analysis, Lemma 9.17 and Corollary 9.18, we reach the conclusions of the theorem after a finite number of blow-ups. □

Remark 11.3. *In (111) of Theorem 11.2, we must have $\beta_j < j$ for some j.*

Proof.

$$u \in \hat{\mathcal{I}}_{\mathrm{sing}(\Phi_X),q} \subset \hat{\mathcal{I}}_{\overline{S_r(X)},q}$$

by Lemma 6.11. Thus $\overline{x} \in \hat{\mathcal{I}}_{\overline{S}_r(X),q}$. $\beta_j \geq j$ for all j implies $F_q \in (y,z)^r$, so that $\hat{\mathcal{I}}_{\overline{S}_r(X),q} \subset (y,z)$ by Lemma 6.22, a contradiction. $\qquad\square$

Theorem 11.4. *Suppose that $r \geq 2$ and $A_r(X)$ holds. Suppose that $p \in X$ is a 1 point or a 2 point with $\nu(p) = \gamma(p) = r$. Let $R = \mathcal{O}_{X,p}$. Suppose that $\pi : Y \to \mathrm{Spec}(\hat{R})$ is the sequence of monoidal transforms of sections over the curve C with local equations $x = y = 0$ at p of Theorem 11.2. Suppose that $t > r$ is a positive integer. Then there exists a sequence of permissible monoidal transforms $\overline{\pi} : \overline{Y} \to \mathrm{spec}(R)$ of sections over C such that for all $q \in \overline{\pi}^{-1}(p)$, F_q is equivalent mod $(\overline{x}, z)^t$ to a form (109) or (111) or F_q is equivalent mod $(\overline{xy}, z)^t$ to a form (110), where $(\overline{x}, \overline{y}, z)$ are permissible parameters for u, v at q, and $z = \sigma\tilde{z}$ for some $\tilde{z} \in R$ and unit $\sigma \in \hat{R}$.*

$\overline{\pi}$ extends to a sequence of permissible monoidal transforms $\overline{U} \to U$ over an affine neighborhood U of p. $\overline{S}_r(\overline{U})$ is the union of the curves in $\overline{\pi}^{-1}(p)$ and the strict transforms of the curves D or D_1, D_2 (if they exist) in the notation of Lemma 8.5. $\overline{S}_r(\overline{U})$ makes SNCs with $\overline{B}_2(\overline{U})$.

Proof. Let m_0 be the maximal ideal of $\hat{R}$. We can after possibly replacing

$$\tilde{x} \text{ with } \tilde{x}\omega, \tag{112}$$

where ω is a unit in R, assume that $x = \gamma\tilde{x}$ with $\gamma \equiv 1 \bmod m_0^t$. We can factor

$$Y = Y_{n'} \to \cdots \to Y_2 \to Y_1 \to \mathrm{spec}(\hat{R}) = Y_0$$

so that each map is a permissible monoidal transform. In fact, if $S_0 = \mathrm{spec}(k[x,y])$, there exists a sequence of quadratic transforms

$$S_{n'} \to \cdots \to S_2 \to S_1 \to \mathrm{spec}(k[x,y]) = S_0$$

centered over (x,y) such that $Y_i = S_i \times_{S_0} Y_0$ for all i. Set $\overline{S}_0 = \mathrm{spec}(k[\tilde{x},y])$. $\tilde{x} \to x$ induces an isomorphism $\overline{S}_0 \cong S_0$. We have then a sequence of quadratic transforms

$$\overline{S}_{n'} \to \cdots \to \overline{S}_1 \to \overline{S}_0$$

where $\overline{S}_i = S_i \times_{S_0} \overline{S}_0$ and isomorphisms $S_i \cong \overline{S}_i$. Set $\overline{Y}_0 = \mathrm{spec}(R)$. We have a natural map $\overline{Y}_0 \to \overline{S}_0$. Define a sequence of permissible monoidal transforms

$$\overline{Y}_{n'} \to \cdots \to \overline{Y}_1 \to \overline{Y}_0$$

by $\overline{Y}_i = \overline{S}_i \times_{\overline{S}_0} \overline{Y}_0$. We have a commutative diagram

$$
\begin{array}{ccccccccc}
Y_{n'} & \to & Y_{n'-1} & \to & \cdots & \to & Y_1 & \to & Y_0 \\
\downarrow & & \downarrow & & & & \downarrow & & \downarrow \\
S_{n'} & \to & S_{n'-1} & \to & \cdots & \to & S_1 & \to & S_0 \\
\uparrow & & \uparrow & & & & \uparrow & & \uparrow \\
\overline{S}_{n'} & \to & \overline{S}_{n'-1} & \to & \cdots & \to & \overline{S}_1 & \to & \overline{S}_0 \\
\uparrow & & \uparrow & & & & \uparrow & & \uparrow \\
\overline{Y}_{n'} & \to & \overline{Y}_{n'-1} & \to & \cdots & \to & \overline{Y}_1 & \to & \overline{Y}_0
\end{array}
\tag{113}
$$

The maps $\overline{S}_i \to S_i$ are isomorphisms, and we have maps $S_i \times_{S_0} \hat{S}_0 \to Y_i$, $\overline{S}_i \times_{\overline{S}_0} \hat{S}_0 \to \overline{Y}_i$ induced by the natural projections

$$k[[x,y,z]] \to k[[x,y]] \text{ and } k[[\tilde{x},y,z]] \to k[[\tilde{x},y]]$$

so that the diagrams

$$
\begin{array}{c}
\qquad\qquad Y_i \\
\qquad\quad \swarrow \quad\ \uparrow \\
S_i \qquad\qquad \\
\qquad\quad \nwarrow \\
\qquad\qquad S_i \times_{S_0} \hat{S}_0
\end{array}
\tag{114}
$$

and

$$
\begin{array}{c}
\qquad\qquad \overline{Y}_i \\
\qquad\quad \swarrow \quad\ \uparrow \\
\overline{S}_i \qquad\qquad \\
\qquad\quad \nwarrow \\
\qquad\qquad \overline{S}_i \times_{\overline{S}_0} \hat{S}_0
\end{array}
$$

commute.

Suppose that $\tilde{q} \in \overline{Y}_{n'}$ is a closed point. (113) and (114) identifies $\tilde{q}$ with a closed point $\tilde{p} \in Y_{n'}$, and closed points $\overline{q} \in \overline{S}_{n'}$, $\overline{p} \in S_{n'}$. We have commutative diagrams:

$$
\begin{array}{ccc}
\hat{\mathcal{O}}_{S_{n'},\overline{p}} & \cong & \hat{\mathcal{O}}_{\overline{S}_{n'},\overline{q}} \\
\uparrow & & \uparrow \\
k[[x,y]] & \cong & k[[\tilde{x},y]]
\end{array}
\tag{115}
$$

induced by $x \to \tilde{x}$. This induces commutative diagrams:

$$
\begin{array}{ccccc}
\hat{\mathcal{O}}_{Y_{n'},\tilde{p}} = \hat{\mathcal{O}}_{S_{n'},\overline{p}}[[z]] & \cong & \hat{\mathcal{O}}_{\overline{S}_{n'},\overline{q}}[[z]] = & \hat{\mathcal{O}}_{\overline{Y}_{n'},\tilde{q}} \\
\uparrow & & & \uparrow \\
\hat{R} = k[[x,y,z]] & & \cong & k[[\tilde{x},y,z]]
\end{array}.
$$

Suppose that $F(x,y,z) \in \hat{R}$. If $(\overline{x},\overline{y})$ are regular parameters in $\hat{\mathcal{O}}_{S_{n'},\overline{p}}$, which are identified with regular parameters $\hat{x},\hat{y}$ in $\hat{\mathcal{O}}_{\overline{S}_{n'},\overline{q}}$ by (115), and $F(x,y,z) = G(\overline{x},\overline{y},z) \in \hat{\mathcal{O}}_{Y_{n'},\tilde{p}}$, then $F(\tilde{x},y,z) = G(\hat{x},\hat{y},z) \in \hat{\mathcal{O}}_{\overline{Y}_{n'},\tilde{q}}$.

Since $x = \gamma\tilde{x}$, $\gamma \equiv 1 \bmod m_0^t$, and

$$F(x,y,z) = F(\gamma\tilde{x},y,z) \equiv F(\tilde{x},y,z) \bmod m_0^t$$

implies

$$F(x,y,z) \equiv G(\hat{x},\hat{y},z) \bmod m_0^t \hat{O}_{\overline{Y}_{n'},\bar{q}}.$$

Let m be the maximal ideal of R.

Suppose that $\tilde{p} \in Y_{n'}$ is a 1 point, $u = \overline{x}^a$ in $\hat{O}_{Y_{n'},\tilde{p}}$. Then (since we can assume that $Y \neq \text{spec}(\hat{R})$) $\overline{x} \mid x$ and $\overline{x} \mid y$ in $\hat{O}_{Y_{n'},\tilde{p}}$ implies $m_0^t \subset (\overline{x},z)^t$, so that $m^t \subset (\hat{x},z)^t \hat{O}_{\overline{Y}_{n'},\bar{q}}.$

Suppose that $\tilde{p} \in Y_{n'}$ is a 2 point, $u = (\overline{x}^a \overline{y}^b)^m$ in $\hat{O}_{Y_{n'},\tilde{p}}$. If $\overline{x} = 0, \overline{y} = 0$ are both local equations of components of the exceptional locus of $Y_{n'} \to Y_0$, we have $\overline{xy} \mid x$, $\overline{xy} \mid y$ which implies that $m_0^t \subset (\overline{xy},z)^t$, so that $m^t \subset (\hat{x}\hat{y},z)^t$.

If one of $\overline{x} = 0, \overline{y} = 0$ is not a local equation of the exceptional locus, then we have regular parameters (x',y') in $\hat{O}_{S_{n'},\tilde{p}}$ such that $x = x'(y')^{\overline{b}}, y = y'$ (or $x = x', y = y'(x')^{\overline{b}}$). In the first case we have

$$
\begin{aligned}
\hat{O}_{Y_{n'},\tilde{p}} &= k[[\tfrac{x}{y^{\overline{b}}}, y, z]] \\
&= k[[\tfrac{\overline{\gamma}\overline{x}}{y^{\overline{b}}}, y, z]] \\
&= k[[\tfrac{\hat{x}}{y^{\overline{b}}}, y, z]] = \hat{O}_{\overline{Y}_{n'},\bar{q}}.
\end{aligned}
$$

Thus we have $\hat{O}_{Y_{n'},\tilde{p}} = \hat{O}_{\overline{Y}_{n'},\bar{q}}$. In the second case, we also have $\hat{O}_{Y_{n'},\tilde{p}} = \hat{O}_{\overline{Y}_{n'},\bar{q}}$.

Let $\overline{\pi} : \overline{Y} = \overline{Y}_{n'} \to \overline{Y}_0$ be the morphism of the bottom row of (113). Suppose that $p_0 \in \overline{Y}_0$ is a 1 point, so that in $\hat{O}_{\overline{Y}_0,p_0}$,

$$u = x^a, v = P_{p_0}(x) + x^{c_0} F_{p_0}.$$

Suppose that $p' \in \overline{\pi}^{-1}(p_0)$. Let q be the corresponding closed point of $Y = Y_{n'}$. Suppose that we have permissible parameters $(\overline{x}, \overline{y}, z)$ in $\hat{O}_{Y,q}$ such that

$$u = \overline{x}^{\overline{a}}, v = P_q(\overline{x}) + \overline{x}^c F_q(\overline{x},\overline{y},z) \tag{116}$$

of the form (109) or (111) with $\frac{\overline{a}}{a}, \frac{c}{c_0} \in \mathbf{N}$, $x = \overline{x}^{\frac{\overline{a}}{a}} = \overline{x}^{\frac{c}{c_0}}$. Let $(\tilde{x}_*, \tilde{y}_*, z)$ be the corresponding regular parameters at p' (by the identification (115)).

$$u = x^a = \gamma^a \tilde{x}^a = \gamma^a \tilde{x}_*^a = (\tilde{x}_*')^{\overline{a}}$$

where we define

$$\tilde{x}_* = \tilde{x}_*' \gamma^{-\frac{a}{\overline{a}}} \equiv \tilde{x}_*' \bmod m_0^t \hat{O}_{\overline{Y},p'}.$$

Thus $(\tilde{x}_*', \tilde{y}_*, z)$ are permissible parameters for (u,v) in $\hat{O}_{\overline{Y},p'}$

There exists a series $\tilde{P}_q(\overline{x})$ such that

$$F_{p_0}(x,y,z) = F_q(\overline{x},\overline{y},z) + \tilde{P}_q(\overline{x}).$$

Thus

$$
\begin{aligned}
F_{p_0}(x,y,z) &\equiv F_q(\tilde{x}_*, \tilde{y}_*, z) + \tilde{P}_q(\tilde{x}_*) \bmod m_0^t \hat{O}_{\overline{Y},p'} \\
&\equiv F_q(\tilde{x}_*', \tilde{y}_*, z) + \tilde{P}_q(\tilde{x}_*') \bmod m_0^t \hat{O}_{\overline{Y},p'}
\end{aligned}
$$

$$P_q(\overline{x}) = P_{p_0}(\overline{x}^{\frac{c}{c_0}}) + \overline{x}^c \tilde{P}_q(\overline{x})$$

implies

$$\begin{aligned}
v &= P_{p_0}(x) + x^{c_0} F_{p_0}(x, y, z) \\
&= P_{p_0}((\tilde{x}'_*)^{\frac{c}{c_0}}) + (\tilde{x}'_*)^c F_{p_0}(x, y, z) \\
&= P_q(\tilde{x}'_*) + (\tilde{x}'_*)^c (F_q(\tilde{x}'_*, \tilde{y}_*, z) + h)
\end{aligned}$$

with $h \in m_0^t \mathcal{O}_{\overline{Y}, p'}$.

The case when $p_0 \in \overline{Y}_0$ is a 1 point and (110) holds in $\hat{\mathcal{O}}_{Y,q}$ is a combination of the case when p_0 is a 1 point and the form (109) or (111) holds in $\hat{\mathcal{O}}_{Y,q}$, and the following case.

Suppose that $p_0 \in \overline{Y}_0$ is a 2 point so that in $\hat{\mathcal{O}}_{\overline{Y}_0, p_0}$,

$$\begin{aligned}
u &= (x^a y^b)^{m_0} \\
v &= P_{p_0}(x^a y^b) + x^{c_0} y^{d_0} F_{p_0}
\end{aligned}$$

Suppose that $p' \in \overline{\pi}^{-1}(p_0)$. Let q be the corresponding closed point in $Y = Y_{n'}$.

Suppose that we have permissible parameters $(\overline{x}, \overline{y}, z)$ at q such that

$$\begin{aligned}
u &= (\overline{x}^{\overline{a}} \overline{y}^{\overline{b}})^{m_1} \\
v &= P_q(\overline{x}^{\overline{a}} \overline{y}^{\overline{b}}) + \overline{x}^c \overline{y}^d F_q(\overline{x}, \overline{y}, z)
\end{aligned}$$

of the form of (110). We have

$$\begin{aligned}
x &= \overline{x}^{a_1} \overline{y}^{b_1} \gamma_1(\overline{x}, \overline{y}) \\
y &= \overline{x}^{a_2} \overline{y}^{b_2} \gamma_2(\overline{x}, \overline{y})
\end{aligned}$$

where γ_1, γ_2 are units in $\hat{\mathcal{O}}_{Y,q}$, such that

$$(aa_1 + ba_2)m_0 = \overline{a}m_1, (ab_1 + bb_2)m_0 = \overline{b}m_1$$

where $m_0 \mid m_1$, $c_0 a_1 + d_0 a_2 = c$, $c_0 b_1 + d_0 b_2 = d$.

We have

$$x^a y^b = (\overline{x}^{\overline{a}} \overline{y}^{\overline{b}})^{\frac{m_1}{m_0}}$$

and

$$x^{c_0} y^{d_0} = \overline{x}^c \overline{y}^d \phi(\overline{x}, \overline{y})$$

where $\phi = \gamma_1^{c_0} \gamma_2^{d_0}$. There exists a series $\tilde{P}_q(\overline{x}^{\overline{a}} \overline{y}^{\overline{b}})$ such that

$$F_q(\overline{x}, \overline{y}, z) = \phi(\overline{x}, \overline{y}) F_{p_0}(x, y, z) - \frac{\tilde{P}_q(\overline{x}^{\overline{a}} \overline{y}^{\overline{b}})}{\overline{x}^c \overline{y}^d}.$$

$$v = P_{p_0}((\overline{x}^{\overline{a}} \overline{y}^{\overline{b}})^{\frac{m_1}{m_0}}) + \tilde{P}_q(\overline{x}^{\overline{a}} \overline{y}^{\overline{b}}) + \overline{x}^c \overline{y}^d F_q(\overline{x}, \overline{y}, z).$$

implies

$$P_q(\overline{x}^{\overline{a}} \overline{y}^{\overline{b}}) = P_{p_0}((\overline{x}^{\overline{a}} \overline{y}^{\overline{b}})^{\frac{m_1}{m_0}}) + \tilde{P}_q(\overline{x}^{\overline{a}} \overline{y}^{\overline{b}}).$$

Let $(\tilde{x}_*, \tilde{y}_*, z)$ be the corresponding regular parameters at p' to $(\overline{x}, \overline{y}, \overline{z})$, by the identification of (115). Define $\tilde{x}_*$ by

$$\tilde{x}_* = \tilde{x}'_* \gamma^{-\frac{m_0 a}{m_1 \overline{a}}} \equiv \tilde{x}'_* \bmod m_0^t \mathcal{O}_{\overline{Y}, p'}.$$

$$x^a y^b = (\overline{x}^{\overline{a}} \overline{y}^{\overline{b}})^{\frac{m_1}{m_0}}$$

implies

$$x^a y^b = \gamma^a \tilde{x}^a y^b = \gamma^a (\tilde{x}^{\bar{a}}_* \tilde{y}^{\bar{b}}_*)^{\frac{m_1}{m_0}} = ((\tilde{x}'_*)^{\bar{a}} \tilde{y}^{\bar{b}}_*)^{\frac{m_1}{m_0}}$$

Thus $(\tilde{x}'_*, \tilde{y}_*, z)$ are permissible parameters for (u, v) in $\hat{\mathcal{O}}_{\overline{Y}, p'}$. $x^{c_0} y^{d_0} = \overline{x}^c \overline{y}^d \phi$ implies

$$x^{c_0} y^{d_0} = \gamma^{c_0} \tilde{x}^{c_0} y^{d_0} = \gamma^{c_0} \tilde{x}^c_* \tilde{y}^d_* \phi(\tilde{x}_*, \tilde{y}_*) = (\tilde{x}'_*)^c \tilde{y}^d_* \tilde{\phi}$$

with

$$\tilde{\phi} \equiv \phi(\tilde{x}'_*, \tilde{y}_*) \mod m_0^t \hat{\mathcal{O}}_{\overline{Y}, p'}.$$

$$
\begin{aligned}
F_{p_0}(x, y, z) &\equiv \phi(\tilde{x}_*, \tilde{y}_*)^{-1}(F_q(\tilde{x}_*, \tilde{y}_*, z) + \frac{\tilde{P}_q(\tilde{x}^{\bar{a}}_* \tilde{y}^{\bar{b}}_*)}{\tilde{x}^c_* \tilde{y}^d_*}) \mod m_0^t \hat{\mathcal{O}}_{\overline{Y}, p'} \\
&\equiv \tilde{\phi}(\tilde{x}'_*, \tilde{y}_*)^{-1}(F_q(\tilde{x}'_*, \tilde{y}_*, z) + \frac{\tilde{P}_q((\tilde{x}'_*)^{\bar{a}} \tilde{y}^{\bar{b}}_*)}{(\tilde{x}'_*)^c \tilde{y}^d_*}) \mod m_0^t \hat{\mathcal{O}}_{\overline{Y}, p'}
\end{aligned}
$$

$$
\begin{aligned}
u &= ((\tilde{x}'_*)^{\bar{a}} \tilde{y}^{\bar{b}}_*)^{m_1} \\
v &= P_{p_0}(((\tilde{x}'_*)^{\bar{a}} \tilde{y}^{\bar{b}}_*)^{\frac{m_1}{m_0}}) + (\tilde{x}'_*)^c (\tilde{y}^d_*) \tilde{\phi} F_{p_0}(x, y, z) \\
&= P_{p_0}(((\tilde{x}'_*)^{\bar{a}} \tilde{y}^{\bar{b}}_*)^{\frac{m_1}{m_0}}) + \tilde{P}_q((\tilde{x}'_*)^{\bar{a}} \tilde{y}^{\bar{b}}_*) + (\tilde{x}'_*)^c (\tilde{y}^d_*)[F_q(\tilde{x}'_*, \tilde{y}_*, z) + h] \\
&= P_q((\tilde{x}'_*)^{\bar{a}} \tilde{y}^{\bar{b}}_*) + (\tilde{x}'_*)^c (\tilde{y}^d_*)[F_q(\tilde{x}'_*, \tilde{y}_*, z) + h]
\end{aligned}
$$

with $h \in m_0^t \hat{\mathcal{O}}_{\overline{Y}, p'}$.

The case when $p_0 \in \overline{Y}_0$ is a 2 point and (109) or (111) holds in $\hat{\mathcal{O}}_{Y, q}$ is similar to the case when (110) holds in $\hat{\mathcal{O}}_{Y, q}$.

Suppose that p' is a generic point of C. If (x', y', z') are permissible parameters at p' such that $x' = y' = 0$ are local equations of C at p', then $\nu(F_{p'}(0, 0, z')) \leq 1$, so that after extending $\overline{\pi}$ to a sequence of permissible blow-ups $\overline{U} \to U$ over a small affine neighborhood U of p, $\gamma(p*) \leq 1$ at all points $p*$ of $\overline{\pi}^{-1}(p')$. Thus the curves in $\overline{S}_r(\overline{U})$ must be components of $\overline{\pi}^{-1}(p)$, and the strict transforms of the curves D or D_1, D_2 in $\overline{S}_r(\overline{U})$, (if they exist), with the notation of Lemma 8.5.

Thus a curve E in $\overline{S}_r(\overline{U})$ must have local equations as asserted by the Theorem. □

Theorem 11.5. *Suppose that $r \geq 3$ and $A_r(X)$ holds. Suppose that $p \in X$ is a 1 point or a 2 point with $\nu(p) = \gamma(p) = r$. Let $R = \mathcal{O}_{X, p}$. Suppose that $\pi : Y_p \to Spec(\hat{R})$ is the sequence of monoidal transforms of sections over the curve $\overline{C}$ with local equations $\tilde{x} = y = 0$ of Theorem 11.2. For $q \in \pi^{-1}(p)$, define*

$$
l_q = \begin{cases}
\left(min_{2 \leq i \leq r}\{[\frac{\alpha_i}{i}]\} + 3\right) r & \text{if } F_q \text{ is a form} \\
& (109) \text{ or } (111). \\
\left(min_{2 \leq i \leq r}\{[\frac{\alpha_i}{i}]\} + min_{2 \leq i \leq r}\{[\frac{\beta_i}{i}]\} + 3\right) r & \text{if } F_q \text{ is a form } (110).
\end{cases}
$$

let $l = max\{l_q \mid q \in \pi^{-1}(p)\}$.

Suppose that $t \geq l$. Let $\overline{\pi} : \overline{Y}_p \to spec(R)$ be the sequence of monoidal transforms of Theorem 11.4. Let

$$\cdots \to Y_n \to \cdots \to Y_1 \to \overline{Y}_p$$

be a sequence of permissible monoidal transforms centered at curves C in $\overline{S}_r$ such that C is r big. Then there exists $n_0 < \infty$ such that

$$V_p = Y_{n_0} \overset{\pi_1}{\to} \overline{Y}_p \to spec(R)$$

extends to a permissible sequence of monoidal transforms

$$\overline{U}_1 \to \overline{U} \to U$$

over an affine neighborhood U of p, in the notation of Theorem 11.4, such that $\overline{S}_r(\overline{U}_1)$ contains no curves C such that C is r big. Let

$$\cdots \to Z_n \to \cdots \to V_p$$

be a permissible sequence of monoidal transforms centered at curves C in $\overline{S}_r$ such that C is r small. Then there exists $n_1 < \infty$ such that $\pi_2 : Z_p = Z_{n_1} \to V_p$ extends to a permissible sequence of monoidal transforms

$$\overline{U}_2 \to \overline{U}_1 \to \overline{U} \to U$$

over an affine neighborhood U of p such that $\overline{S}_r(\overline{U}_2) = \emptyset$.

Finally, there exists a sequence of quadratic transforms $\pi_3 : W_p \to Z_p$ which extends to a permissible sequence of monoidal transforms

$$\overline{U}_3 \to \overline{U}_2 \to \overline{U}_1 \to \overline{U} \to U$$

over an affine neighborhood U of p such that $\overline{S}_r(\overline{U}_3) = \emptyset$, and if

$$q \in (\overline{\pi} \circ \pi_1 \circ \pi_2 \circ \pi_3)^{-1}(p),$$

1. $\nu(q) \leq r - 1$ if q is a 1 or 2 point.
2. If q is a 2 point and $\nu(q) = r - 1$, then $\tau(q) > 0$.
3. $\nu(q) \leq r - 2$ if q is a 3 point.

Proof. Let $\overline{Y} = \overline{Y}_p$. If $q \in \overline{\pi}^{-1}(p)$, we have permissible parameters (x, y, z) in $\hat{\mathcal{O}}_{\overline{Y}, q}$ for (u, v) with forms obtained from those of (109), (110), (111) by modifying F_q by adding an appropriate series h to F_q.

(109) is modified by changing F_q to

$$F_q = \tau z^r + \sum_{j=2}^{r-1} \overline{a}_j(x, y) x^{\alpha_j} z^{r-j} + \epsilon x^{\alpha_r} y + h \tag{117}$$

with $h \in (x, z)^t$. By assumption, there exists $\tilde{z} \in \mathcal{O}_{\overline{Y}, q}$ and a unit $\sigma \in \hat{\mathcal{O}}_{\overline{Y}, q}$ such that $z = \sigma \tilde{z}$. Then $x = z = 0$ defines a germ of an algebraic curve D at q. We also assume that $\nu(q) = r$.

Given a form (117) at q, suppose that $D \subset \overline{S}_r(\overline{Y})$ is a curve such that $q \in D$. By assumption, $\overline{S}_r(\overline{Y})$ makes SNCs with $\overline{B}_2(\overline{Y})$. Since D is nonsingular at q, $x \in \hat{\mathcal{I}}_{D, q}$ and $F_q \in \hat{\mathcal{I}}_{D, q}^r + (x)^{r-1}$ by Lemma 6.24 implies

$$\frac{\partial^{r-1} F_q}{\partial z^{r-1}} \in \hat{\mathcal{I}}_{D, q},$$

so that $z \in \hat{\mathcal{I}}_{D, q}$ and $x = z = 0$ are local equations of D at q.

(110) is modified by changing F_q to

$$F_q = \tau z^r + \sum_{j=2}^{r-1} \bar{a}_j(x,y) x^{\alpha_j} y^{\beta_j} z^{r-j} + \epsilon x^{\alpha_r} y^{\beta_r} + h \qquad (118)$$

with $h \in (xy,z)^t$. By assumption, there exists $\tilde{z} \in \mathcal{O}_{\overline{Y},q}$ and a unit $\sigma \in \hat{\mathcal{O}}_{\overline{Y},q}$ such that $z = \sigma \tilde{z}$. Then $x = z = 0$ and $x = y = 0$ define germs of algebraic curves D_1 and D_2 at q. We also assume $\nu(q) = r$.

Given a form (118) at q, suppose that $D \subset \overline{S}_r(\overline{Y})$ is a curve such that $q \in D$. Since (by assumption) $\overline{S}_r(\overline{Y})$ makes SNCs with $\overline{B}_2(\overline{Y})$, either x or $y \in \hat{\mathcal{I}}_{D,q}$, and by Lemma 6.26, there exist $d_i \in k$ such that

$$F_q - \frac{1}{x^c y^d} \sum d_i (x^a y^b)^i \in \hat{\mathcal{I}}_{D,q} + (x)^{r-1}$$

or

$$F_q - \frac{1}{x^c y^d} \sum d_i (x^a y^b)^i \in \hat{\mathcal{I}}_{D,q} + (y)^{r-1}.$$

$$\frac{\partial^{r-1} F_q}{\partial z^{r-1}} \in \hat{\mathcal{I}}_{D,q}$$

implies $z \in \hat{\mathcal{I}}_{D,q}$ so that either $x = z = 0$ or $y = z = 0$ are local equations of D at q.

(111) is modified by changing F_q to

$$F_q = \tau z^r + \sum_{j=2}^{r-1} \bar{a}_j(x,y) x^{\alpha_j} y^{\beta_j} z^{r-j} + \epsilon x^{\alpha_r} y^{\beta_r} + h \qquad (119)$$

with $h \in (x,z)^t$. By assumption, there exists $\tilde{z} \in \mathcal{O}_{\overline{Y},q}$ and a unit $\sigma \in \hat{\mathcal{O}}_{\overline{Y},q}$ such that $z = \sigma \tilde{z}$. Then $x = z = 0$ defines a germ of an algebraic curve D at q. We also assume $\nu(q) = r$.

Given a form (119) at q, suppose that $D \subset \overline{S}_r(\overline{Y})$ is a curve such that $q \in D$. By assumption, $\overline{S}_r(\overline{Y})$ makes SNCs with $\overline{B}_2(\overline{Y})$. As in the analysis of the case when (117) holds, we conclude that $x = z = 0$ are local equations of D at q.

Suppose that $D \subset \overline{S}_r(\overline{Y})$ is a curve such that D is r big. Let $\pi' : Y' \to \overline{Y}$ be the blow-up of D. By assumption, π' is a permissible monoidal transform. $\overline{S}_r(Y')$ makes SNCs with $\overline{B}_2(Y')$ by Lemma 8.8.

First suppose that $q \in \overline{\pi}^{-1}(p) \cap D$, and that u, v have the form of (117). Then $x = z = 0$ are local equations of D at q. $F_q \in (x,z)^r$ implies $\alpha_r \geq r$ if $\epsilon = 1$. Suppose that $q' \in (\pi')^{-1}(q)$. First suppose that q' has permissible parameters (x_1, y, z_1) where $x = x_1$, $z = x_1(z_1 + \alpha)$ for some $\alpha \neq 0$. Substituting into F_q, we get $\nu(F_{q'}(0,0,z_1)) \leq r-1$. Suppose that q' has regular parameters (x_1, y, z_1) where $x = x_1 z_1$, $z = z_1$. Then $F_{q'}$ is a unit. The remaining case is when q' has permissible parameters (x_1, y, z_1) where $x = x_1$, $z = x_1 z_1$. Then

$$\begin{aligned} u &= x_1^a \\ F_{q'} &= \tau z_1^r + \sum_{j=2}^{r-1} \bar{a}_j(x_1, y) x_1^{\alpha_j'} z_1^{r-j} + \epsilon x_1^{\alpha_r'} y + h_1 \end{aligned} \qquad (120)$$

where $\alpha_j' = \alpha_j - j$ for $2 \leq j \leq r$ and $h_1 \in (x_1, z_1)^{t-r}$. We either have a reduction in multiplicity $\nu(q') < r$, or $\nu(q') = r$ and we are back in the form of (117) with a reduction in the α_j by j, and a decrease of t by r. By Lemma 8.8, $\overline{S}_r(Y') \cup \overline{B}_2(Y')$ makes SNCs in a neighborhood of $(\pi')^{-1}(q)$. Since $S_r(Y')$ is closed in the open set of 1 points of $E_{Y'}$, and $r \geq 2$, by Lemma 7.7, $\overline{S}_r(Y') \cap (\pi')^{-1}(q) = \emptyset$ or it is the point q' of (120), if $\nu(q') = r$. Suppose that $\nu(q') = r$ in (120). Since there exists a unit series σ such that $\sigma z \in \mathcal{O}_{\overline{Y},q}$, there exists a unit series σ' such that $\sigma' z_1 \in \mathcal{O}_{Y',q'}$.

Now suppose that $q \in \overline{\pi}^{-1}(p) \cap D$, and that u, v have the form of (119). Then we have that $x = z = 0$ are local equations of D at q. Since $F_q \in \hat{\mathcal{I}}^r_{D,q}$, we have $\alpha_j \geq j$ for all j.

Suppose that $q' \in (\pi')^{-1}(q)$. First suppose that q' has permissible parameters (x_1, y, z_1) where $x = x_1$, $z = x_1(z_1 + \alpha)$ for some $\alpha \neq 0$. substituting into F_q, we get $\nu(F_{q'}(0,0,z_1)) \leq r-1$. Suppose that q' has regular parameters (x_1, y, z_1) where $x = x_1 z_1$, $z = z_1$. Then $F_{q'}$ is a unit. The remaining case is when q' has regular parameters (x_1, y, z_1) where $x = x_1$, $z = x_1 z_1$. Then

$$
\begin{aligned}
u &= x_1^a \\
F_{q'} &= \tau z_1^r + \sum_{j=2}^{r-1} \overline{a}_j(x_1,y) x_1^{\alpha'_j} y^{\beta_j} z_1^{r-j} + \epsilon x_1^{\alpha'_r} y^{\beta_r} + h_1
\end{aligned}
\tag{121}
$$

where $\alpha'_j = \alpha_j - j$ for $2 \leq j \leq r$ and $h_1 \in (x_1, z_1)^{t-r}$. We either have a reduction in multiplicity $\nu(q') < r$, or we are back in the form of (119) with a reduction in α_i by i, and a decrease of t by r. As in the analysis of (117), we either have $\overline{S}_r(Y') \cap (\pi')^{-1}(q) = \emptyset$, or $\overline{S}_r(Y') \cap (\pi')^{-1}(q)$ is the single point q' of (121). In this case there exists a unit series σ' such that $\sigma' z_1 \in \mathcal{O}_{Y',q'}$.

Now suppose that $q \in \overline{\pi}^{-1}(p) \cap D$, and that u, v have the form of (118). Then either $x = z = 0$ or $y = z = 0$ are local equations of D at q. We may suppose that $x = z = 0$ are local equations of D at q, so that $F_q \in (x, z)^r$.

Suppose that $q' \in (\pi')^{-1}(q)$. First suppose that q' has permissible parameters (x_1, y, z_1) where $x = x_1$, $z = x_1(z_1 + \alpha)$ for some $\alpha \neq 0$. substituting into F_q, we get $u = (x_1^a y^b)^m$ and $\nu(F_{q'}(0,0,z_1)) \leq r-1$. Suppose that q' has permissible parameters (x_1, y, z_1) where $x = x_1 z_1$, $z = z_1$. Then $F_{q'}$ is a unit. The remaining case is when q' has permissible parameters (x_1, y, z_1) where $x = x_1$, $z = x_1 z_1$. Then

$$
\begin{aligned}
u &= (x_1^a y^b)^m \\
v &= P(x_1^a y^b) + x_1^{c+r} y^d F_{q'} \\
F_{q'} &= \tau z_1^r + \sum_{j=2}^{r-1} \overline{a}_j(x_1,y) x_1^{\alpha'_j} y^{\beta_j} z_1^{r-j} + \epsilon x_1^{\alpha'_r} y^{\beta_r} + h_1
\end{aligned}
\tag{122}
$$

where $\alpha'_j = \alpha_j - j$ for $2 \leq j \leq r$ and $h_1 \in (x_1 y_1, z_1)^{t-r}$. We either have a reduction in multiplicity $\nu(q') < r$, or we are back in the form of (118) with a reduction of α_i by i and a decrease of t by r. In this case, there exists a unit series σ' such that $\sigma' z_1 \in \mathcal{O}_{Y',q'}$.

Suppose that $q' \in \overline{S}_r(Y') \cap (\pi')^{-1}(q)$ is a 2 point with $\nu(q') = r - 1$, and q' lies on a curve E in $\overline{S}_r(Y'')$. E is transversal to the 2 curve at q' by Lemma 8.8. By Lemma 6.26, there exist $b_t \in k$ such that

$$
F_{q'} + \frac{1}{x_1^{c+r} y^d} \sum b_t (x_1^a y^b)^t \in \hat{\mathcal{I}}^r_{E,q'} + (x_1)^{r-1},
$$

if $x_1 \in \hat{\mathcal{I}}_{E,q'}$ or the series is in $\hat{\mathcal{I}}^r_{E,q'} + (y)^{r-1}$ if $y \in \hat{\mathcal{I}}_{E,q'}$. Then $\frac{\partial^{r-2} F_{q'}}{\partial z^{r-2}} \in \hat{\mathcal{I}}^2_{E,q'}$.
Suppose that q' has permissible parameters (x_1, y, z_1) such that $x = x_1, z = x_1(z_1 + \alpha)$ with $\alpha \neq 0$.

$$F_{q'} = \Lambda z_1^{r-1} + \sum_{i=1}^{r-1} \tilde{a}_i(x_1, y) z_1^{r-1-i}$$

where Λ is a unit. $x_1^i \mid \tilde{a}_i$ (or $y^i \mid \tilde{a}_i$) for $1 \leq i \leq r-1$ since $F_{q'} \in \hat{\mathcal{I}}^{r-1}_{E,q'}$. Then $z_1 \in \hat{\mathcal{I}}^2_{E,q'} + (x_1)$ which is impossible.

Thus, by Lemma 7.7, the only possible point in $\overline{S}_r(Y') \cap (\pi')^{-1}(q)$ is the point q' of (122). If there is a curve $E \subset \overline{S}_r(Y')$ containing q', then we must have $z_1 \in \hat{\mathcal{I}}_{E,q'}$ since

$$\frac{\partial^{r-1} F_{q'}}{\partial z^{r-1}} \in \hat{\mathcal{I}}_{E,q'}.$$

Thus E has local equations $x_1 = z_1 = 0$ or $y = z_1 = 0$.

After any sequence of permissible monoidal transforms, centered at r big curves $C \subset \overline{S}_r$, we eventually obtain $\pi_1 : V_p \to Y$ where there are no r big curves C in $\overline{S}_r(V_p)$ and $\overline{S}_r(V_p)$ makes SNCs with $\overline{B}_2(V_p)$.

Further, if $q \in (\overline{\pi} \circ \pi_1)^{-1}(p)$, and either $q \in \overline{S}_r(V_p)$ or one of 1. - 3. of the conclusions of W_p fail at q then q must satisfy one of (117), (118) or (119) (with $\nu(q) = r$ or $\nu(q) = r - 1$).

Suppose that (118) holds at $q \in (\overline{\pi} \circ \pi_1)^{-1}(p)$, and $\nu(q) = r$. Then we either have $\alpha_j \geq j$ for all j or $\beta_j \geq j$ for all j by Remark 9.19. If $\alpha_j \geq j$ for all j, then $F_q \in (x, z)^r$. If $\beta_j \geq j$ for all j, then $F_q \in (y, z)^r$. Since $x = z = 0$ and $y = z = 0$ are local equations of curves on V_p, in either case we have a curve $D \subset \overline{S}_r(V_p)$ such that D is r big by Lemma 8.2. Thus (118) cannot hold on V_p with $\nu(q) = r$.

Suppose that (119) holds at $q \in (\overline{\pi} \circ \pi_1)^{-1}(p)$, and $\nu(q) = r$. Then we either have $\alpha_j \geq j$ for all j or $\beta_j \geq j$ for all j by Remark 9.19. By Remark 11.3, $\beta_j < j$ for some j. Thus $F_q \in (x, z)^r$. Since $x = z = 0$ are local equations of a curve in V_p, By Lemma 8.2, we must have a curve $D \subset \overline{S}_r(V_p)$ such that D is r big. Thus (119) cannot hold on V_p, with $\nu(q) = r$.

The only points on $(\overline{\pi} \circ \pi_1)^{-1}(p)$ where the conclusions of the Theorem do not hold are at points q' over p where one of (123) or (124) following hold.

$$\begin{aligned}
u &= x^a \\
v &= P(x) + x^c F_q \\
F_q &= \tau z^r + \sum_{j=2}^{r-1} \overline{a}_j(x, y) x^{\alpha_j} z^{r-j} + \epsilon x^{\alpha_r} y + h
\end{aligned} \tag{123}$$

where $\nu(q') = r$, some $\alpha_j < j$, and $h \in (x, z)^{3r}$. Further, there exists a series σ such that $\sigma z \in \mathcal{O}_{V_p, q'}$. The other possiblity is that q' has permissible parameters (x_1, y_1, z_1) of the form of (122), with

$$\begin{aligned}
u &= (x_1^a y_1^b)^m \\
v &= P(x_1^a y_1^b) + x_1^c y_1^d F_{q'} \\
F_{q'} &= \tau z_1^r + \sum_{j=2}^{r-1} \overline{a}_j(x_1, y_1) x_1^{\alpha'_j} y_1^{\beta'_j} z_1^{r-j} + \epsilon x_1^{\alpha'_r} y_1^{\beta'_r} + h_1
\end{aligned} \tag{124}$$

with $\nu(q') = r - 1$,

$$h_1 \in (x_1 y_1, z_1)^{3r}$$

and there exists i such that

$$\frac{\alpha'_i}{i} \leq \frac{\alpha'_j}{j}, \quad \frac{\beta'_i}{i} \leq \frac{\beta'_j}{j}$$

for $2 \leq j \leq r$ and

$$\left\{\frac{\alpha'_i}{i}\right\} + \left\{\frac{\beta'_i}{i}\right\} < 1.$$

Further, there exists a series σ such that $\sigma z_1 \in \mathcal{O}_{V_p, q'}$.

Suppose that $D \subset \bar{S}_r(V_p)$ is a curve (which is necessarily r small). Let $\pi' : Z_1 \to V_p$ be the blow-up of D.

Suppose that $q' \in D \cap (\bar{\pi} \circ \pi_1)^{-1}(p)$. q' can only be a point of the form of (123) or (124).

Suppose that q' satisfies (123). Then $x = z = 0$ are local equations of D at q', and by Lemma 6.24, $\epsilon = 1$, $\alpha_r = r - 1$ and $\alpha_j \geq j$ if $j \neq r - 1$.

Suppose that $q'' \in (\pi')^{-1}(q')$. Suppose that $\hat{\mathcal{O}}_{Z_1, q''}$ has regular parameters (x_1, y, z_1) where $x = x_1$, $z = x_1(z_1 + \alpha)$ for some $\alpha \in k$.

$$u = x_1^a$$
$$F_{q''} = \tau x_1 (z_1 + \alpha)^r + \sum_{j=2}^{r-1} \bar{a}_j(x_1, y) x_1^{\alpha_j - j + 1}(z_1 + \alpha)^{r-j} + y - g(x_1) + h_1$$

with $h_1 \in (x_1, z_1)^{2r}$ for some series $g(x_1)$. q'' is resolved, since $\nu(F_{q'}(0, y, 0)) = 1$. If q'' has permissible parameters (x_1, y, z_1) where $x = x_1 z_1$, $z = z_1$,

$$u = x_1^a z_1^a$$
$$F_{q''} = \tau z_1 + \sum_{j=2}^{r-1} \bar{a}_j(x_1 z_1, y) x_1^{\alpha_j} z_1^{\alpha_j - j + 1} + x_1^{r-1} y + h_1 \tag{125}$$

with $h_1 \in (x_1, z_1)^{2r}$. $\nu(q'') \leq r - 2$ since $\nu(F_{q'}(0, 0, z_1)) = 1$ and $r \geq 3$. By Lemma 7.6, $(\pi')^{-1}(q') \cap \bar{S}_r(Z_1) = \emptyset$, and the conclusions of 1. - 3. of the Theorem hold on $(\pi')^{-1}(q')$.

Suppose that $q' \in (\bar{\pi} \circ \pi_1)^{-1}(p)$ satisfies (124). Then either $x_1 = z_1 = 0$ or $y_1 = z_1 = 0$ are local equations of D at q'. Without loss of generality, assume that $y_1 = z_1 = 0$ are local equations of D at q'. Then we have $\beta'_j \geq j$ if $2 \leq j \leq r - 1$ and $\epsilon = 1$, $\beta'_r = r - 1$ by Lemma 6.26. $\nu(q') = r - 1$ implies $\alpha'_r = 0$.

We have

$$u = (x_1^a y_1^b)^m$$
$$v = P(x_1^a y_1^b) + x_1^c y_1^d F_{q'} \text{ where} \tag{126}$$
$$F_{q'} = \tau z_1^r + \sum_{j=2}^{r-1} \bar{a}_j(x_1, y_1) x_1^{\alpha'_j} y_1^{\beta'_j} z_1^{r-j} + y_1^{r-1} + h_1$$

with $\beta'_j \geq j$ for all j,

$$h_1 \in (x_1 y_1, z_1)^{3r}$$

Suppose that $q'' \in (\pi')^{-1}(q')$, and q'' has regular parameters (x_1, y_2, z_2) defined by

$$y_1 = y_2, z_1 = y_2(z_2 + \alpha)$$

Then

$$
\begin{aligned}
u &= (x_1^a y_2^b)^m \\
v &= P_{q''}(x_1^a y_2^b) + x_1^c y_2^{d+r-1} F_{q''} \\
\frac{F_{q'}}{y_2^{r-1}} &= \tau(z_2 + \alpha)^r y_2 + \sum_{j=2}^{r-1} \bar{a}_j x_1^{\alpha'_j} y_2^{\beta'_j - j + 1}(z_2 + \alpha)^{r-j} + 1 + h_2 \\
&= 1 + y_2 \Omega
\end{aligned}
$$

since

$$
h_2 \in (y_2)^{2r}.
$$

$a(d + r - 1) - bc \neq 0$ since $F_{q'}$ is normalized. Thus $F_{q''} = 1 + y_2 \Omega'$ is a unit.

Suppose that $q'' \in (\pi')^{-1}(q')$, and q'' has regular parameters (x_1, y_2, z_2) defined by

$$
y_1 = y_2 z_2, z_1 = z_2
$$

Then

$$
\begin{aligned}
F_{q''} = \frac{F_{q'}}{z_2^{r-1}} &= \tau z_2 + \sum_{j=2}^{r-1} \bar{a}_j x_1^{\alpha'_j} y_2^{\beta'_j} z_2^{\beta'_j - j + 1} + y_2^{r-1} + h_2 \\
&= \tau z_2 + + y_2 + z_2^2 \Omega'
\end{aligned}
\tag{127}
$$

since

$$
h_2 \in (z_2)^{2r}
$$

Thus $\nu(q'') = 1$. $\nu(q'') \leq r - 2$ since $r \geq 3$. Thus $(\pi')^{-1}(q') \cap \overline{S}_r(Z_1) = \emptyset$ by Lemma 7.7, and the conclusions 1. - 3. of the Theorem hold on $(\pi')^{-1}(q')$.

We thus construct a permissible sequence of monoidal transforms $\pi_2 : Z_p \to V_p$ centered at the strict transforms of curves $C \subset \overline{S}_r(V_p)$ which are r small so that $\overline{S}_r(Z_p)$ contains no curves. $Z_p \to \overline{Y}_p$ extends to $\overline{U}_2 \to U$ in the notation of the Theorem.

Suppose that $q' \in (\overline{\pi} \circ \pi_1 \circ \pi_2)^{-1}(p)$ does not satisfy the conclusions of the Theorem. Then q must either satisfy (123) or (124).

We cannot have that (123) holds at q', since then $\nu(q') = r$, which implies that $\alpha_j \geq j$ for $j \geq 2$ and $\alpha_{r-1} \geq r - 1$, so that $x = z = 0$ are local equations of a curve D in $\overline{S}_r(\overline{U}_2)$. We further see that $\overline{S}_r(\overline{U}_2) = \emptyset$.

Suppose that (124) holds at q'. Then $\nu(q') = r - 1$ and 2. of the conclusions of the Theorem does not hold, so that $\tau(q') = 0$. Thus $\alpha'_j + \beta'_j \geq j$ for $j \neq r$ and $\epsilon = 1$, $\alpha'_r + \beta'_r = r - 1$.

First suppose that (124) holds, with $\tau(q') = 0$, α'_r and $\beta'_r \neq 0$. Let $\pi'' : W_1 \to Z_p$ be the quadratic transform with center q'. Suppose that $q'' \in (\pi'')^{-1}(q')$ and $\hat{O}_{W_1,q''}$ has regular parameters (x_2, y_2, z_2) such that

$$
x_1 = x_2, y_1 = x_2(y_2 + \alpha), z_1 = x_2(z_2 + \beta).
$$

with $\alpha \neq 0$. Set $x_2 = \overline{x}_2(y_2 + \alpha)^{\frac{-b}{a+b}}$. there exists

$$
h_2 \in (x_2)^{2r}
$$

such that

$$
\begin{aligned}
\frac{F_{q'}}{x_2^{r-1}} &= \tau x_2(z_2 + \beta)^r + \sum_{j=2}^{r-1} \bar{a}_j x_2^{\alpha'_j + \beta'_j - j + 1}(y_2 + \alpha)^{\beta'_j}(z_2 + \beta)^{r-j} \\
&\quad + (y_2 + \alpha)^{\beta'_r} + h_2 \\
&= (y_2 + \alpha)^{\beta'_r} + \overline{x}_2 \Omega
\end{aligned}
$$

Thus

$$u = (\bar{x}_2^{a+b})^m$$
$$v = P_{q''}(\bar{x}_2) + \bar{x}_2^{c+d+r-1} F_{q''}$$
$$F_{q''} = (y_2 + \alpha)^{\lambda+\beta'_r} - \alpha^{\lambda+\beta'_r} + \bar{x}_2 \Omega'$$

where $\lambda = d - \frac{b(c+d+r-1)}{a+b}$. Since $F_{q'}$ is normalized,

$$(a+b)(\beta'_r + d) - b(c+d+r-1) = a(d+\beta'_r) - b(c+r-1-\beta'_r)$$
$$= a(d+\beta'_r) - b(c+\alpha'_r) \neq 0$$

Thus $\lambda + \beta'_r \neq 0$. q'' is thus a resolved point.

Suppose that $q'' \in (\pi'')^{-1}(q')$ has permissible parameters (x_2, y_2, z_2) such that

$$x_1 = x_2, y_1 = x_2 y_2, z_1 = x_2(z_2 + \beta).$$

There exists

$$h_2 \in (x_2)^{2r}$$

such that

$$\frac{F_{q'}}{x_2^{r-1}} = \tau x_2(z_2 + \beta)^r + \sum_{j=2}^{r-1} \bar{a}_j x_2^{\alpha'_j + \beta'_j - j + 1} y_2^{\beta'_j}(z_2 + \beta)^{r-j} + y_2^{\beta'_r} + h_2$$
$$= y_2^{\beta'_r} + x_2 \Omega$$

$$b(c+d+r-1) - (a+b)(d+\beta'_r) \neq 0$$

since $F_{q'}$ is normalized.

Thus

$$u = (x_2^{a+b} y_2^b)^m$$
$$v = P_{q''}(x_2^{a+b} y_2^b) + x_2^{c+d+r-1} y_2^d F_{q''}$$
$$F_{q''} = y_2^{\beta'_r} + x_2 \Omega'$$

$\nu(F_{q''}) \leq \beta'_r < r - 1$.

Suppose that $q'' \in (\pi'')^{-1}(q')$ has regular parameters (x_2, y_2, z_2) such that

$$x_1 = x_2 y_2, y_1 = y_2, z_1 = y_2(z_2 + \beta).$$

Then

$$u = (x_2^a y_2^{a+b})^m$$
$$v = P_{q''}(x_2^a y_2^{a+b}) + x_2^c y_2^{c+d+r-1} F_{q''}$$
$$F_{q''} = x_2^{\alpha'_r} + y_2 \Omega'$$

since $F_{q'}$ is normalized. $\nu(F_{q''}) \leq \alpha'_r < r - 1$.

The remaining point in $q'' \in (\pi'')^{-1}(q')$ has regular parameters (x_2, y_2, z_2) such that

$$x_1 = x_2 z_2, y_1 = y_2 z_2, z_1 = z_2.$$

There exists

$$h_2 \in (z_2)^{2r}$$

such that

$$F_{q''} = \frac{F_{q'}}{z_1^{r-1}} = \tau z_2 + \sum_{j=2}^{r-1} \bar{a}_j x_2^{\alpha'_j} y_2^{\beta'_j} z_2^{\alpha'_j + \beta'_j - j + 1} + x_1^{\alpha'_r} y_2^{\beta'_r} + h_2$$
$$\equiv \tau z_2 \bmod (x_2, y_2, z_2^2) \tag{128}$$

Thus q'' is a 3 point with $\nu(F_{q''}) = 1 \leq r - 2$, since $r \geq 3$.

$\nu(q'') < r - 1$ for $q'' \in (\pi'')^{-1}(q')$, so that $(\pi'')^{-1}(q') \cap \overline{S}_r(Y'') = \emptyset$, and the conclusions of 1.-3. of Theorem 11.5 hold on $(\pi'')^{-1}(q')$.

Now suppose that (124) holds, with $\tau(q') = 0$ and $\alpha'_r = 0$ or $\beta'_r = 0$. Since the 2 cases are symmetric, we may assume that $\alpha'_r = 0$.

We thus have $\beta'_r = r - 1$ (and $\alpha'_j + \beta'_j \geq j$ for $j \neq m$). Suppose that $i \neq r$. Then $\frac{\alpha'_i}{i} \leq \frac{\alpha'_r}{r}$ implies $\alpha'_i = 0$ and $\frac{\beta'_i}{i} \leq \frac{r-1}{r} < 1$ implies $\beta'_i < i$, so that $\tau(q') > 0$, a contradiction. We thus have $i = r$.

$$\frac{r-1}{r} = \frac{\beta'_r}{r} \leq \frac{\beta'_j}{j}$$

for all j implies $\beta'_j \geq j - \frac{j}{r}$ for $2 \leq j < r$. Since $\beta'_j \in \mathbf{N}$, we have $\beta'_j \geq j$, and the curve D with local equations $y_1 = z_1 = 0$ at q' is such that $D \subset \overline{S}_r(Z_p)$, a contradiction.

$\square$

Theorem 11.6. *Suppose that $r = 2$ and $A_2(X)$ holds. Suppose that $p \in X$ is a 1 point or a 2 point with $\nu(p) = \gamma(p) = 2$. Let $R = \mathcal{O}_{X,p}$. Suppose that $\pi : Y_p \to \operatorname{Spec}(\hat{R})$ is the sequence of monoidal transforms of sections over the curve $\overline{C}$ with local equations $\tilde{x} = y = 0$ of Theorem 11.2. For $q \in \pi^{-1}(p)$, define*

$$l_q = \begin{cases} ([\frac{\alpha_2}{2}] + 3)2 & \text{if } F_q \text{ is a form (109) or (111).} \\ ([\frac{\alpha_2}{2}] + [\frac{\beta_2}{2}] + 3)2 & \text{if } F_q \text{ is a form (110).} \end{cases}$$

let $l = max\{l_q \mid q \in \pi^{-1}(p)\}$.

Suppose that $t \geq l$. Let $\overline{\pi} : \overline{Y}_p \to \operatorname{spec}(R)$ be the sequence of monoidal transforms of Theorem 11.4. Let

$$\cdots \to Y_n \to \cdots \to Y_1 \to \overline{Y}_p$$

be a sequence of permissible monoidal transforms centered at curves C in $\overline{S}_2$ such that C is 2 big. Then there exists $n_0 < \infty$ such that

$$V_p = Y_{n_0} \xrightarrow{\pi_1} \overline{Y}_p \to \operatorname{spec}(R)$$

extends to a permissible sequence of monoidal transforms

$$\overline{U}_1 \to \overline{U} \to U$$

over an affine neighborhood U of p, in the notation of Theorem 11.4, such that $\overline{S}_2(\overline{U}_1)$ contains no curves C such that C is 2 big. Let

$$\cdots \to Z_n \to \cdots \to V_p$$

be a permissible sequence of monoidal transforms centered at curves C in $\overline{S}_2$ such that C is 2 small. Then there exists $n_1 < \infty$ such that $\pi_2 : Z_p = Z_{n_1} \to V_p$ extends to a permissible sequence of monoidal transforms

$$\overline{U}_2 \to \overline{U}_1 \to \overline{U} \to U$$

over an affine neighborhood U of p such that $\overline{S}_2(\overline{U}_2) = \emptyset$.

Finally, there exists a sequence of quadratic transforms and monoidal transforms centered at strict transforms of 2 curves C on Z_p such that C is 1 big and

C is a section over a 2 small curve blown up in $Z_p \to V_p$, $\pi_3 : W_p \to Z_p$ which extends to a permissible sequence of monoidal transforms

$$\overline{U}_3 \to \overline{U}_2 \to \overline{U}_1 \to \overline{U} \to U$$

over an affine neighborhood U of p such that $\overline{S}_2(\overline{U}_3) = \emptyset$, and if

$$q \in (\overline{\pi} \circ \pi_1 \circ \pi_2 \circ \pi_3)^{-1}(p)$$

then q is resolved.

Proof. The analysis of Theorem 11.5 is valid for $r = 2$, except in (125), (127) and (128).

The situation of (128) cannot occur when $r = 2$, since this comes from the case when (124) holds, with $\tau(q') = 0$, α'_r and $\beta'_r \neq 0$. Since $\alpha'_r + \beta'_r = r - 1 = 1$, this case cannot occur.

Suppose that a case (125) occurs in

$$(\overline{\pi} \circ \pi_1 \circ \pi_2) : Z_p \to \mathrm{spec}(R).$$

Then we have a 2 point $q'' \in (\overline{\pi} \circ \pi_1 \circ \pi_2)^{-1}(p)$ such that

$$
\begin{aligned}
u &= (x_1^a z_1^b)^m \\
v &= P_{q''}(x_1^a z_1^b) + x_1^c z_1^d F_{q''} \\
F_{q''} &= z_1 + x_1 y_1 + h_1
\end{aligned}
$$

with $h_1 \in (x_1, z_1)^4$.

Let C be the 2 curve on Z_p with local equations $x_1 = z_1 = 0$. C is 1 big by Lemma 8.1.

Let $\pi' : W_1 \to Z_p$ be the blow-up of C. Suppose that $\overline{q} \in (\pi')^{-1}(q'')$ is a 1 point. Then there exist regular parameters (x_2, y_1, z_2) in $\hat{O}_{W_1, \overline{q}}$ such that

$$x_1 = x_2, \; z_1 = x_2(z_2 + \alpha)$$

with $\alpha \neq 0$. Set

$$x_2 = \overline{x}_2(z_1 + \alpha)^{-\frac{b}{a+b}}.$$

$$
\begin{aligned}
u &= \overline{x}_2^{(a+b)m} \\
v &= P_{q''}(\overline{x}_2^{a+b}) + \overline{x}_2^{c+d+1}(z_2 + \alpha)^{d - \frac{b(c+d+1)}{a+b}}(z_2 + \alpha + y_1 + x_1\Omega)
\end{aligned}
$$

Thus $\nu(F_{\overline{q}}(0, y_1, z_2)) = 1$ and $\overline{q}$ is resolved.

Suppose that $\overline{q} \in (\pi')^{-1}(q'')$ is the 2 point with permissible parameters

$$x_1 = x_2, \; z_1 = x_2 z_2.$$

Then $F_{\overline{q}} = \frac{F_{q''}}{x_2} = z_2 + y_1 + x_1\Omega$ and $\overline{q}$ is resolved.

If $\overline{q} \in (\pi')^{-1}(q'')$ is the 2 point with permissible parameters

$$x_1 = x_2 z_2, \; z_1 = z_2$$

then

$$F_{\overline{q}} = \frac{F_{q''}}{z_2} = 1 + x_2 y_1 + z_2\Omega$$

and is resolved.

Suppose that a case (127) occurs in

$$(\overline{\pi} \circ \pi_1 \circ \pi_2) : Z_p \to \mathrm{spec}(R).$$

Then we have a 3 point $q'' \in (\overline{\pi} \circ \pi_1 \circ \pi_2)^{-1}(p)$ such that

$$
\begin{aligned}
u &= (x_2^a y_2^b z_2^c)^m \\
v &= P_{q''}(x_2^a y_2^b z_2^c) + x_2^d y_2^e z_2^f F_{q''} \\
F_{q''} &= z_2 + y_2 + h_1
\end{aligned}
$$

with $h_1 \in (z_2)^4$.

Let C be the 2 curve on Z_p with local equations $y_2 = z_2 = 0$. C is 1 big by Lemma 8.1. Let $\pi' : W_1 \to Z_p$ be the blow-up of C. Suppose that $\overline{q} \in (\pi')^{-1}(q'')$ is a 2 point. Then there exist regular parameters (x_2, y_3, z_3) in $\hat{\mathcal{O}}_{W_1, \overline{q}}$ such that

$$
y_2 = y_3, z_2 = y_3(z_3 + \alpha)
$$

with $\alpha \neq 0$. Set $y_3 = \overline{y}_3(z_3 + \alpha)^{-\frac{c}{b+c}}$.

$$
\begin{aligned}
u &= (x_2^a \overline{y}_3^{b+c})^m = (x_2^{\overline{a}} \overline{y}_3^{\overline{b}})^{\overline{m}} \\
v &= P_{q''}(x_2^a \overline{y}_3^{b+c}) + x_2^d \overline{y}_3^{e+f+1}(z_3 + \alpha)^{f - \frac{c(e+f+1)}{b+c}}(z_3 + \alpha + 1 + y_3^3 \Omega)
\end{aligned}
$$

with $(\overline{a}, \overline{b}) = 1$. Thus $\nu(F_{\overline{q}}(0, 0, z_3)) = 1$ and $\overline{q}$ is resolved.

Suppose that $\overline{q} \in (\pi')^{-1}(q'')$ is the 3 point with permissible parameters $y_2 = y_3, z_2 = y_3 z_3$. Then

$$
F_{\overline{q}} = \frac{F_{q''}}{y_3} = z_3 + 1 + y_3^3 \Omega
$$

and $\overline{q}$ is resolved.

Suppose that $\overline{q} \in (\pi')^{-1}(q'')$ is the 3 point with permissible parameters $y_2 = y_3 z_3, z_2 = z_3$. Then

$$
F_{\overline{q}} = \frac{F_{q''}}{z_3} = 1 + y_3 + z_3^3 \Omega
$$

and $\overline{q}$ is resolved.

$\square$

Theorem 11.7. *Suppose that $r \geq 3$ and $A_r(X)$ holds. Suppose that $p \in X$ is a 2 point such that $\nu(p) = r - 1$, $\tau(p) = 0$ and $\gamma(p) = r$. Let $\overline{C}$ be the 2 curve containing p. Let $R = \mathcal{O}_{X, p}$.*

There exists a sequence of permissible monoidal transforms centered at sections over $\overline{C}$, $\overline{Y}_p \to \operatorname{spec}(R)$, which extends to a sequence of permissible monoidal transforms $\overline{U} \to U$ where U is an affine neighborhood of p, with the following property.

Let

$$
\cdots \to Y_n \to \cdots \to Y_1 \to \overline{Y}_p
$$

be a sequence of permissible monoidal transforms centered at curves C in $\overline{S}_r$ such that C is r big. Then there exists $n_0 < \infty$ such that

$$
V_p = Y_{n_0} \xrightarrow{\pi_1} \overline{Y} \to \operatorname{spec}(R)
$$

extends to a permissible sequence of monoidal transforms

$$
\overline{U}_1 \to \overline{U} \to U
$$

over an affine neighborhood U of p, in the notation of Theorem 11.4, such that $\overline{S}_r(\overline{U}_1)$ contains no curves C such that C is r big. Let

$$\cdots \to Z_n \to \cdots \to V_p$$

be a permissible sequence of monoidal transforms centered at curves C in $\overline{S}_r$ such that C is r small. Then there exists $n_1 < \infty$ such that $\pi_2 : Z_p = Z_{n_0} \to V_p$ extends to a permissible sequence of monoidal transforms

$$\overline{U}_2 \to \overline{U}_1 \to \overline{U} \to U$$

over an affine neighborhood U of p such that $\overline{S}_r(\overline{U}_2) = \emptyset$.

Finally, there exists a sequence of quadratic transforms $\pi_3 : W_p \to Z_p$ which extends to a permissible sequence of monoidal transforms

$$\overline{U}_3 \to \overline{U}_2 \to \overline{U}_1 \to \overline{U} \to U$$

over an affine neighborhood U of p such that $\overline{S}_r(\overline{U}_3) = \emptyset$, and if

$$q \in (\overline{\pi} \circ \pi_1 \circ \pi_2 \circ \pi_3)^{-1}(p),$$

1. *$\nu(q) \leq r - 1$ if q is a 1 or 2 point.*
2. *If q is a 2 point and $\nu(q) = r - 1$, then $\tau(q) > 0$.*
3. *$\nu(q) \leq r - 2$ if q is a 3 point.*

Theorem 11.8. *Suppose that $r = 2$ and $A_2(X)$ holds Suppose that $p \in X$ is a 2 point such that $\nu(p) = 1$, $\tau(p) = 0$ and $\gamma(p) = 2$. Let $R = \mathcal{O}_{X,p}$.*

There exists a sequence of permissible monoidal transforms centered at sections over $\overline{C}$, $\overline{Y}_p \to spec(R)$ which extends to a sequence of permissible monoidal transforms $\overline{U} \to U$, where U is an affine neighborhood of p, with the following property.

Let

$$\cdots \to Y_n \to \cdots \to Y_1 \to \overline{Y}_p$$

be a sequence of permissible monoidal transforms centered at curves C in $\overline{S}_2$ such that C is 2 big. Then there exists $n_0 < \infty$ such that

$$V_p = Y_{n_0} \xrightarrow{\pi_1} \overline{Y}_p \to spec(R)$$

extends to a permissible sequence of monoidal transforms

$$\overline{U}_1 \to \overline{U} \to U$$

over an affine neighborhood U of p, in the notation of Theorem 11.4, such that $\overline{S}_2(\overline{U}_1)$ contains no curves C such that C is 2 big. Let

$$\cdots \to Z_n \to \cdots \to V_p$$

be a permissible sequence of monoidal transforms centered at curves C in $\overline{S}_2$ such that C is 2 small. Then there exists $n_1 < \infty$ such that $\pi_2 : Z_p = Z_{n_1} \to Y_{n_0}$ extends to a permissible sequence of monoidal transforms

$$\overline{U}_2 \to \overline{U}_1 \to \overline{U} \to U$$

over an affine neighborhood U of p such that $\overline{S}_2(\overline{U}_2) = \emptyset$.

Finally, there exists a sequence of quadratic transforms and monoidal transforms centered at strict transforms of 2 curves C on Z_p such that C is 1 big and

C is a section over a 2 small curve blown up in $Z_p \to V_p$, $\pi_3 : W_p \to V_p$ which extends to a permissible sequence of monoidal transforms

$$\overline{U}_3 \to \overline{U}_2 \to \overline{U}_1 \to \overline{U} \to U$$

over an affine neighborhood U of p such that $\overline{S}_2(\overline{U}_3) = \emptyset$, and if

$$q \in (\overline{\pi} \circ \pi_1 \circ \pi_2 \circ \pi_3)^{-1}(p)$$

then q is resolved.

Proof. (of Theorems 11.7 and 11.8) The conclusions of Lemma 8.5 hold at p.

The conclusions of Lemma 11.1 must be modified to: $\alpha_j + \beta_j \geq j$ for $2 \leq j \leq r - 2$, $\alpha_r + \beta_r \geq r - 1$.

In the conclusions of Theorem 11.2, we must add a fourth case:

$$\begin{aligned}
u &= (x^a y^b)^m \\
v &= P(x^a y^b) + x^c y^d F_q \\
F_q &= \tau z^r + \sum_{i=2}^{r-1} \overline{a}_i(x, y) x^{\alpha_i} y^{\beta_i} z^{r-i} + y^{r-1}
\end{aligned} \tag{129}$$

with $\beta_i \geq i$ for all i.

Theorem 11.4 must be modified by adding the case F_q equivalent mod $(\overline{xy}, z)^t$ to a form (129). The proof of Theorem 11.5 must be modified by adding an analysis of (129). Such q are not effected by blowing up r big curves, so the construction of $\pi_1 : V_p \to \overline{Y}_p$ is as in the proof of Theorem 11.5. Suppose that $q \in (\overline{\pi} \circ \pi_1)^{-1}(p)$ satisfies (129). There is a unique r small curve $D \subset \overline{S}_r(V_p)$ containing q, which has local equations $y = z = 0$. Let $\pi' : Z_1 \to \overline{Y}_p$ be the blow-up of D. Then if $r \geq 3$, all points of $(\pi')^{-1}(q)$ satisfy 1. - 3. of the conclusions of the Theorem, and $\overline{S}_r(Z_1) \cap (\pi')^{-1}(q) = \emptyset$.

If $r = 2$, and $q' \in (\pi')^{-1}(q)$ is the 3 point, then there exist permissible parameters (x, y_1, z_1) at q' such that

$$F_{q'} = z_1 + y_1 + z_1^2 \Omega.$$

If C is the 2 curve with local equations $y_1 = z_1 = 0$, then C is 1 big, and if $\pi' : W_1 \to Z_p$ is the blow-up of C, then all points of $(\pi')^{-1}(q')$ are resolved. $\qquad \square$

12. REDUCTION OF ν IN A SECOND SPECIAL CASE

Throughout this section, we will assume that $\Phi_X : X \to S$ is weakly prepared.

Theorem 12.1. *Suppose that $r \geq 2$, $A_r(X)$ holds, $p \in X$ is a 2 point with $\nu(p) = r-1$, C is a generic curve through p, and there are permissible parameters (x, y, z) at p for (u, v) (with $y, z \in \mathcal{O}_{X,p}$) such that $L_p(x, 0, 0) \neq 0$, and C has local equations $y = z = 0$ at p. Let $R = \mathcal{O}_{X,p}$. We have an expression at p*

$$
\begin{aligned}
u &= (x^a y^b)^m \\
v &= P(x^a y^b) + x^c y^d F_p \\
F_p &= \tau x^{r-1} + \sum_{i=1}^{r-1} \tilde{a}_i(y, z) x^{r-i-1}
\end{aligned}
\tag{130}
$$

where τ is a unit and $\nu(\tilde{a}_i) \geq i$ for all i. Then there exists a finite sequence of permissible monoidal transforms $\pi : Y \to \operatorname{Spec}(R)$ centered at sections over C, such that for $q \in \pi^{-1}(p)$, there exist permissible parameters $(\bar{x}, \bar{y}, \bar{z})$ at q such that F_q has one of the following forms.

1.

$$
\begin{aligned}
u &= (\bar{x}^a \bar{y}^b)^m \\
v &= P(\bar{x}^a \bar{y}^b) + \bar{x}^c \bar{y}^d F_q \text{ with} \\
F_q &= \tau \bar{x}^{r-1} + \sum_{j=1}^{r-1} \bar{y}^{d_j} \Lambda_j(\bar{y}, \bar{z}) \bar{x}^{r-1-j}
\end{aligned}
\tag{131}
$$

where τ is a unit, $\Lambda_j(\bar{y}, \bar{z}) = 0$ or $e_j = \nu(\Lambda_j(0, \bar{z})) = 0$ or 1 and $d_j + e_j \geq j$ for all j.

2.

$$
\begin{aligned}
u &= (\bar{x}^a \bar{y}^b \bar{z}^c)^m \\
v &= P(\bar{x}^a \bar{y}^b \bar{z}^c) + \bar{x}^d \bar{y}^e \bar{z}^f F_q \text{ with} \\
F_q &= \tau \bar{x}^{r-1} + \sum_{j=1}^{r-1} \bar{a}_j(\bar{y}, \bar{z}) \bar{y}^{d_j} \bar{z}^{e_j} \bar{x}^{r-1-j}
\end{aligned}
\tag{132}
$$

where τ is a unit, $d_j + e_j \geq j$, $\bar{a}_j$ are units (or zero) for all j and there exists an i such that $1 \leq i \leq r-1$, $\bar{a}_i \neq 0$ and

$$
\frac{d_i}{i} \leq \frac{d_j}{j}, \frac{e_i}{i} \leq \frac{e_j}{j}
$$

for $1 \leq j \leq r-1$. We further have

$$
\left\{ \frac{d_i}{i} \right\} + \left\{ \frac{e_i}{i} \right\} < 1.
$$

3.

$$
\begin{aligned}
u &= (\bar{x}^a \bar{y}^b)^m \\
v &= P(\bar{x}^a \bar{y}^b) + \bar{x}^c \bar{y}^d F_q \text{ with} \\
F_q &= \tau \bar{x}^{r-1} + \sum_{j=1}^{r-1} \bar{a}_j(\bar{y}, \bar{z}) \bar{y}^{d_j} \bar{z}^{e_j} \bar{x}^{r-1-j}
\end{aligned}
\tag{133}
$$

where τ is a unit, $d_j + e_j \geq j$, $\bar{a}_j$ are units (or zero) for all j and there exists an i such that $1 \leq i \leq r-1$, $\bar{a}_i \neq 0$ and

$$
\frac{d_i}{i} \leq \frac{d_j}{j}, \frac{e_i}{i} \leq \frac{e_j}{j}
$$

for $1 \leq j \leq r - 1$. *We further have*

$$\left\{ \frac{d_i}{i} \right\} + \left\{ \frac{e_i}{i} \right\} < 1.$$

In all these cases $\overline{x} = x$ *and* $\overline{x} = 0$ *is a local equation at* q *of the strict transform of the component of* E_X *with local equation* $x = 0$ *at* p.

There exists an affine neighborhood U *of* p *such that* $Y \to spec(R)$ *extends to a sequence of permissible monoidal transforms* $\overline{U} \to U$ *such that* $A_r(\overline{U})$ *holds.*

Suppose that $r \geq 3$. *Let* D_i *be the curves in* $\overline{S}_{r-1}(X)$ *which contain* p, *and such that* $x \in \hat{I}_{D_i,p}$. *We further have that the strict transforms* $\overline{D}_i$ *of the* D_i *on* $\overline{U}$ *are nonsingular, disjoint and make SNCs with* $\overline{B}_2(\overline{U})$.

Proof. Set $S_0 = \mathrm{Spec}(k[y, z])$, $Y_0 = \mathrm{Spec}(R)$. Consider the sequence of monoidal transforms centered at sections over C

$$\cdots \to Y_n \to Y_{n-1} \to Y_{n-2} \to \cdots \to Y_1 \to Y_0 \tag{134}$$

where the sequence is obtained from a sequence of quadratic transforms

$$\cdots \to S_n \to S_{n-1} \to \cdots \to S_1 \to S_0 \tag{135}$$

over the closed point p_0 with local equations $y = z = 0$ in S_0, and (134) is obtained from (135) by base change with $Y_0 \to S_0$, so that $Y_i \cong S_i \times_{S_0} Y_0$.

The map $\hat{S}_0 \to \hat{Y}_0$ obtained from the natural projection

$$k[[x, y, z]] \to k[[y, z]]$$

induces maps $S_i \times_{S_0} \hat{S}_0 \to Y_i \times_{Y_0} \hat{Y}_0$ such that the composed map

$$S_i \times_{S_0} \hat{S}_0 \to Y_i \times_{Y_0} \hat{Y}_0 \to S_i \times_{S_0} \hat{S}_0$$

is an isomorphism for all i.

We can thus identify the center of the quadratic transform $S_{i+1} \to S_i$ with a point $p_i \in Y_i$ over p. A section over C through p_i is blown up in (134) only if none of the forms (131), (132) or (133) hold at p_i.

We will show that (134) is finite, so that there exists n such that Y_n satisfies the conclusions of the theorem.

Suppose that (134) is not finite. Then we may assume that there exists an infinite sequence of points $p_0, p_1, p_2, \ldots, p_n, \ldots$ such that $Y_{i+1} \to Y_i$ is a permissible monoidal transform, centered at a section C_i over C, containing p_i, such that p_i maps to p_{i-1} for all i, and F_{p_i} does not satisfy (131), (132) or (133) for any i.

Each point p_i has permissible parameters (x, y_i, z_i) for (u, v), such that one of the following cases hold.

Case 1 p_i is a 2 point

$$u = (x^{a_i} y_i^{b_i})^{m_i}, v = P_i(x^{a_i} y_i^{b_i}) + x^c y_i^{d_i} F_i$$

with $a_i m_i = am$, and permissible parameters at p_{i+1} are as in one of the following cases.

Case 1a

$$y_i = y_{i+1}, z_i = y_{i+1}(z_{i+1} + \alpha_{i+1})$$

Case 1b

$$y_i = y_{i+1} z_{i+1}, z_i = z_{i+1}$$

Case 2 p_i is a 3 point

$$u = (x^{a_i} \omega_i^{k_i})^{m_i}, \omega_i = y_i^{\bar{b}_i} z_i^{\bar{c}_i}$$
$$v = P_i(x^{a_i} \omega_i^{k_i}) + x^c \omega_i^{d_i} F_i$$

with $a_i m_i = am$, $(\bar{b}_i, \bar{c}_i) = 1$, $(a_i, k_i) = 1$, and permissible parameters at p_{i+1} are as in one of the following cases.

Case 2a

$$y_i = y_{i+1}, z_i = y_{i+1} z_{i+1}$$

Case 2b

$$y_i = y_{i+1} z_{i+1}, z_i = z_{i+1}$$

Case 2c

$$y_i = y_{i+1}(z_{i+1} + \alpha_{i+1})^{-\frac{\bar{c}_i}{\bar{b}_i + \bar{c}_i}}, z_i = y_{i+1}(z_{i+1} + \alpha_{i+1})^{\frac{\bar{b}_i}{\bar{b}_i + \bar{c}_i}}$$

with $\alpha_{i+1} \neq 0$. In Case 2c, y_{i+1}, z_{i+1} are constructed from the monoidal transform

$$y_i = \bar{y}_{i+1}, z_i = \bar{y}_{i+1}(\bar{z}_{i+1} + \alpha_{i+1}).$$

Then define

$$\bar{y}_{i+1} = y_{i+1}(z_{i+1} + \alpha_{i+1})^{-\frac{\bar{c}_i}{\bar{b}_i + \bar{c}_i}}, \bar{z}_{i+1} = z_{i+1}.$$

If p_i is a 2 point, then y is a power of y_i, and if q_i is a 3 point, then y is a monomial in y_i and z_i. If p_i is a 2 point, then there is a series g_i such that

$$F_i = F_p - \frac{g_i(x^{a_i} y_i^{b_i})}{x^c y_i^{d_i}}.$$

If p_i is a 3 point, then there is a series g_i such that

$$F_i = F_p - \frac{g_i(x^{a_i} \omega_i^{k_i})}{x^c \omega_i^{d_i}}.$$

In either case, we have an expression

$$F_i = F_{p_i} = \tau' x^{r-1} + \sum_{j=1}^{r-1} a_j'(y_j, z_j) x^{r-j-1}.$$

We will show that τ' is a unit. Suppose not. First suppose that p_i is a 2 point. Then $a_i d_i - b_i(c + r - 1) = 0$.

$$(x^a y^b)^m = (x^{a_i} y^{b_i})^{m_i}$$

implies $y = y_i^{\frac{b_i m_i}{bm}}$. $x^{c+r-1} y^d = x^{c+r-1} y_i^{d_i}$ implies

$$d_i = \frac{b_i m_i d}{mb}.$$

Thus $ad - b(c+r-1) = 0$, a contradiction to the assumption that F_p is normalized.

Now suppose that p_i is a 3 point and τ' is not a unit. Then $a_i d_i - (c+r-1)k_i = 0$.

$$(x^a y^b)^m = (x^{a_i} \omega_i^{k_i})^{m_i}$$

implies $y = \omega_i^{\frac{k_i m_i}{mb}}$. $x^{c+r-1} y^d = x^{c+r-1} \omega_i^{d_i}$ implies

$$d_i = \frac{d k_i m_i}{mb}.$$

Thus $ad - (c+r-1)b = 0$, a contradiction to the assumption that F_p is normalized.

If p_i is a 2 point,

$$a'_j = \begin{cases} \tilde{a}_j - \tilde{c} y_i^{t_j(i)} & \text{if } a_i(d_i + t_j(i)) - b_i(c + r - j - 1) = 0 \text{ with } t_j(i) \in \mathbf{N} \\ & \text{and } \tilde{c} \text{ is the coefficient of } y_i^{t_j(i)} \text{ in the expansion of } \tilde{a}_j \\ & \text{in terms of } y_i, z_i \\ \tilde{a}_j & \text{if } a_i(d_i + t) - b_i(c + r - j - 1) \neq 0 \text{ for any } t \in \mathbf{N}. \end{cases}$$

If p_i is a 3 point,

$$a'_j = \begin{cases} \tilde{a}_j - \tilde{c} \omega_i^{t_j(i)} & \text{if } a_i(d_i + t_j(i)) - k_i(c + r - j - 1) = 0 \text{ with } t_j(i) \in \mathbf{N} \\ & \text{and } \tilde{c} \text{ is the coefficient of } \omega_i^{t_j(i)} \text{ in the expansion of } \tilde{a}_j \\ & \text{in terms of } y_i, z_i \\ \tilde{a}_j & \text{if } a_i(d_i + t) - k_i(c + r - j - 1) \neq 0 \text{ for any } t \in \mathbf{N}. \end{cases}$$

In particular, there exists at most one value of $t_j(i)$ such that a term can be removed from any $\tilde{a}_j$.

Set $\bar{u} = y^b$, $\bar{v}_j = y^d \tilde{a}_j(y, z)$. We have

$$\begin{aligned} u &= (x^a \bar{u})^m \\ v &= P(x^a \bar{u}) + \tau x^{c+r-1} y^d + \sum_{j=1}^{r-1} \bar{v}_j x^{c+r-1-j}. \end{aligned}$$

By Theorem 9.4, Lemma 9.2 and Theorem 9.8, there exists i_0 such that for $i \geq i_0$ in (135), one of the following forms holds at p_i, for $1 \leq j \leq r - 1$.

If p_i is a 2 point,

$$\begin{aligned} \bar{u} &= y_i^{b_i m_i} \\ \bar{v}_j &= P_{ji}(y_i) + y_i^{d_j(i)} \psi_{ji}(y_i, z_i)^{e_j(i)} \end{aligned} \tag{136}$$

where $e_j(i) \geq 1$, $\nu(\psi_{ji}(0, z_i)) = 1$ or $\psi_{ji} = 0$ for $1 \leq j \leq r - 1$.

If p_i is a 3 point,

$$\begin{aligned} \bar{u} &= (\omega_i^{k_i})^{m_i} = (y_i^{\bar{b}_i} z_i^{\bar{c}_i})^{k_i m_i} \\ \bar{v}_j &= P_{ji}(\omega_i) + y_i^{d_j(i)} z_i^{e_j(i)} \phi_{ji}(y_i, z_i) \end{aligned} \tag{137}$$

where $\phi_{ji}(y_i, z_i)$ is a unit and $d_j(i)\bar{c}_i - \bar{b}_i e_j(i) \neq 0$, or $\phi_{ji} = 0$ for $1 \leq j \leq r - 1$. For i sufficiently large, we have that $\bar{u}\bar{v}_j = 0$ are SNC divisors for $1 \leq j \leq r - 1$ (by Lemma 9.16).

Suppose that (136) holds at p_i with $e_j(i) = 1$ or $\phi_{ji} = 0$ for all j. If there exists $t_j(i) \in \mathbf{N}$ such that

$$a_i(d_i + t_j(i)) - b_i(c + r - j - 1) = 0,$$

then the normalized form of $x^{c+r-1-j}\bar{v}_j$ at p_i is

$$x^{c+r-1-j}\bar{v}_j - \tilde{c}x^{c+r-1-j}y_i^{t_j(i)+d_i} = x^{c+r-1-j}y_i^{\lambda_j(i)}\Lambda_{ji}(y_i, z_i)$$

where Λ_{ji} is a unit, zero, or $\nu(\Lambda_{ji}(0, z_i)) = 1$.

If there does not exist $t \in \mathbf{N}$ such that

$$a_i(d_i + t) - b_i(c + r - j - 1) = 0,$$

then

$$x^{c+r-1-j}\bar{v}_j = x^{c+r-1-j}y_i^{\lambda_j(i)}\Lambda_{ji}(y_i, z_i)$$

where Λ_{ji} is a unit, 0 or $\nu(\Lambda_{ji}(0, z_i)) = 1$.

Thus if (136) holds at p_i, with $e_j(i) = 1$ or $\phi_{ji} = 0$ for all j, (131) holds at p_i.

If p_i is a 3 point, so that (137) holds, and p_{i+1} is a 2 point, then (136) holds at p_{i+1}, with $e_j(i) = 1$ or $\psi_{ji} = 0$ for $1 \leq j \leq r - 1$, so that (131) holds at p_{i+1}.

We are reduced to the 2 cases where either for all $i \geq i_0$ in (135) all monoidal transforms are of the forms 2a or 2b, or for all $i \geq i_0$ in (135), all monoidal transforms are of the form 1a with some $e_j(i) > 1$.

If all monoidal transforms are of the form 2a or 2b for $i \geq i_0$, then $F_i = F_{i_0}$ for all $i \geq i_0$. If

$$F_{i_0} = \tau_0 x^{r-1} + \sum_{j=1}^{r-1} \tilde{a}_j(y_{i_0}, z_{i_0})x^{r-1-j}$$

then for $i >> i_0$, $\tilde{a}_j(y_{i_0}, z_{i_0})$ is a monomial in y_i and z_i times a unit for all j by Lemma 9.16. By Lemmas 9.17 and Corollary 9.18, there exists $i_1 > i_0$ such that (132) holds at p_{i_1}.

Suppose that all monoidal transforms are of the form 1a for $i \geq i_0$ and some $e_j(i) > 1$. There exists a permissible change of parameters $(x, y_{i_0}, \bar{z}_{i_0})$ such that

$$\bar{z}_{i_0} = z_{i_0} - \tilde{p}(y_{i_0})$$

for some series $\tilde{p}$, such that $(x, y_i, \bar{z}_i)$ are permissible parameters at p_i for all $i \geq i_0$ with $y_i = y_{i_0}$, $\bar{z}_{i_0} = y_i^{i-i_0}\bar{z}_i$. Let

$$F = \bar{P}(y_i) + \bar{F}_i$$

be the normalized form of F with respect to the parameters $(x, y_i, \bar{z}_i)$. Then $\bar{F}_i = \bar{F}_{i_0}$ for $i \geq i_0$. If

$$\bar{F}_{i_0} = \tau_0 x^{r-1} + \sum_{j=1}^{r-1} \tilde{a}_j(y_{i_0}, \bar{z}_{i_0})x^{r-1-j}$$

then for $i >> i_0$, $\tilde{a}_j(y_{i_0}, \bar{z}_{i_0})$ is a monomial in y_i and $\bar{z}_i$ times a unit for all j by Lemma 9.16. By Lemma 9.17 and Corollary 9.18 there exists $i_1 > i_0$ such that (133) holds at p_{i_1}.

C generic implies F_q is resolved for $q \in C$ a generic point. There exists an affine neighborhood U of p such that $Y \to \mathrm{spec}(R)$ extends to a permissible sequence of monoidal transforms of sections over C, $\bar{U} \to U$ such that $\bar{S}_r(\bar{U}) \cup \bar{B}_2(\bar{U})$ is contained in the union of $\bar{B}_2(\bar{U})$ and the strict transform of $\bar{S}_r(U)$. Thus $A_r(\bar{U})$ holds.

If $r \geq 3$, we can choose i_0 sufficiently large in obtaining the forms of (136) and (137) so that the strict transforms of the curves D_i in $\overline{S}_{r-1}(U)$ such that $x \in \hat{I}_{D_i,p}$ are disjoint and make SNCs with $\overline{B}_2(\overline{U})$.

$\square$

Theorem 12.2. *Suppose that $r \geq 3$, $A_r(X)$ holds, p is a 2 point with $\nu(p) = r - 1$, and $L(x, 0, 0) \neq 0$, as in the assumptions of Theorem 12.1. Let $R = \mathcal{O}_{X,p}$. Suppose that $\pi : Y_p \to Spec(R)$ is the morphism of Theorem 12.1.*
    Let

$$\cdots \to Y_n \to \cdots \to Y_1 \to Y_p$$

be a sequence of permissible monoidal transforms centered at 2 curves D such that D is r-1 big. Then there exists $n_0 < \infty$ such that

$$V_p = Y_{n_0} \overset{\pi_1}{\to} Y_p \to spec(R)$$

extends to a permissible sequence of monoidal transforms

$$\overline{U}_1 \to \overline{U} \to U$$

over an affine neighborhood U of p (with the notation of Theorem 12.1) such that $\overline{U}_1$ contains no 2 curves D such that D is r-1 big or r small, and for $q \in \overline{U}_1$,

1. *If q is a 1 or a 2 point then $\nu(q) \leq r$. $\nu(q) = r$ implies $\gamma(q) = r$.*
2. *If q is a 3 point then $\nu(q) \leq r - 2$.*
3. *$\overline{S}_r(\overline{U}_1)$ makes SNCs with $\overline{B}_2(\overline{U}_1)$.*

There exists a sequence of quadratic transforms $W_p \to V_p$ such that if $Z_p \to W_p$ is the sequence of monoidal transforms (in any order) centered at the strict transforms of curves C in $\overline{S}_r(X)$ then

$$Z_p \to W_p \to V_p \to Y_p \to spec(R)$$

extends to a permissible sequence of monoidal transforms

$$\overline{\pi} : \overline{U}_2 \to \overline{U}_1 \to \overline{U} \to U$$

over an affine neighborhood of p such that $\overline{U}_2$ contains no 2 curves D such that D is r-1 big or r small. $\overline{S}_r(\overline{U}_2)$ makes SNCs with $\overline{B}_2(\overline{U}_2)$, and if $q \in \overline{\pi}^{-1}(p)$,

1'.: *$\nu(q) \leq r$ if q is a 1 or 2 point. $\nu(q) = r$ implies $\gamma(q) = r$.*
2'.: *If q is a 2 point and $\nu(q) = r - 1$, then either*
    a.: *$\tau(q) > 0$ or*
    b.: *$\gamma(q) = r$ or*
    c.: *$\tau(q) = 0$ and (133) holds at q with $0 < d_i < i$, $e_i = i$ and $\overline{S}_{r-1}(Y_1)$ contains a single curve D containing q, and containing a 1 point, which has local equations $x = z = 0$ at q.*
3'.: *$\nu(q) \leq r - 2$ if q is a 3 point.*

Proof. Suppose that there exists a 2 curve $D \subset Y = Y_p$ such that D is r-1 big. Let $\pi_1 : Y_1 \to Y$ be the blow-up of D. Then $A_r(Y_1)$ holds by Lemmas 8.6 and 8.7, since $\nu(q) = r - 1$ for all $q \in D$.

Suppose that $q \in D$ and (131) holds at q. Then $d_j \geq j$ for all j. Suppose that $q' \in \pi_1^{-1}(q)$ and $\hat{\mathcal{O}}_{Y_1,q'}$ has regular parameters $(x_1, y_1, \bar{z})$ such that

$$\bar{x} = x_1, \bar{y} = x_1(y_1 + \alpha)$$

with $\alpha \neq 0$. Then q' is a 1 point, so that $\nu(q') \leq r$ and $\nu(q') = r$ implies $\gamma(q') = r$ by Lemma 8.6.

Suppose that $q' \in \pi_1^{-1}(q)$ and q' has permissible parameters $(x_1, y_1, \bar{z})$ such that

$$\bar{x} = x_1, \bar{y} = x_1 y_1$$

Then

$$
\begin{aligned}
u &= (x_1^{a+b} y_1^b)^m \\
v &= P(x_1^{a+b} y_1^b) + x_1^{c+d+r-1} y_1^d F_{q'} \\
F_{q'} &= \frac{F_q}{x_1^{r-1}} = \tau + \sum_{j=1}^{r-1} y_1^{d_j} \Lambda_j(x_1 y_1, \bar{z}) x_1^{d_j - j}
\end{aligned}
$$

so that $\nu(F_{q'}) = 0$.

Suppose that $q' \in \pi_1^{-1}(q)$ and q' has permissible parameters $(x_1, y_1, \bar{z})$ such that

$$\bar{x} = x_1 y_1, \bar{y} = y_1$$

Then

$$
\begin{aligned}
u &= (x_1^a y_1^{a+b})^m \\
v &= P(x_1^a y_1^{a+b}) + x_1^c y_1^{c+d+r-1} F_{q'} \\
F_{q'} &= \frac{F_q}{y_1^{r-1}} = \tau x_1^{r-1} + \sum_{j=1}^{r-1} y_1^{d_j - j} \Lambda_j(y_1, \bar{z}) x_1^{r-1-j}
\end{aligned}
$$

so that either $\nu(F_{q'}) < r - 1$, or we are back in the form (131) but the d_j have decreased by j.

Suppose that $q \in D$ and (132) holds at q. Without loss of generality, we may assume that D has local equations $\bar{x} = \bar{z} = 0$ at q. Then $e_j \geq j$ for all j. Suppose that $q' \in \pi_1^{-1}(q)$ and $\hat{\mathcal{O}}_{Y_1,q'}$ has regular parameters $(x_1, \bar{y}, z_1)$ such that

$$\bar{x} = x_1, \bar{z} = x_1(z_1 + \alpha)$$

with $\alpha \neq 0$. Set

$$x_1 = \bar{x}_1(z_1 + \alpha)^{-\frac{c}{a+c}}$$

Set $\lambda = f + (d + f + r - 1)(\frac{-c}{a+c})$,

$$G(x_1, \bar{y}, z_1) = \frac{(z_1 + \alpha)^\lambda F_q}{x_1^{r-1}} = (z_1 + \alpha)^\lambda \tau + \sum_{j=1}^{r-1} (z_1 + \alpha)^{\lambda + e_j} \bar{a}_j(\bar{y}, x_1(z_1 + \alpha)) \bar{y}^{d_j} x_1^{e_j - j}$$

Then

$$
\begin{aligned}
u &= (\bar{x}_1^{a'} \bar{y}^{b'})^{m'} \\
v &= P((\bar{x}_1^{a'} \bar{y}^{b'})^{\frac{m'}{m}}) + \bar{x}_1^{d+f+r-1} \bar{y}^e G
\end{aligned}
$$

$$G(0, 0, z_1) = (z_1 + \alpha)^\lambda \left[\tau_0 + \sum_{d_j = 0, e_j = j} \bar{a}_j(0,0)(z_1 + \alpha)^j \right]$$

where $\tau_0 = \tau(0,0,0)$, $(a + c)m = a'm'$, $bm = b'm'$, $(a', b') = 1$.

$$F_{q'}(0, 0, z_1) = \begin{cases} G(0, 0, z_1) & \text{if } a'e - b'(d + f + r - 1) \neq 0 \\ G(0, 0, z_1) - G(0, 0, 0) & \text{if } a'e - b'(d + f + r - 1) = 0 \end{cases}$$

Thus $\nu(F_{q'}(0,0,z_1)) \leq r$, except possibly if $\lambda = 0$ and $a'e - b'(d+f+r-1) = 0$. Then we have

$$af - c(d+r-1) = 0 \tag{138}$$

and

$$[ae - b(d+r-1)] + [ce - fb] = 0 \tag{139}$$

with $a, b, c > 0$. Substituting $d + r - 1 = \frac{af}{c}$ into (139), we get

$$(\frac{a}{c} + 1)(ce - bf) = 0$$

so that

$$ce - bf = 0 \tag{140}$$

and

$$ae - b(d+r-1) = 0 \tag{141}$$

(138), (140) and (141) cannot all hold since F_q is normalized. Thus

$$\nu(F_{q'}(0,0,z_1)) \leq r$$

and $\nu(q') \leq r$, $\gamma(q') \leq r$.

Suppose that $q' \in \pi_1^{-1}(q)$ and q' has permissible parameters $(x_1, \bar{y}, z_1)$ such that

$$\bar{x} = x_1, \bar{z} = x_1 z_1$$

Then $\nu(q') = 0$.

Suppose that $q' \in \pi_1^{-1}(q)$ and q' has permissible parameters $(x_1, y_1, \bar{z})$ such that

$$\bar{x} = x_1 z_1, \bar{z} = z_1$$

Then we either have a 3 point with $\nu(q') < r - 1$, or we are back in the form of (132) with e_i decreased by i.

Suppose that $q \in D$ and (133) holds at q. Then $d_j \geq j$ for all j. Suppose that $q' \in \pi_1^{-1}(q)$ and $\hat{\mathcal{O}}_{Y_1, q'}$ has regular parameters $(x_1, y_1, \bar{z})$ such that

$$\bar{x} = x_1, \bar{y} = x_1(y_1 + \alpha)$$

with $\alpha \neq 0$. Then q' is a 1 point so that $\nu(q') \leq r$ and $\gamma(q') \leq r$ by Lemma 8.6.

Suppose that $q' \in \pi_1^{-1}(q)$ and q' has permissible parameters $(x_1, y_1, \bar{z})$ such that

$$\bar{x} = x_1, \bar{y} = x_1 y_1$$

Then $\nu(q') = 0$.

Suppose that $q' \in \pi_1^{-1}(q)$ and q' has permissible parameters $(x_1, y_1, \bar{z})$ such that

$$\bar{x} = x_1 y_1, \bar{y} = y_1$$

Then we either have a 2 point with $\nu(q') < r - 1$, or we are back in the form of (133) with d_i decreased by i.

After a finite number of blow-ups of 2 curves $\pi_1 : Y_{n_0} \to Y$ we have that there are no 2 curves D on Y_{n_0} such that D is r-1 big. By Lemmas 8.6, 8.7, and since all 3 points $q \in (\pi \circ \pi_1)^{-1}(p)$ have $\nu(q) \leq r - 2$ (if q is a 3 point and $\nu(q) = r - 1$

so that q satisfies (132), then either $d_j \geq j$ for all j or $e_j \geq j$ for all j), there exists a neighborhood $\overline{U}_1$ with the properties asserted by the statement of the Theorem.

The only 2 points $q \in (\pi \circ \pi_1)^{-1}(p)$ where $\nu(q) = r - 1$ and $\gamma(q) > r$ either satisfy (131) with some $d_j = j - 1$ and $e_j = 1$ (so that $\tau(q) \geq 1$) or satisfy (133) with $\nu(q) = r - 1$ and $d_i < i$ so that $e_i \geq i$. Furthermore, $\overline{x} = 0$ is a local equation at q of the strict transform of the surface with local equation $x = 0$ at p.

Suppose that $D \subset \overline{S}_r(X)$ is a curve containing p, and $\overline{D}$ is the strict transform of D on $\overline{U}_1$. $\overline{D}$ can only intersect $(\pi \circ \pi_1)^{-1}(p)$ at points q such that q is a 2 point and $\nu(q) = r$ or $\nu(q) = r - 1$ by Lemma 7.7 and Lemma 7.6. We must either have $\gamma(q) \leq r$ or q satisfies (131) or (133).

Suppose that $q \in \overline{D} \cap (\pi \circ \pi_1)^{-1}(p)$ satisfies (131) or (133) and $\overline{y} \in \hat{I}_{\overline{D},q}$. Then there exist $a_j \in k$ such that

$$F_q - \sum a_j \frac{(\overline{x}^a \overline{y}^b)^j}{\overline{x}^c \overline{y}^d} \in (\overline{y})^{r-1} + (\overline{y}, f(\overline{x}, \overline{z}))^r$$

where $\overline{y} = f(\overline{x}, \overline{z}) = 0$ are local equations of $\overline{D}$ at q (by Lemma 6.26). This is impossible by the form of F_q. Thus $\overline{x} \in \hat{I}_{\overline{D},q}$.

Suppose that q satisfies (131). We have

$$\hat{I}_{\overline{D},q} = (\overline{x}, \overline{z} - \phi(\overline{y}))$$

for some series ϕ. There exist $a_j \in k$ such that, when renormalizing with respect to these new parameters,

$$F_q - \sum a_j \frac{(\overline{x}^a \overline{y}^b)^j}{\overline{x}^c \overline{y}^d} \in (\overline{x})^{r-1} + (\overline{x}, \overline{z} - \phi(\overline{y}))^r$$

by Lemma 6.26. Setting $\overline{x} = 0$ in

$$F_q - \sum a_j \frac{(\overline{x}^a \overline{y}^b)^j}{\overline{x}^c \overline{y}^d},$$

we get $\overline{y}^{d_r-1} \Lambda_{r-1}(\overline{y}, \overline{z})$ or $\overline{y}^{d_r-1} \Lambda_{r-1}(\overline{y}, \overline{z}) + \tilde{c}\overline{y}^{\overline{n}}$ for some $\tilde{c} \in k, \overline{n} \in \mathbf{N}$. Thus

$$(\overline{z} - \phi(\overline{y}))^r \mid \overline{y}^{d_r-1} \Lambda_{r-1}(\overline{y}, \overline{z})$$

or

$$(\overline{z} - \phi(\overline{y}))^r \mid \overline{y}^{d_r-1} \Lambda_{r-1}(\overline{y}, \overline{z}) + \tilde{c}\overline{y}^{\overline{n}}$$

which is nonzero since $\overline{x} \nmid F_q$. As $\nu(\Lambda_{r-1}(\overline{y}, \overline{z})) \leq 1$, this is a contradiction. Thus q cannot have the form of (131).

Suppose that q satisfies (133). We have $\hat{I}_{\overline{D},q} = (\overline{x}, \overline{z} - \phi(\overline{y}))$ for some series ϕ. There exists $a_j \in k$ such that

$$F_q - \sum a_j \frac{(\overline{x}^a \overline{y}^b)^j}{\overline{x}^c \overline{y}^d} \in (\overline{x})^{r-1} + (\overline{x}, \overline{z} - \phi(\overline{y}))^r$$

by Lemma 6.26. Setting $\overline{x} = 0$ in

$$F_q - \sum a_j \frac{(\overline{x}^a \overline{y}^b)^j}{\overline{x}^c \overline{y}^d},$$

we get $\bar{a}_{r-1}(\bar{y},\bar{z})\bar{y}^{d_{r-1}}\bar{z}^{e_{r-1}}$ or $\bar{a}_{r-1}(\bar{y},\bar{z})\bar{y}^{d_{r-1}}\bar{z}^{e_{r-1}} + \tilde{c}\bar{y}^{\bar{n}}$ for some $\tilde{c} \in k$, $\bar{n} \in \mathbf{N}$. Thus

$$(\bar{z} - \phi(\bar{y}))^r \mid \bar{y}^{d_{r-1}}\bar{z}^{e_{r-1}}$$

or

$$(\bar{z} - \phi(\bar{y}))^r \mid \bar{a}_{r-1}(\bar{y},\bar{z})\bar{y}^{d_{r-1}}\bar{z}^{e_{r-1}} + \tilde{c}\bar{y}^{\bar{n}}$$

which is nonzero since $\bar{x} \nmid F_q$. In either case, we have

$$(\bar{z} - \phi(\bar{y}))^{r-1} \mid \bar{y}^{d_{r-1}}\bar{z}^{e_{r-1}-1}$$

since

$$\frac{\partial}{\partial \bar{z}}(\bar{a}_{r-1}\bar{y}^{d_{r-1}}\bar{z}^{e_{r-1}} + \tilde{c}\bar{y}^{\bar{n}}) = \bar{z}^{e_{r-1}-1}\bar{y}^{d_{r-1}}\left(e_{r-1}\bar{a}_{r-1} + \frac{\partial \bar{a}_{r-1}}{\partial \bar{z}}\bar{z}\right),$$

which implies that $e_{r-1} \geq r$ and $\phi(\bar{y}) = 0$. Thus $\bar{x} = \bar{z} = 0$ are local equations of $\overline{D}$ at q.

Suppose that $\overline{D}$ is such that $\overline{D}$ is r small and $q \in (\pi \circ \pi_1)^{-1}(p) \cap \overline{D}$ satisfies $\nu(q) = r$ and $\gamma(q) = r$. By Lemma 8.10, there exists a sequence of quadratic transforms $\sigma_1 : W_1 \to Y_{n_0}$ such that the strict transform $\tilde{D}$ of $\overline{D}$ intersects $\sigma_1^{-1}(q)$ in a 2 point q' such that $\nu(q') = r-1$ and $\gamma(q') = r$. Furthermore, there are no 2 curves C in $\sigma_1^{-1}(q)$ such that C is r-1 big, and 1'. - 3'. of the conclusions of the Theorem hold at all points of $\sigma_1^{-1}(q)$. Thus there exists a sequence of quadratic transforms $\sigma : W \to Y_{n_0}$, centered at 2 points $\{q_1, \ldots, q_m\}$ such that $\nu(q_i) = r$ and $\gamma(q_i) = r$ on the strict transform $\overline{D}$ of curves D in $\overline{S}_r(X)$ containing p such that if $\overline{D} \subset \overline{S}_r(W)$ is the strict transform of a curve $D \subset \overline{S}_r(X)$ containing p, and $\overline{D}$ is r small then $\overline{D}$ intersects $(\pi \circ \pi_1 \circ \sigma)^{-1}(p)$ in 2 points q of the form of (133), and in 2 points q such that $\nu(q) = r-1$ and $\gamma(q) = r$. W contains no 2 curves C such that C is r-1 big, $\overline{S}_r(W)$ makes SNCs with $\overline{B}_2(W)$ and $\gamma(q) \leq r$ for all exceptional 1 and 2 points of σ, $\nu(q) = 0$ for all exceptional 3 points of σ.

Suppose that $\overline{D} \subset W$ is the strict transform of a curve D in $\overline{S}_r(X)$ containing p. First suppose that $\overline{D}$ is r big. If $q \in \overline{D} \cap (\pi \circ \pi_1 \circ \sigma)^{-1}(p)$, then q must be a 2 point with $\nu(q) = \gamma(q) = r$. Suppose that $\lambda_1 : Z_1 \to W$ is the blow-up of $\overline{D}$. By Lemma 8.8, 1'. - 3'. of the conclusions of the Theorem hold on Z_1, and the conclusions of $\overline{U}_2$ hold in a neighborhood of $\lambda_1^{-1}(q)$.

If $\overline{D}$ is r small, then if $q \in \overline{D} \cap (\pi \circ \pi_1 \circ \sigma)^{-1}(p)$, q must be either a 2 point where $\nu(q) = r-1$ and $\gamma(q) = r$ or q satisfies (133) and $\bar{x} = \bar{z} = 0$ are local equations of $\overline{D}$ at q.

Let $\lambda_1 : Z_1 \to W$ be the blow-up of $\overline{D}$. If $q \in \overline{D} \cap (\pi \circ \pi_1 \circ \sigma)^{-1}(p)$ is a 2 point such that $\nu(p) = r-1$ and $\gamma(p) = r$, then 1.' - 3.' of the conclusions of the Theorem hold and the conclusions of $\overline{U}_2$ hold in a neighborhood of $\lambda_1^{-1}(q)$ (since $r \geq 3$) by Lemma 8.10.

Suppose that $q \in \overline{D} \cap (\pi \circ \pi_1 \circ \sigma)^{-1}(p)$ satisfies (133). $\bar{x} = \bar{z} = 0$ are local equations of $\overline{D}$ at q and $d_i < i$. Since $\overline{D} \subset \overline{S}_r(W)$, we have $e_i > i$.

Since $A_r(X)$ holds, $\gamma(q') = r$ if $q' \in \overline{D}$ is a 1 point. Then $e_{r-1} = r$ in (133). Suppose that $q' \in \lambda_1^{-1}(q)$, q' has permissible parameters $(x_1, \bar{y}, z_1)$ such that

$$\bar{x} = x_1, \bar{z} = x_1(z_1 + \alpha)$$

q' is a 2 point. $e_i > i$ implies

$$\frac{F_p}{x_1^{r-1}} = \tau + x_1 \Omega$$

so that $\nu(q) = 0$.

Suppose that $q' \in \lambda_1^{-1}(q)$ has permissible parameters $(x_1, \bar{y}, z_1)$ such that

$$\bar{x} = x_1 z_1, \bar{z} = z_1$$

Then we either have a 3 point with $\nu(q') < r - 1$, or we are in the form of (132) with e_i decreased by i, $d_i < i$, and $\bar{x} = \bar{z} = 0$ is a local equation of a 2 curve which is a section over D. Since we have $e_{r-1} = 1$ in (132), we must have $e_i < i$ (since $r \geq 2$), $d_i < i$ so that $\nu(q') < r - 1$ by Remark 9.19.

Thus the conclusions of $\bar{U}_2$ hold in a neighborhood of $\lambda_1^{-1}(q)$ by Lemma 8.10.

Then if $\bar{\lambda} : Z \rightarrow W$ is the sequence of monoidal transforms (in any order) centered at the strict transforms of curves C in $\bar{S}_r(X)$, the conclusions of 1'. - 3'. of the Theorem hold, except possibly at a finite number of points q of the form of (133) with $d_i < i$ and $e_i \geq i$. There are no 2 curves $C \subset Z$ such that C is r-1 big.

If $q \in (\pi \circ \pi_1 \circ \sigma \circ \bar{\lambda})^{-1}(p)$ does not satisfy one of 1' - 3' of the conclusions of the Theorem then q satisfies (133), $\tau(q) = 0$ and $d_i < 0$ so that $e_i \geq i$. $\bar{x} = 0$ is then a local equation of the surface with local equation $x = 0$ at p.

$e_i \geq i$ implies $F_q \in (\bar{x}, \bar{z})^{r-1}$. Since $r \geq 3$, this implies (by Lemma 6.22) that there exists an algebraic curve $\bar{D} \subset \bar{S}_{r-1}(Z)$ such that $\bar{x} = \bar{z} = 0$ are local equations of a formal branch of $\bar{D}$. $\bar{D}$ is necessarily the strict transform of a curve $D \subset \bar{S}_{r-1}(U)$, since $\bar{x} \in \hat{I}_{\bar{D},q}$ and $\bar{y} \notin \hat{I}_{\bar{D},q}$. $\bar{D}$ is thus nonsingular at q, by the conclusions of Theorem 12.1. Thus $\bar{x} = \bar{z} = 0$ are local equations of an algebraic curve $\bar{D}$ at q. If $e_i > i$, then $\bar{D} \subset \bar{S}_r(Z)$ and if $e_i = i$, then $\bar{D} \subset \bar{S}_{r-1}(Z)$.

Since the strict transforms of all curves in $\bar{S}_r(X)$ have been blown up in the map $Z \rightarrow W$, we must have $e_i = i$.

Thus at q, (133) holds, $e_i = i$, $d_i < i$ and $\tau(q) = 0$. Let T be the component of E_X with local equation $x = 0$ at p. $\tau(q) = 0$ implies $d_i > 0$, which implies $d_j > 0$ for all j, so that $F_q = \tau \bar{x}^{r-1} + \bar{y}\Omega$. Since $\bar{x} = 0$ is a local equation of the strict transform T' of T, the only curve in $\bar{S}_{r-1}(Z) \cap T'$ containing q is the curve $\bar{D}$ with local equations $\bar{x} = \bar{z} = 0$, since a curve in $\bar{S}_{r-1}(Z) \cap T'$ containing q must be the strict transform of a curve in $\bar{S}_{r-1}(U) \cap T$.

$$u = (\bar{x}^a \bar{y}^b)^m$$
$$v = P(\bar{x}^a \bar{y}^b) + \bar{x}^c \bar{y}^d F$$

and $F_q = \tau \bar{x}^{r-1} + \bar{y}\Omega$, where τ is a unit.

We will show that there does not exist a curve $C \subset \bar{S}_{r-1}(Z)$ containing q (and a 1 point) such that $\bar{y} \in \hat{I}_{C,q}$.

After a permissible change of parameters, we may assume that $\bar{y}, \bar{z} \in \mathcal{O}_{X,q}$ with

$$F_q = \tau \bar{x}^{r-1} + \bar{y}\Omega.$$

$\mathcal{I}_{C,q} = (\overline{y}, g(\overline{x}, \overline{z}))$. Set $\overline{x} = \tilde{x}^b$. By Lemma 6.29, either there exists a series f such that

$$\tilde{x}^{bc-ad}F_q - f(\tilde{x}^a\overline{y}) \in \left((\overline{y}, g(\tilde{x}^b, \overline{z}))^{r-1} + (\overline{y})^{r-2}\right)k[[\tilde{x}, \overline{y}, \overline{z}]] \qquad (142)$$

if $bc - ad \geq 0$ or

$$F_q - f(\tilde{x}^a\overline{y})\tilde{x}^{ad-bc} \in \left((\overline{y}, g(\tilde{x}^b, \overline{z}))^{r-1} + (\overline{y})^{r-2}\right)k[[\tilde{x}, \overline{y}, \overline{z}]] \qquad (143)$$

if $ad - bc > 0$.

If (142) holds, since $\nu(q) > 0$, we have $\nu(f) > 0$, which implies

$$g(\tilde{x}^b, \overline{z})^{r-1} \mid \tilde{x}^{b(r-1)+bc-ad},$$

and $g = \overline{x}$, a contradiction since C is then a 2 curve.

Suppose that (143) holds. Let $\overline{c} = f(0)$.

$$\tau(\overline{x}, 0, \overline{z})\tilde{x}^{b(r-1)} - \overline{c}\tilde{x}^{ad-bc} = h(\tilde{x}, \overline{z})g(\tilde{x}^b, \overline{z})^{r-1}$$

for some series h. If $ad - bc = b(r-1)$ then $ad - b(c+r-1) = 0$, a contradiction to the assumption that F_q is normalized.

Let

$$\overline{d} = \begin{cases} \min\{b(r-1), ad-bc\} & \text{if } \overline{c} \neq 0 \\ b(r-1) & \text{if } \overline{c} = 0. \end{cases}$$

$$\tau(\overline{x}, 0, \overline{z})\tilde{x}^{b(r-1)} - \overline{c}\tilde{x}^{ad-bc} = \Lambda\tilde{x}^{\overline{d}}$$

where Λ is a unit. Then $g(\tilde{x}^b, \overline{z})^{r-1}$ is a power of $\tilde{x}$, and $g = \tilde{x}$, a contradiction since C is then a 2 curve.

Thus the conclusions of 1'. - 3'. of the Theorem hold and the conclusions of $\overline{U}_2$ hold on Z. $\qquad \square$

Theorem 12.3. *Suppose that $r = 2$, $A_2(X)$ holds, p is a 2 point with $\nu(p) = r - 1 = 1$, and $L(x,0,0) \neq 0$, as in the assumptions of Theorem 12.1. Let $R = \mathcal{O}_{X,p}$. Suppose that $\pi : Y_p \to Spec(R)$ is the morphism of Theorem 12.1.*

Let

$$\cdots \to Y_n \to \cdots \to Y_1 \to Y_p$$

be a sequence of permissible monoidal transforms centered at 2 curves D such that D is 1 big. Then there exists $n_0 < \infty$ such that

$$V_p = Y_{n_0} \xrightarrow{\pi_1} Y_p \to spec(R)$$

extends to a permissible sequence of monoidal transforms

$$\overline{U}_1 \to \overline{U} \to U$$

over an affine neighborhood U of p (with the notation of Theorem 12.1) such that $\overline{U}_1$ contains no 2 curves D such that D is 1 big or 2 small, and for $q \in \overline{U}_1$,

1. *If q is a 1 or a 2 point then $\nu(q) \leq 2$. $\nu(q) = 2$ implies $\gamma(q) = 2$.*
2. *If q is a 3 point then $\nu(q) = 0$.*
3. *$\overline{S}_2(\overline{U}_1)$ makes SNCs with $\overline{B}_2(\overline{U}_1)$.*

There exists a sequence of quadratic transforms $W_p \to V_p$ such that if $Z_p \to W_p$ is the sequence of monoidal transforms (in any order) centered at the strict transforms C' of curves C in $\overline{S}_2(X)$, followed by monoidial transforms centered at any 2 curves $\overline{C}$ which are sections over C' such that $\overline{C}$ is 1 big, then

$$Z_p \to W_p \to V_p \to Y_p \to spec(R)$$

extends to a permissible sequence of monoidal transforms

$$\overline{\pi} : \overline{U}_2 \to \overline{U}_1 \to \overline{U} \to U$$

over an affine neighborhood of p such that $\overline{U}_2$ contains no 2 curves D such that D is 1 big or 2 small. $\overline{S}_2(\overline{U}_2)$ makes SNCs with $\overline{B}_2(\overline{U}_2)$, and if $q \in \overline{\pi}^{-1}(p)$,

1'.: $\nu(q) \leq 2$ *if q is a 1 or 2 point. $\nu(q) = 2$ implies $\gamma(q) = 2$.*
2'.: *If q is a 2 point and $\nu(q) = 1$, then either q is resolved or $\gamma(q) = 2$.*
3'.: $\nu(q) = 0$ *if q is a 3 point.*

Proof. We can construct

$$W = W_p \overset{\sigma}{\to} V_p = Y_{n_0} \overset{\pi_1}{\to} Y = Y_p \overset{\pi}{\to} spec(R)$$

exactly as in the proof of Theorem 12.2.
   If $q \in (\pi \circ \pi_1 \circ \sigma)^{-1}(p)$ satisfies 133), then

$$\begin{aligned} u &= (\overline{x}^a \overline{y}^b)^m \\ F_q &= \overline{x} + \overline{z}^{e_1}. \end{aligned} \tag{144}$$

   If $D \subset \overline{S}_2(X)$ contains p, and $\overline{D}$ is the strict transform of D on W, and $q \in \overline{D} \cap (\pi \circ \pi_1 \circ \sigma)^{-1}(p)$, then either q satisfies (144),

$$F_q = \overline{x} + \overline{z}^{e_1}$$

with $e_1 \geq 2$ and $\overline{x} = \overline{z} = 0$ are local equations of $\overline{D}$ at q, or q is a 2 point with $\nu(q) = 1$, $\gamma(q) = 2$, or $\overline{D}$ is 2 big and q is a 2 point with $\nu(q) = \gamma(q) = 2$.
   Suppose that $\overline{D} \subset \overline{S}_2(W)$ is the strict transform of $D \subset \overline{S}_2(X)$ such that $p \in D$. Let $\lambda_1 : Z_1 \to W$ be the blow-up of $\overline{D}$.
   First suppose that $\overline{D}$ is 2 big. Suppose that $q \in \overline{D} \cap (\pi \circ \pi_1 \circ \sigma)^{-1}(p)$. Then q is a 2 point with $\nu(q) = \gamma(q) = 2$. By Lemma 8.8, 1' - 3' of the conclusions of the Theorem hold on Z_1, and the conclusions of $\overline{U}_2$ hold in a neighborhood of $\lambda_1^{-1}(q)$.
   Suppose that $\overline{D}$ is 2 small. If $q \in \overline{D} \cap (\pi \circ \pi_1 \circ \sigma)^{-1}(p)$, then either q is a 2 point with $\nu(q) = 1$, $\gamma(q) = 2$, or q satisfies (144) with $e_1 \geq 2$, $\overline{x} = \overline{z} = 0$ are local equations of $\overline{D}$ at q.
   If $q \in \overline{D} \cap (\pi \circ \pi_1 \circ \sigma)^{-1}(p)$ is a 2 point with $\nu(q) = 1$, $\gamma(q) = 2$, then by Lemma 8.10 and Lemma 8.11, either 1' - 3' of the conclusions of the Theorem and the conclusions of $\overline{U}_2$ hold in a neighborhood of $\lambda_1^{-1}(q)$ or there exists a 2 curve $\overline{C}$ which is a section over $\overline{D}$, such that if $\lambda_2 : Z_2 \to Z_1$ is the blow-up of $\overline{C}$, then 1' - 3' of the conclusions of the Theorem hold and the conclusions of $\overline{U}_2$ hold in a neighborhood of $(\lambda_1 \circ \lambda_2)^{-1}(q)$.
   Suppose that $q \in \overline{D} \cap (\pi \circ \pi_1 \circ \sigma)^{-1}(p)$ satisfies (144) with $e_1 \geq 2$, $\overline{x} = \overline{z} = 0$ are local equations of $\overline{D}$ at q. Since $A_2(X)$ holds, $\gamma(q') = 2$ if $q' \in \overline{D}$ is a 1 point, so that $e_1 = 2$. If $q' \in \lambda_1^{-1}(q)$ we have $\nu(q') = 0$ except if q' is the 3

point with permissible parameters $(x_1, \overline{y}, z_1)$ such that $\overline{x} = x_1 z_1$, $\overline{z} = z_1$. Then $F_{q'} = x_1 + z_1$.

Let $\overline{C}$ be the 2 curve through q' with local equations $x_1 = z_1 = 0$ at q'. $\overline{C}$ is a section over $\overline{D}$. By Lemma 8.1, $F_a \in \hat{\mathcal{I}}_{\overline{C},a}$ for all $a \in \overline{C}$, so that $\overline{C}$ is 1 big.

Let $\lambda_2 : Z_2 \to Z_1$ be the blow-up of $\overline{C}$. Then 1' - 3' of the conclusions of the Theorem, and the conclusions of $\overline{U}_2$ hold in a neighborhood of $(\lambda_1 \circ \lambda_2)^{-1}(q)$.

Then if $\overline{\lambda} : Z \to W$ is the sequence of monoidal transforms (in any order) centered at the strict transform of curves C in $\overline{S}_2(X)$, followed by the monoidal transforms centered at 2 curves $\overline{C}$ which are sections over C' such that $\overline{C}$ is 1 big, we have that there are no 2 curves $C \subset Z$ such that $\overline{C}$ is 1 big.

The only points of Z which may not satisfy the conclusions of 1' - 3' of the Theorem are the 2 points $q \in (\pi \circ \pi_1 \circ \sigma \circ \overline{\lambda})^{-1}(p)$ which satisfy (144) with $\tau(q) = 0$. Then (after a permissible change of parameters)

$$
\begin{aligned}
u &= (\overline{x}^a \overline{y}^b)^m \\
F_q &= \overline{x} + \overline{z}^{e_r}
\end{aligned}
$$

with $e_r \geq 2$.

By Lemma 6.22, there exists an algebraic curve $\overline{D} \subset \overline{S}_2(Z)$ such that $\overline{x} = \overline{z} = 0$ are local equations of D at q. Since $\overline{x} = 0$ is a local equation of the strict transform of the component of E_X with local equation $x = 0$, and $\overline{D}$ is not contained in the component of E_X with local equation $\overline{y} = 0$, $\overline{D}$ is the strict transform of a curve in $\overline{S}_r(X)$ containing p, a contradiction to the construction of Z. Thus q is a resolved point.

$\square$

Theorem 12.4. *Suppose that $r \geq 2$, $A_r(X)$ holds, $p \in X$ is a 3 point with $\nu(p) = r - 1$, and we have permissible parameters (x, y, z) at p for u, v (with $y, z \in \mathcal{O}_{X,p}$) such that*

$$
\begin{aligned}
u &= (x^b y^{a+nb} z^{a+(n+1)b})^m \\
v &= P(x^b y^{a+nb} z^{a+(n+1)b}) + x^d y^{c+n(d+r)} z^{c+(n+1)(d+r)} F_p \\
F_p &= \tau x^{r-1} + \sum_{i=1}^{r-1} a_i(y, z) x^{r-i-1}
\end{aligned}
\tag{145}
$$

with $n \geq 0$, $a, b > 0$, $L(x, 0, 0) \neq 0$, so that τ is a unit, and (eq510)

$$
a(d + r - 1) - bc = 0
\tag{146}
$$

or

$$
\begin{aligned}
u &= (x^b y^a z^b)^m \\
v &= P(x^b y^a z^b) + x^d y^c z^{d+r} F_p \\
F_p &= \tau x^{r-1} + \sum_{i=1}^{r-1} a_i(y, z) x^{r-i-1}
\end{aligned}
\tag{147}
$$

with $n \geq 0$, $a, b > 0$, $L(x, 0, 0) \neq 0$, so that τ is a unit, and

$$
a(d + r - 1) - bc = 0
\tag{148}
$$

Let $R = \mathcal{O}_{X,p}$. Let C be the 2 curve with local equations $y = z = 0$ at p. Then there exists a finite sequence of permissible monoidal transforms $\pi : Y \to \text{Spec}(R)$ centered at sections over $C = V(y, z)$, such that for $q \in \pi^{-1}(p)$, F_q has one of the forms (131), (132) or (133) of Theorem 12.1.

In all these cases $\bar{x} = x$ and $\bar{x} = 0$ is a local equation at q of the strict transform of the component of E_X with local equation $x = 0$ at p.

There exists an affine neighborhood U of p such that $Y \to \operatorname{spec}(R)$ extends to a sequence of permissible monoidal transforms $\bar{U} \to U$ such that $A_r(\bar{U})$ holds.

$\gamma(q) \leq 1$ at a generic point q of C, so that all points q' on the fiber of the blow-up of C over q are resolved.

Suppose that $r \geq 3$. Let D_i be the curves in $\overline{S}_{r-1}(X)$ which contain p, and $x \in \hat{I}_{D_i,p}$. We further have that the strict transforms $\overline{D}_i$ of the D_i on $\bar{U}$ are nonsingular, disjoint, and make SNCs with $\overline{B}_2(\bar{U})$.

Proof. We modify the proof of Theorem 12.1 to prove this Theorem. In the sequence of (134) we must add a new case,

Case 0:

$$u = (x^{a_i}(y_i^{\bar{b}_i} z_i^{\bar{c}_i})^{k_i})^{m_i}$$
$$v = P_i(x^{a_i}(y_i^{\bar{b}_i} z_i^{\bar{c}_i})^{k_i}) + x^c y_i^{\bar{d}_i} z_i^{\bar{e}_i} F_i$$

Case 0a:

$$y_i = y_{i+1}, z_i = y_{i+1} z_{i+1}$$

Case 0b:

$$y_i = y_{i+1} z_{i+1}, z_i = z_{i+1}$$

Case 0c:

$$y_i = y_{i+1}(z_{i+1} + \alpha_{i+1})^{-\frac{\bar{c}_i}{\bar{b}_i + \bar{c}_i}},$$

$$z_i = y_{i+1}(z_{i+1} + \alpha_{i+1})^{\frac{\bar{b}_i}{\bar{b}_i + \bar{c}_i}}$$

In the sequence (134), Y_0 has the form Case 0 (and not Case 2). The transformations of type 0a and type 0b produce a p_{i+1} of the type of Case 0, and $F_i = F_{i+1}$.

If all transformations in (134) are of types 0a or 0b, then we eventually get a p_i of type (132) by Lemmas 9.16, 9.17 and Corollary 9.18.

Otherwise, we eventually reach a first p_i where p_{i+1} is obtained by a transformation of type 0c. We have $F_i = F_p$.

$$u = (x^{a_i} y_{i+1}^{(\bar{b}_i + \bar{c}_i)k_i})^{m_i} = (x^{a_{i+1}} y_{i+1}^{b_{i+1}})^{m_{i+1}}$$
$$v = P_i(x^{a_i} y_{i+1}^{(\bar{b}_i + \bar{c}_i)k_i}) + x^c y_{i+1}^{d_{i+1}}(z_{i+1} + \alpha_{i+1})^{\lambda} F_p$$

$$\lambda = \frac{\bar{e}_i \bar{b}_i - \bar{d}_i \bar{c}_i}{\bar{b}_i + \bar{c}_i},$$

$d_{i+1} = \bar{d}_i + \bar{e}_i$. Thus p_{i+1} has the form of Case 1, with

$$F_{i+1} = (z_{i+1} + \alpha_{i+1})^{\lambda} F_p - \frac{g_{i+1}(x^{a_{i+1}} y_{i+1}^{b_{i+1}})}{x^c y_{i+1}^{d_{i+1}}}$$

Thus p_{i+1} satisfies the assumptions of Theorem 12.1, provided $\nu(F_{i+1}(x,0,0)) = r - 1$.

p_i has permissible parameters (x, y_1, z_1) with $y_1 = y_i, z_1 = z_i$ such that

$$y = y_1^{\alpha} z_1^{\beta}, z = y_1^{\gamma} z_1^{\delta}$$

with $\alpha\delta - \beta\gamma = \pm 1$,

p_{i+1} has permissible parameters (x, y_2, z_2) with $y_2 = y_{i+1}$, $z_2 = z_{i+1}$ such that

$$y_1 = y_2, z_1 = y_2(z_2 + \overline{\alpha})$$

with $\overline{\alpha} \neq 0$.

First suppose that we are in the situation of (145) and (146). Set

$$\lambda_1 = \alpha(a + nb) + \gamma(a + (n+1)b) + \beta(a + nb) + \delta(a + (n+1)b),$$

$$\lambda_2 = \beta(\alpha + nb) + \delta(a + (n+1)b).$$

$$y_2 = \overline{y}_2(z_2 + \overline{\alpha})^{-\frac{\lambda_2}{\lambda_1}}$$

$$\begin{aligned}
u &= (x^b y_1^{\alpha(a+nb)+\gamma(a+(n+1)b)} z_1^{\beta(a+nb)+\delta(a+(n+1)b)})^m \\
&= (x^b y_2^{\lambda_1}(z_2 + \overline{\alpha})^{\lambda_2})^m \\
&= (x^b \overline{y}_2^{\lambda_1})^m
\end{aligned}$$

$$\begin{aligned}
x^{d+r-1} y^{c+n(d+r)} & z^{c+(n+1)(d+r)} \\
&= x^{d+r-1} y_1^{\alpha(c+n(d+r))+\gamma(c+(n+1)(d+r))} z_1^{\beta(c+n(d+r))+\delta(c+(n+1)(d+r))} \\
&= x^{d+r-1} y_2^{\lambda_3}(z_2 + \overline{\alpha})^{\lambda_4} \\
&= x^{d+r-1} \overline{y}_2^{\lambda_3}(z_2 + \overline{\alpha})^{\lambda_4 - \frac{\lambda_2\lambda_3}{\lambda_1}}
\end{aligned}$$

where

$$\lambda_3 = \alpha(c+n(d+r)) + \gamma(c+(n+1)(d+r)) + \beta(c+n(d+r)) + \delta(c+(n+1)(d+r)),$$

$$\lambda_4 = \beta(c + n(d + r)) + \delta(c + (n+1)(d + r)).$$

$$\begin{aligned}
b\lambda_3 &- (d+r-1)\lambda_1 \\
&= b[(\alpha + \beta)(c + n(d+r)) + (\gamma + \delta)(c + (n+1)(d+r))] \\
&\quad -(d+r-1)[(\alpha + \beta)(a + nb) + (\gamma + \delta)(a + (n+1)b)] \\
&= (\alpha + \beta + \gamma + \delta)[bc - (d+r-1)a] + nb(\alpha + \beta + \gamma + \delta) + b(\gamma + \delta) \\
&= n(\alpha + \beta + \gamma + \delta)b + (\gamma + \delta)b > 0.
\end{aligned}$$

Since we cannot remove

$$\tau_0 \overline{\alpha}^{\lambda_4 - \frac{\lambda_2\lambda_3}{\lambda_1}} x^{d+r-1} \overline{y}_2^{\lambda_3}$$

from $x^d y^{c+n(d+r)} z^{c+(n+1)(d+r)} F_p$, where $\tau_0 = \tau(0,0,0)$, when normalizing to obtain $F_{p_{i+1}}$, we must have $\nu(F_{p_{i+1}}(x,0,0)) = r - 1$.

Now suppose that we are in the situation of (147) and (148). Set

$$\lambda_1 = \alpha a + \gamma b + \beta a + \delta b,$$

$$\lambda_2 = \beta a + \delta b.$$

$$y_2 = \overline{y}_2(z_2 + \overline{\alpha})^{-\frac{\lambda_2}{\lambda_1}}$$

$$\begin{aligned}
u &= (x^b y_1^{\alpha a+\gamma b} z_1^{\beta a+\delta b})^m \\
&= (x^b y_2^{\lambda_1}(z_2 + \overline{\alpha})^{\lambda_2})^m \\
&= (x^b \overline{y}_2^{\lambda_1})^m
\end{aligned}$$

Set

$$\lambda_3 = \alpha c + \gamma(d + r) + \beta c + \delta(d + r),$$

$$\lambda_4 = \beta c + \delta(d + r).$$

$$x^{d+r-1}y^c z^{d+r} = x^{d+r-1}y_1^{ac+\gamma(d+r)}z_1^{\beta c+\delta(d+r)}$$
$$= x^{d+r-1}y_2^{\lambda_3}(z_2+\overline{\alpha})^{\lambda_4}$$
$$= x^{d+r-1}\overline{y}_2^{\lambda_3}(z_2+\overline{\alpha})^{\lambda_4-\frac{\lambda_2\lambda_3}{\lambda_1}}$$

$b\lambda_3 - \lambda_1(d+r-1)$
$= b(ac+\gamma(d+r)+\beta c+\delta(d+r)) - (\alpha a+\gamma b+\beta a+\delta b)(d+r-1)$
$= (\alpha+\beta)[bc-a(d+r-1)]+(\gamma+\delta)[b(d+r)-b(d+r-1)]$
$= (\gamma+\delta)b \neq 0$

Thus $\nu(F_{p_{i+1}}(x,0,0)) = r-1$ in this case also. The proof now preceeds as in Theorem 12.1. $\square$

Theorem 12.5. *Suppose that* $r \geq 3$, $A_r(X)$ *holds,* p *is a 3 point with* $\nu(p) = r-1$, *and we have permissible parameters* (x,y,z) *at* p *for* u,v *(with* $y,z \in \mathcal{O}_{X,p}$*) such that*

$$u = (x^b y^{a+nb}z^{a+(n+1)b})^m$$
$$v = P(x^b y^{a+nb}z^{a+(n+1)b}) + x^d y^{c+n(d+r)}z^{c+(n+1)(d+r)}F_p$$
$$F_p = \tau x^{r-1} + \sum_{i=1}^{r-1} a_i(y,z)x^{r-i-1}$$

with $n \geq 0$, $a,b > 0$, $\nu(p) = r-1$, $L(x,0,0) \neq 0$, *so that* τ *is a unit, and*

$$a(d+r-1) - bc = 0$$

or

$$u = (x^b y^a z^b)^m$$
$$v = P(x^b y^a z^b) + x^d y^c z^{d+r}F_q$$
$$F_p = \tau x^{r-1} + \sum a_i(y,z)x^{r-i-1}$$

with $n \geq 0$, $a,b > 0$, $\nu(p) = r-1$, $L(x,0,0) \neq 0$, *so that* τ *is a unit, and*

$$a(d+r-1) - bc = 0$$

Let $R = \mathcal{O}_{X,p}$.
Suppose that $\pi : Y_p \to \operatorname{Spec}(R)$ *is the morphism of Theorem 12.4.*
Let

$$\cdots \to Y_n \to \cdots \to Y_1 \to Y_p$$

be a sequence of permissible monoidal transforms centered at 2 curves D *such that* D *is r-1 big. Then there exists* $n_0 < \infty$ *such that*

$$V_p = Y_{n_0} \xrightarrow{\pi_1} Y \to \operatorname{spec}(R)$$

extends to a permissible sequence of monoidal transforms

$$\overline{U}_1 \to \overline{U} \to U$$

over an affine neighborhood U *of* p *(with the notation of Theorem 12.4) such that* $\overline{U}_1$ *contains no 2 curves* D *such that* D *is r-1 big or r small, and for* $q \in \overline{U}_1$,

1. *If* q *is a 1 or a 2 point then* $\nu(q) \leq r$. $\nu(q) = r$ *implies* $\gamma(q) = r$.
2. *If* q *is a 3 point then* $\nu(q) \leq r-2$.
3. $\overline{S}_r(\overline{U}_1)$ *makes SNCs with* $\overline{B}_2(\overline{U}_1)$.

There exists a sequence of quadratic transforms $W_p \to V_p$ such that if $Z_p \to W_p$ is the sequence of monoidal transforms (in any order) centered at the strict transforms of curves C in $\overline{S}_r(X)$ then

$$Z_p \to W_p \to V_p \to Y_p \to \text{spec}(R)$$

extends to a permissible sequence of monoidal transforms

$$\overline{\pi} : \overline{U}_2 \to \overline{U}_1 \to \overline{U} \to U$$

over an affine neighborhood of p such that $\overline{U}_2$ contains no 2 curves D such that D is r-1 big or r small. $\overline{S}_r(\overline{U}_2)$ makes SNCs with $\overline{B}_2(\overline{U}_2)$, and if $q \in \overline{\pi}^{-1}(p)$,

    1'.: $\nu(q) \le r$ *if q is a 1 or 2 point. $\nu(q) = r$ implies $\gamma(q) = r$.*
    2'.: *If q is a 2 point and $\nu(q) = r - 1$, then either $\tau(q) > 0$ or $\gamma(q) = r$ or $\tau(q) = 0$ and (133) holds at q with $0 < d_i < i$, $e_i = i$ and $\overline{S}_{r-1}(Y_1)$ contains a single curve D containing q, and containing a 1 point, which has local equations $x = z = 0$ at q.*
    3'.: $\nu(q) \le r - 2$ *if q is a 3 point.*

If $p \notin \overline{S}_r(X)$, then $Z = Y_{n_0}$.

Proof. The proof of Theorem 12.2 applied to the conclusions of Theorem 12.4 proves this theorem. □

Theorem 12.6. *Suppose that $r = 2$, $A_2(X)$ holds, p is a 3 point with $\nu(p) = r - 1 = 1$, and we have permissible parameters (x, y, z) at p for u, v (with $y, z \in \mathcal{O}_{X,p}$) such that*

$$
\begin{aligned}
u &= (x^b y^{a+nb} z^{a+(n+1)b})^m \\
v &= P(x^b y^{a+nb} z^{a+(n+1)b}) + x^d y^{c+n(d+2)} z^{c+(n+1)(d+2)} F_p \\
F_p &= \tau x + a_1(y, z)
\end{aligned}
$$

with $n \ge 0$, $a, b > 0$, $L(x, 0, 0) \ne 0$ so that τ is a unit and

$$a(d + r - 1) - bc = a(d + 1) - bc = 0$$

or

$$
\begin{aligned}
u &= (x^b y^a z^b)^m \\
v &= P(x^b y^a z^b) + x^d y^c z^{d+2} F_q \\
F_p &= \tau x + a_1(y, z)
\end{aligned}
$$

with $n \ge 0$, $a, b > 0$, $L(x, 0, 0) \ne 0$ so that τ is a unit and

$$a(d + r - 1) - bc = a(d + 1) - bc = 0$$

Let $R = \mathcal{O}_{X,p}$.
Suppose that $\pi : Y_p \to \text{Spec}(R)$ is the morphism of Theorem 12.4.
Let

$$\cdots \to Y_n \to \cdots \to Y_1 \to Y_p$$

be a sequence of permissible monoidal transforms centered at 2 curves D such that D is 1 big. Then there exists $n_0 < \infty$ such that

$$V_p = Y_{n_0} \xrightarrow{\pi_1} Y_p \to \text{spec}(R)$$

extends to a permissible sequence of monoidal transforms

$$\overline{U}_1 \to \overline{U} \to U$$

over an affine neighborhood U of p (with the notation of Theorem 12.4) such that $\overline{U}_1$ contains no 2 curves D such that D is 1 big or 2 small, and for $q \in \overline{U}_1$,

1. *If q is a 1 or a 2 point then $\nu(q) \leq 2$. $\nu(q) = 2$ implies $\gamma(q) = 2$.*
2. *If q is a 3 point then $\nu(q) = 0$.*
3. *$\overline{S}_2(\overline{U}_1)$ makes SNCs with $\overline{B}_2(\overline{U}_1)$.*

There exists a sequence of quadratic transforms $W_p \to V_p$ such that if $Z_p \to W_p$ is the sequence of monoidal transforms (in any order) centered at the strict transforms C' of curves C in $\overline{S}_2(X)$, followed by monoidal transforms centered at any 2 curves $\overline{C}$ which are sections over C' such that $\overline{C}$ is 1 big, then

$$Z_p \to W_p \to V_p \to Y_p \to \text{spec}(R)$$

extends to a permissible sequence of monoidal transforms

$$\overline{\pi} : \overline{U}_2 \to \overline{U}_1 \to \overline{U} \to U$$

over an affine neighborhood of p such that $\overline{U}_2$ contains no 2 curves D such that D is 1 big or 2 small. $\overline{S}_2(\overline{U}_2)$ makes SNCs with $\overline{B}_2(\overline{U}_2)$, and if $q \in \overline{\pi}^{-1}(p)$,

1'.: *$\nu(q) \leq 2$ if q is a 1 or 2 point. $\nu(q) = 2$ implies $\gamma(q) = 2$.*
2'.: *If q is a 2 point and $\nu(q) = 1$, then either q is resolved or $\gamma(q) = 2$.*
3'.: *$\nu(q) = 0$ if q is a 3 point.*
If $p \notin \overline{S}_2(X)$, then $Z_p = Y_{n_0}$.

Proof. The proof of Theorem 12.3 applied to the conclusions of Theorem 12.4 proves the Theorem. □

13. Resolution 1

Throughout this section we will assume that $\Phi_X : X \to S$ is weakly prepared.

In this chapter we will need to consider the following condition on a 2 point $p \in X$ such that $\nu(p) = r$ and $\tau(p) = 1$. The condition is that $\Phi_X(p)$ has permissible parameters (u, v) such that $u = 0$ is a local equation of E_X at p and p has permissible parameters (x, y, z) for (u, v) such that(eq998)

$$
\begin{aligned}
u &= (x^a y^b)^m \\
v &= P(x^a y^b) + x^c y^d F_p
\end{aligned}
\tag{149}
$$

and L_p contains a nonzero $y^{r-1}z$ term with $a(d + r - 1) - bc = 0$. Up to interchanging x and y, this condition is independent of permissible parameters at p for (u, v).

Lemma 13.1. *Suppose that X satisfies $A_r(X)$, with $r \geq 2$, and C is a 2 curve on X such that $C \subset \bar{S}_r(X)$. Suppose that*

$$
\cdots \to X_n \to \cdots \to X_1 \to X
$$

is a sequence of permissible monoidal transforms centered at 2 curves C_i such that $C_i \subset \bar{S}_r(X_i)$ are sections over C. Then this sequence is finite. That is, there exists $n < \infty$ such that X_n contains no 2 curve C_n with this property.

Proof. Since $A_r(X)$ holds, C must be r small. Suppose that $q \in C$ is a 2 point and the sequence has infinite length. Let q_n be the point on X_n which is the intersection of the fiber over q and C_n. With the notations of (70) in the proof of Lemma 8.6, there are permissible parameters (x, y, z) at q such that

$$
\begin{aligned}
u &= (x^a y^b)^m \\
F_q &= \bar{c}zy^{r-1} + \Sigma_{i+j \geq r, k \geq 0} c_{ijk} x^i y^j z^k.
\end{aligned}
$$

For all n there are permissible parameters (x_n, y_n, z) at q_n such that

$$
x = x_n, y = x_n^n y_n
$$

$$
F_{q_n} = \frac{F_q}{x_1^{n(r-1)}}
$$

If the sequence has infinite length, then $c_{ijk} = 0$ if $j < r - 1$, so that $y \mid F_q$, a contradiction to the assumption that F_q is normalized. $\qquad\square$

Lemma 13.2. *Suppose that X satisfies $A_r(X)$ with $r \geq 2$ and C is a 2 curve on X such that C is r-1 big. Suppose that*

$$
\cdots \to X_n \to \cdots \to X_1 \to X
$$

is a sequence of permissible monoidal transforms, centered at 2 curves C_i such that C_i is a section over C and C_i is r-1 big. Then the sequence is finite. That is, there exists $n < \infty$ such that X_n contains no 2 curves C_n with this property.

Proof. Suppose that $p \in C$ is a 2 point such that $\nu(p) = r - 1$. p has permissible parameters (x, y, z) such that

$$
\begin{aligned}
u &= (x^a y^b)^m \\
v &= P(x^a y^b) + x^c y^d F_p \\
F_p &= \sum_{i+j \geq r-1} a_{ij}(z) x^i y^j
\end{aligned}
$$

Let $\pi : Y \to \operatorname{spec}(\mathcal{O}_{X,p})$ be the blow-up of C.

Suppose that $q \in \pi^{-1}(p)$ is the 2 point with permissible parameters (x_1, y_1, z_1)

$$
x = x_1 y_1, y = y_1
$$

$$
\begin{aligned}
u &= (x_1^a y_1^{a+b})^m \\
v &= P(x_1^a y_1^{a+b}) + x_1^c y_1^{c+d+r-1}\left(\sum_{i+j=r-1} a_{ij}(z) x_1^i + y_1 \Omega\right)
\end{aligned}
\tag{150}
$$

$\nu(q) \leq r - 2$ unless $L_p = x^{r-1}$. Then $\nu(q) = r - 1$.

Suppose that $L_p = x^{r-1}$.

$$
F_p = \tau x^{r-1} + \sum_{i=1}^{r-1} y^{b_i} a_i(y, z) x^{r-1-i}
$$

where τ is a unit and $x \nmid a_i$, $y \nmid a_i$, $b_i \geq i$ for all i.

Suppose that the 2 curve $C_1 \subset Y$ containing q is such that C_1 is r-1 big. $C_1 = V(x_1, y_1)$ in (150).

$$
F_q = \tau x_1^{r-1} + \sum_{i=1}^{r-1} y_1^{b_i - i} a_i(y_1, z) x_1^{r-1-i}.
$$

By induction on b_i, after a finite number of blow-ups of 2 curves, we reach $\lambda : Z \to \operatorname{spec}(\mathcal{O}_{X,p})$ such that if D is a 2 curve in Z which is a section over C, then D is not r-1 big. $\qquad \square$

Definition 13.3. *Suppose that X satisfies $A_r(X)$ with $r \geq 2$. A 2 point $p \in X$ contained in a 2 curve C is called bad if $\nu(p) = r$, $\tau(p) = 1$ and one of the following holds.*

1. *$C \not\subset \overline{S}_r(X)$.*
2. *$C \subset \overline{S}_r(X)$ is r small and there exists a sequence of monoidal transforms*

$$
X_n \to X_{n-1} \to \cdots \to X_1 \to \operatorname{spec}(\mathcal{O}_{X,p})
$$

   and 2 curves $C_i \subset X_i$ which are sections over C such that $C_i \subset \overline{S}_r(X_i)$ is r small for $i < n$, $C_n \not\subset \overline{S}_r(X_i)$, $X_{i+1} \to X_i$ is centered at C_i if $i < n$, and if p_n is the point on C_n over p then $\nu(p_n) = r$.
3. *There exists a curve $D \subset \overline{S}_r(X)$ such that D contains a 1 point $p \in D$, and D is r big at p.*

Suppose that $r \geq 2$ and $A_r(X)$ holds. Then there are only finitely many bad 2 points on X.

Lemma 13.4. *Suppose that $A_r(X)$ holds with $r \geq 2$ and $p \in X$ is a bad 2 point. Suppose that there does not exist a curve $D \subset \overline{S}_r(X)$ such that $p \in D$, D contains a 1 point and D is r big at p. Then there exists a sequence of quadratic transforms $\pi : X_1 \to X$ centered at 2 points over p such that $A_r(X_1)$ holds and all 2 points $q \in \pi^{-1}(p)$ are good.*

Proof. There exist permissible parameters (x, y, z) at p such that

$$
\begin{aligned}
u &= (x^a y^b)^m \\
v &= P(x^a y^b) + x^c y^d F_p \\
F_p &= \textstyle\sum_{i+j+k \geq r} a_{ijk} x^i y^j z^k
\end{aligned}
$$

Let $\pi_1 : X_1 \to X$ be the blow-up of p. By Theorems 7.1, 7.3 and Lemma 7.9, $A_r(X_1)$ holds. Suppose that $q \in \pi^{-1}(p)$ is a 2 point such that $\nu(q) = r$ and $\tau(q) = 1$. After a permissible change of parameters at p, we may assume that q has permissible parameters $(\overline{x}_1, \overline{y}_1, \overline{z}_1)$ such that $x = \overline{x}_1 \overline{y}_1, y = \overline{y}_1, z = \overline{y}_1 \overline{z}_1$. $\tau(q) = 1$ and $\nu(q) = r$ implies that, after replacing z by a constant times z, that

$$
L_p = L(x, z) = \overline{d} x^r + x^{r-1} z
$$

for some $\overline{d} \in k$.

Suppose that there exists a 2 point $q' \in \pi^{-1}(p)$ such that $\nu(q') = r$ and q' has permissible parameters x', y', z' such that $x = x', y = x'y', z = x'(z' + \alpha)$ for some $\alpha \in k$. Then there exists a form L_1 and $\overline{c} \in k$ such that

$$
L_p = \begin{cases}
L_1 + \overline{c} x^{\overline{a}} y^{\overline{b}} & \text{if there exists } \overline{a}, \overline{b} \in \mathbf{N} \text{ such that } \overline{a} + \overline{b} = r \\
& \text{and } a(d + \overline{b}) - b(c + \overline{a}) = 0 \\
L_1 & \text{otherwise.}
\end{cases}
$$

where

$$
L_1 = L_1(y, z - \alpha x) = e y^r + f(z - \alpha x) y^{r-1}
$$

for some $e, f \in k$, with $f \neq 0$. This is not possible, since $r \geq 2$. Thus all 2 points $q_1 \in \pi^{-1}(p)$ with $\nu(q_1) = r$ have permissible parameters (x_1, y_1, z_1) such that

$$
x = x_1 y_1, y = y_1, z = y_1(z_1 + \alpha)
$$

for some $\alpha \in k$. There exist at most finitely many bad 2 points $q_1 \in \pi_1^{-1}(p)$.

Consider the following sequence of quadratic transforms

$$
\cdots \to X_n \to X_{n-1} \to \cdots \to X_1 \to X
$$

with maps $\lambda_n : X_n \to X$, where $\pi_i : X_i \to X_{i-1}$ is the blow-up of all bad 2 points in $\lambda_{i-1}^{-1}(p)$. We will show that there exists $n < \infty$ such that $\lambda_n^{-1}(p)$ contains no bad 2 points. Suppose not. Then there exist bad 2 points $q_i \in X_i$ such that $\pi_i(q_i) = q_{i-1}$ for all i.

q_1 has permissible parameters (x_1, y_1, z_1) such that

$$
x = x_1 y_1, y = y_1, z = y_1(z_1 + \alpha_1)
$$

for some $\alpha_1 \in k$. $\nu(q_1) = r$ implies

$$
L_{q_1} = \overline{d}_1 x_1^r + x_1^{r-1} z_1 + y_1 \Omega_1
$$

for some $\bar{d}_1 \in k$ and series Ω_1. Since $\nu(q_2) = r$, and because of the existence of the $x_1^{r-1} z_1$ term in L_{q_1}, q_2 must have permissible parameters (x_2, y_2, z_2) such that

$$x_1 = x_2 y_2, y_1 = y_2, z_1 = y_2(z_2 + \alpha_2)$$

for some $\alpha_2 \in k$. $\nu(q_2) = r$ implies

$$L_{q_2} = \bar{d}_2 x_2^r + x_2^{r-1} z_2 + y_2 \Omega_2.$$

We see that there exists a series $\sigma(y) = \sum_{i=1}^{\infty} \alpha_i y^i$ such that if we replace z with $\tilde{z} = z - \sigma(y)$, we have permissible parameters $(x_n, y_n, \tilde{z}_n)$ at q_n such that

$$x = x_n y_n^n, y = y_n, \tilde{z} = \tilde{z}_n y_n^n.$$

Then

$$F_{q_n} = \frac{F_p(x_n y_n^n, y_n, y_n^n \tilde{z}_n)}{y_n^{rn}}$$

for all n, so that $a_{ijk} = 0$ if $i + k < r$ and $F_p \in (x, \tilde{z})^r$. By Lemma 6.22, $\hat{\mathcal{I}}_{\bar{S}_r(X),p} \subset (x, \tilde{z})$, so that $x = \tilde{z} = 0$ are local equations of a curve $D \subset \bar{S}_r(X)$, since $A_r(X)$ holds. D is r big at p by Lemma 8.2. $\qquad\square$

Theorem 13.5. *Suppose that $A_r(X)$ holds with $r \geq 2$. Then there exists a sequence of permissible monoidal transforms $X_1 \to X$ such that the following properties hold:*

1. *$A_r(X_1)$ holds.*
2. *All bad 2 points $p \in X_1$ satisfy (149).*
3. *Suppose that $D \subset \bar{S}_r(X)$ is a curve which is r big at a 1 point. Then there exists at most one 3 point $q \in D$ and D has a tangent direction at q distinct from those of $\bar{B}_2(X)$ at q. Furthermore, if D is not r big, there exists only one 2 point $q \in D$. If C is the 2 curve containing q, then C is not r-1 big or r small.*
4. *If C is a r small or r-1 big 2 curve containing a 2 point p such that $p \in D$ where D is a curve containing a 1 point and D is r big at p, then D is r big.*

Proof. By Theorems 7.1 and 7.8, there exists a sequence of quadratic transforms $\pi_1 : X_1 \to X$ centered at 3 points so that $A_r(X_1)$ holds and if $q \in X_1$ is a bad 2 point such that (149) doesn't hold, and there exists a curve $D \subset \bar{S}_r(X_1)$ such that $q \in D$ and D is r big at q, then D is r big. All exceptional 2 points for π_1 which are bad must satisfy (149).

Let $\pi_2 : X_2 \to X_1$ be the blow-up of such a D. By Lemma 8.8, $A_r(X_2)$ holds and all 2 points in $\pi_2^{-1}(D)$ are good. We have that if $q \in X_2$ is a bad 2 point such that (149) doesn't hold, and there exists a curve $D \subset \bar{S}_r(X_2)$ such that $q \in D$ and D is r big at q, then D is r big. If such a D exists, let $\pi_3 : X_3 \to X_2$ be the blow-up of D.

After a finite sequence of blow-ups, we then obtain $\lambda_1 : Z_1 \to X$ such that $A_r(Z_1)$ holds, and if $q \in Z_1$ is a bad 2 point which doesn't satisfy (149), then there doesn't exist a curve $D \subset \bar{S}_r(Z_1)$ such that D is r big at q. By Lemma

13.4, there exists a sequence of quadratic transforms $\lambda_2 : Z_2 \to Z_1$ such that $A_r(Z_2)$ holds, and if $q \in Z_2$ is a bad 2 point, then (149) holds at q.

By Theorems 7.1 and 7.8, there exists a sequence of quadratic transforms $\lambda_3 : Z_2 \to Z_2$ centered at 3 points such that the conclusions of the Theorem hold on Z_3.   □

Theorem 13.6. *Suppose that the conclusions of Theorem 13.5 hold on X. Then there exists a finite sequence of quadratic transforms centered at 3 points $X_1 \to X$ such that*

1. *$A_r(X_1)$ holds.*
2. *All bad 2 points $p \in X_1$ satisfy (149).*
3. *Suppose that $D \subset \overline{S}_r(X)$ is a curve which is r big at a 1 point. Then there exists at most one 3 point $q \in D$ and D has a tangent direction at q distinct from those of $\overline{B}_2(X)$ at q. If D is not r big, there exists only one 2 point $q \in D$. If C is the 2 curve containing q, then C is not r-1 big or r small.*
4. *If C is a r small or r-1 big 2 curve containing a 2 point p such that $p \in D$ where D is a curve r big at p, then D is r big.*
5. *If $q \in X_1$ is a 3 point with $\nu(q) = r - 1$, then either there are permissible parameters (x, y, z) at q such that*

$$L_q \text{ depends on both } y \text{ and } z \text{ and } F_q \in (y, z)^{r-1} \tag{151}$$

or there are permissible parameters (x, y, z) at q such that

$$F_q = \tau y^{r-1} + \sum_{j=1}^{r-1} a_j(x, z) x^{\alpha_j} z^{\beta_j} y^{r-1-j} \tag{152}$$

where τ is a unit, a_j are units (or zero), $\alpha_j + \beta_j \geq j$ for all j, and there exists i such that

$$\frac{\alpha_i}{i} \leq \frac{\alpha_j}{j}, \frac{\beta_i}{i} \leq \frac{\beta_j}{j}$$

for all j, and

$$\left\{ \frac{\alpha_i}{i} \right\} + \left\{ \frac{\beta_i}{i} \right\} < 1.$$

Proof. X satisfies 1. - 4. of the conclusions of the Theorem. By Theorems 7.1 and 7.8, 1. - 4. are stable under quadratic transforms centered at 3 points.

Suppose that $\pi_1 : X_1 \to X$ is the blow-up of a 3 point p.

If L_p depends on all three variables x, y, z then $\nu(q) \leq r - 2$ for all 3 points $q \in \pi^{-1}(p)$.

Suppose that L_p depends on both y and z. Then $\nu(q) \leq r - 2$ for all 3 points $q \in \pi^{-1}(p)$, except possibly under the quadratic transform

$$x = x_1, y = x_1 y_1, z = x_1 z_1.$$

At this 3 point q,

$$F_q = \frac{F_p}{x_1^{r-1}} = L_p(y_1, z_1) + x_1 \Omega.$$

By a sequence of quadratic transforms centered at 3 points, we can get the Theorem to hold above p, except possibly along an infinite sequence

$$R = \mathcal{O}_{X,p} \to R_1 \to \cdots \to R_n \to \cdots$$

where for all n R_n has permissible parameters (x_n, y_n, z_n) with

$$x = x_n, y = x_n^n y_n, z = x_n^n z_n$$

and $\nu(\frac{F_p(x_n, x_n^n y_n, x_n^n z_n)}{x^{n(r-1)}}) = r - 1$. Thus $F_p \in (y, z)^{r-1}$.

Now suppose that L_p depends only on y. Then

$$F_p = \tau y^{r-1} + \sum_{i=1}^{r-1} a_i(x, z) y^{r-1-i} \tag{153}$$

where τ is a unit.

If $p_1 \in \pi_1^{-1}(p)$ is a 3 point with $\nu(q) = r - 1$, then p_1 has permissible parameters (x_1, y_1, z_1) of one of the following 2 forms:

$$x = x_1, y = x_1 y_1, z = x_1 z_1$$

or

$$x = x_1 z_1, y = y_1 z_1, z = z_1$$

and

$$F_{p_1} = \tau y_1^{r-1} + \sum_{i=1}^{r-1} \frac{a_i(x_1, x_1 z_1)}{x_1^i} y_1^{r-1-i} \tag{154}$$

or

$$F_{p_1} = \tau y_1^{r-1} + \sum_{i=1}^{r-1} \frac{a_i(x_1 z_1, z_1)}{z_1^i} y_1^{r-1-i}.$$

Suppose that

$$\cdots \to X_n \to X_{n-1} \to \cdots \to X_1 \to X$$

is a sequence of quadratic transforms, $\pi_i : X_i \to X_{i-1}$ with induced maps $\lambda_i : X_i \to X$ such that for all i, π_i is the blow-up of a 3 point p_{i-1}, with $\nu(p_{i-1}) = r - 1$ and $\pi_{i-1}(p_{i-1}) = p_{i-2}$.

We will show that there exists n such that p_n satisfies (152). Each p_i has permissible parameters $(x_i, y_i.z_i)$ such that either

$$x_{i-1} = x_i, y_{i-1} = x_i y_i, z_{i-1} = x_i z_i$$

or

$$x_{i-1} = x_i z_i, y_{i-1} = y_i z_i, z_{i-1} = z_i.$$

By (154), and resolution of plane curve singularities, there exists n_0 such that $n \geq n_0$ implies F_{p_n} has the form

$$F_{p_n} = \tau y_n^{r-1} + \sum_{i=1}^{r-1} a_{n,i}(x_n, z_n) x_n^{\alpha_{ni}} z_n^{\beta_{ni}} y_n^{r-1-i}$$

where $a_{n,i}(x_n, z_n)$ are units, and either

$$(\alpha_{n+1,i}, \beta_{n+1,i}) = (\alpha_{n,i} + \beta_{n,i} - i, \beta_{n,i})$$

or

$$(\alpha_{n+1,i}, \beta_{n+1,i}) = (\alpha_{n,i}, \alpha_{n,i} + \beta_{n,i} - i)$$

for all i. The proof now follows from Lemmas 9.20 and 9.21.

□

Theorem 13.7. *Suppose that $A_r(X)$ holds with $r \geq 2$ and $p \in X$ is a 2 point such that (149) holds. Then either*

1. *There exists a sequence of quadratic transforms $\pi : Y \to X$ centered at points over p such that*
   (a) *If $q \in \pi^{-1}(p)$ is a 1 point then $\nu(q) \leq r$. $\nu(q) = r$ implies $\gamma(q) = r$.*
   (b) *If $q \in \pi^{-1}(p)$ is a 2 point then $\nu(p) \leq r$. $\nu(p) = r$ implies $\tau(p) \geq 2$.*
   (c) *$q \in \pi^{-1}(p)$ a 3 point implies $\nu(q) \leq r - 1$. $\nu(q) = r - 1$ implies q satisfies the assumptions of (145) and (146) of Theorem 12.4. If D_q is the 2 curve containing q, with local equations $y = z = 0$ at q, in the notation of Theorem 12.4, then $F_{q'}$ is resolved for all $q \neq q' \in D_q$.*
   (d) *$A_r(Y)$ holds.*
   or
2. *There exists a curve $C \subset \overline{S}_r(X)$ which is r big at p. Then there exists an affine neighborhood U of p such that the blow-up of $C \cap U$, $\pi : Z \to U$ is a permissible monoidal transform such that*
   (a) *If $q \in \pi^{-1}(p)$ is a 2 point then $\nu(q) = 0$.*
   (b) *If $q \in \pi^{-1}(p)$ is the 3 point then either $\nu(q) \leq r - 2$ or q satisfies the assumptions of (147) and (148) of Theorem 12.4. $D_q = \pi^{-1}(p)$ is the 2 curve with local equations $y = z = 0$ at q in the notation of Theorem 12.4.*
   (c) *$A_r(Z)$ holds.*

Proof. We first assume that the assumption of 2. doesn't hold. There are permissible parameters (x, y, z) at p such that

$$\begin{aligned} u &= (x^a y^b)^m \\ v &= P(x^a y^b) + x^c y^d F_p. \end{aligned} \tag{155}$$

By assumption, L_p has the form

$$L_p = f(x, y) + z g(x, y) \tag{156}$$

and L_p contains a $y^{r-1} z$ term with

$$a(d + r - 1) - bc = 0. \tag{157}$$

Let $\pi_1 : X_1 \to X$ be the blow-up of p.

Suppose that there exists a 2 point $q \in \pi_1^{-1}(p)$ such that $\nu(q) = r$ and q has permissible parameters (x_1, y_1, z_1) such that

$$x = x_1 y_1, y = y_1, z = y_1(z_1 + \alpha).$$

After a permissible change of parameters, we may assume that $\alpha = 0$. Then $F_q = \frac{F_p}{y_1^r}$ and $\nu(F_q(0, 0, z_1)) \leq 1$, so that q is resolved.

Suppose that $p_1 \in \pi_1^{-1}(p)$ is a 2 point such that $\nu(p_1) = r$ and p_1 has permissible parameters (x_1, y_1, z_1) such that

$$x = x_1, y = x_1 y_1, z = x_1(z_1 + \alpha).$$

After making a permissible change of parameters, we may assume that $\alpha = 0$. Then L_p depends only on y and z, and

$$L_p = \bar{e} y^r + \bar{b} z y^{r-1}$$

for some $\bar{b}, \bar{e} \in k$ with $\bar{b} \neq 0$.

Suppose that $q \in \pi_1^{-1}(p)$ is another 2 point such that $\nu(q) = r$, and q has permissible paramters (x_1, y_1, z_1) such that

$$x = x_1, y = x_1 y_1, z = x_1(z_1 + \alpha)$$

with $\alpha \neq 0$.

Then there exists a form L such that

$$L_p = \begin{cases} L(y, z - \alpha x) \text{ or} \\ L(y, z - \alpha x) + \bar{c} x^{\bar{\alpha}} y^{\bar{\beta}} \\ \quad \text{for some } \bar{c} \in k, \text{ such that } a(d + \bar{\beta}) - b(c + \bar{\alpha}) = 0, \bar{\alpha} + \bar{\beta} = r. \end{cases}$$

$$L_p = \bar{e} y^r + \bar{b} \alpha x y^{r-1} + \bar{b}(z - \alpha x) y^{r-1}$$

implies $L(y, z - \alpha x) = L_p - \bar{b}\alpha x y^{r-1}$, but
$\bar{\alpha} = 1, \bar{\beta} = r - 1$ is not possible, since

$$a(d + r - 1) - b(c + 1) = -b \neq 0 \tag{158}$$

Thus p_1 is the unique 2 point $q \in \pi_1^{-1}(p)$ with $\nu(q) = r$. There are permissible parameters (x_1, y_1, z_1) at p_1 such that

$$x = x_1, y = x_1 y_1, z = x_1 z_1$$

and $L_p = L(y, z) = \bar{e} y^r + \bar{b} y^{r-1} z$, $L_{p_1} = L(y_1, z_1) + x_1 \Omega$.

If $p' \in \pi^{-1}(p)$ is the 3 point, then $\nu(p') \leq r - 1$. If $p' \in \pi^{-1}(p)$ is a 1 point then $\nu(p') \leq r$ and $\nu(p') = r$ implies $\gamma(p') = r$ by Theorem 7.3.

By Theorem 7.3, $\tau(p_1) \geq 1$. Suppose that $\nu(p_1) = r$ and $\tau(p_1) = 1$. Let $\pi_2 : X_2 \to X_1$ be the blow-up of p_1.

We can make an analysis of $\pi_2^{-1}(p_1)$ which is similar to that of $\pi_1^{-1}(p)$. (155) is replaced with

$$\begin{aligned} u &= (x^{a+b} y^b)^m \\ v &= P(x^{a+b} y^b) + x^{c+d+r} y^d F \end{aligned}$$

$\tau(p_1) = 1$ implies L_{p_1} has the form of (156). (157) holds at p_1. (158) is then modified to: $\bar{\alpha} = 1, \bar{\beta} = r - 1$ is not possible since

$$\begin{aligned} (a + b)(d + r - 1) &- b(c + d + r + 1) \\ = a(d + r - 1) + b(d + r) &- b - bc - b(d + r) - b = -2b \neq 0 \end{aligned}$$

We conclude that there is at most one 2 point $p_2 \in \pi_2^{-1}(p_1)$ with $\nu(p_2) = r$, and after replacing z with $z - \alpha x^2$ for some $\alpha \in k$, we have that p_2 has permissible parameters (x_2, y_2, z_2) such that

$$x = x_2, y = x_2^2 y_2, z = x_2^2 z_2.$$

Suppose that we can construct an infinite sequence of quadratic transforms

$$\cdots \to X_n \to \cdots \to X_1 \to X$$

Where $\pi_n : X_{n+1} \to X_n$ is the blow-up of a 2 point $p_n \in X_n$ over p such that $\nu(p_n) = r$ and $\tau(p_n) = 1$. Then there exists a series $\sigma(x)$ such that if we make a formal change of variables, replacing z with $z - \sigma(x)$, we get that there are permissible parameters (x_n, y_n, z_n) at p_n such that

$$x = x_1, y = x_1^n y_1, z = x_1^n z_1$$

p_n must be the only 2 point of $\pi_n^{-1}(p_{n-1})$ such that $\nu(p_n) = r$ and $\tau(p_n) = 1$. To show this, the argument following (155) is modified by replacing (155) with

$$
\begin{aligned}
u &= (x^{a+nb} y^b)^m \\
v &= P(x^{a+nb} y^b) + x^{c+n(d+r)} y^d F
\end{aligned}
$$

where (157) holds at p_n. (158) is modified to:
$\bar{\alpha} = 1, \bar{\beta} = r - 1$ is not possible since

$$
\begin{aligned}
&(a + nb)(d + r - 1) - b(c + n(d + r) + 1) \\
&= a(d + r - 1) + nb(d + r - 1) - bc - nb(d + r) - b \\
&= -nb - b = -(n + 1)b \neq 0
\end{aligned}
$$

$\nu(p_n) = \nu(\frac{F_p}{x_1^{nr}}) = r$ for all n implies that $F_p \in (y, z)^r$. $\hat{\mathcal{I}}_{\overline{S}_r(X),p} \subset (y, z)$ by Lemma 6.22, so that since $A_r(X)$ holds, $y = z = 0$ are local equations at p of a curve $C \subset \overline{S}_r(X)$, which is r big at p by Lemma 8.2, a contradiction to the assumption that the assumption of 2. doesn't hold. Thus there exists a sequence of quadratic transforms $\pi : Y \to X$ such that if $q \in \pi^{-1}(p)$ is a 2 point, then $\nu(q) \leq r$, and $\nu(q) = r$ implies $\tau(q) > 1$. By Theorem 7.3 and Lemma 7.9, $A_r(Y)$ holds.

Suppose that $q \in \pi^{-1}(p)$ is a 3 point such that $\nu(q) = r - 1$. There exist permissible parameters (x_2, y_2, z_2) at q, and there exists a 2 point $p_n \in X_n$ such that $\nu(p_n) = r$, $\tau(p_n) = 1$ and p_n has permissible parameters (x_1, y_1, z_1) such that

$$x = x_1, y = x_1^n y_1, z = x_1^n z_1 \tag{159}$$

$$x_1 = x_2 z_2, y_1 = y_2 z_2, z_1 = z_2$$

$$
\begin{aligned}
u &= (x_1^{a+nb} y_1^b)^m \\
v &= P(x_1^{a+nb} y_1^b) + x_1^{c+n(d+r)} y_1^d (\bar{b} y_1^{r-1} z_1 + \bar{e} y_1^r + x_1 \Omega)
\end{aligned}
\tag{160}
$$

with $\bar{b} \neq 0$.

$$
\begin{aligned}
u &= (x_2^{a+nb} y_2^b z_2^{a+(n+1)b})^m \\
v &= P(x_2^{a+nb} y_2^b z_2^{a+(n+1)b}) + x_2^{c+n(d+r)} y_2^d z_2^{c+(n+1)(d+r)} (\bar{b} y_2^{r-1} + \cdots)
\end{aligned}
\tag{161}
$$

with $a(d + r - 1) - bc = 0$. Thus q satisfies the assumptions of (145) and (146) of Theorem 12.4, with $x = y_2, y = x_2, z = z_2$.

On Y we have:

1. If $q \in \pi^{-1}(p)$ is a 1 point then $\nu(p) \leq r$. $\nu(p) = r$ implies $\gamma(p) = r$.
2. If $q \in \pi^{-1}(p)$ is a 2 point then $\nu(p) \leq r$. $\nu(p) = r$ implies $\tau(p) \geq 2$.

3. $q \in \pi^{-1}(p)$ a 3 point implies $\nu(q) \le r - 1$. $\nu(q) = r - 1$ implies q satisfies the assumptions of (145) and (146) of Theorem 12.4

4. $A_r(Y)$ holds.

Let T be the set of 3 points $q \in \pi^{-1}(p)$ such that $\nu(q) = r - 1$, so that q satisfies (145) and (146) of Theorem 12.4.

Suppose that $q \in T$. In the factorization $Y \to X$ by quadratic transforms there exists a factorization $Y \to X_n \to X$ such that q is an exceptional point on the blow-up of a 2 point p_n of the form of (159) on X_n. Let $\tau : Y \to X_n$ be this map, D_q be the nonsingular curve $D_q \subset \tau^{-1}(p_n)$ such that $\hat{\mathcal{I}}_{D_q,q} = (x_2, z_2)$. The other points $q' \in D_q$ have regular parameters (x', y', z') with the notation of (159) such that

$$x_1 = x'y', y_1 = y', z_1 = y'(z' + \alpha)$$

with $\alpha \in k$. At such a 2 point q', we have $\nu(F_{q'}(0, 0, z')) \le 1$ by (160). Thus q' is resolved.

We now prove 2. There are permissible parameters (x, y, z) at p such that $y = z = 0$ are local equations of C at p and

$$\begin{aligned} u &= (x^a y^b)^m \\ v &= P(x^a y^b) + x^c y^d F_p. \end{aligned}$$

There exists $\bar{a} \in k$ such that

$$L_p = \bar{a} y^r + z y^{r-1}$$

Let $\pi_1 : Y \to \mathrm{spec}(\mathcal{O}_{X,p})$ be the blow-up of C. Suppose that $q \in \pi_1^{-1}(p)$ is a 2 point. q has permissible parameters (x, y_1, z_1) such that

$$y = y_1, z = y_1(z_1 + \alpha)$$

for some $\alpha \in k$.

$$\begin{aligned} u &= (x^a y_1^b)^m \\ v &= p_1(x^a y_1^b) + x^c y_1^{d+r}(z_1 + \alpha' + x\Omega' + y_1\Omega'') \end{aligned}$$

with $\alpha' \in k$, so that q is resolved. At the 3 point $q \in \pi^{-1}(p)$, there are permissible parameters (x, y_1, z_1) such that

$$y = y_1 z_1, z = z_1.$$

$$\begin{aligned} u &= (x^a y_1^b z_1^b)^m \\ v &= P(x^a y_1^b z_1^b) + x^c y_1^d z_1^{d+r} F_q \\ F_q &= y_1^{r-1} + \bar{a} y_1^r + x\Omega' + z\Omega'' \end{aligned}$$

with $a(d + r - 1) - bc = 0$. Either $\nu(q) \le r - 2$ or $\nu(q) = r - 1$ and q satisfies the assumptions of (147) and (148) of Theorem 12.4 (with $x = y_1, y = x_1, z = z_1$).

The curve C blown up in Theorem 12.4 is the fiber $\pi^{-1}(p)$, which is resolved away from q. There exists an affine neighborhood U of q such that if $Z \to U$ is the blow-up of $C \cap U$, then $A_r(Z)$ holds by Lemma 8.8. $\square$

Theorem 13.8. *Suppose that $A_r(X)$ holds with $r \ge 2$. Then there exists a finite sequence of permissible monoidal transforms $X_1 \to X$ such that*

1. $A_r(X_1)$ holds.
2. If $p \in X_1$ is a 3 point, then $\nu(p) \le r - 2$.
3. If $p \in X_1$ is a 2 point such that $\nu(p) = r$ and $\tau(p) = 1$ then p has permissible parameters (x, y, z) such that

$$\begin{aligned} u &= (x^a y^b)^m \\ v &= P(x^a y^b) + x^c y^d F_q \end{aligned} \tag{162}$$

and L_p contains a nonzero $y^{r-1}z$ term with $a(d + r - 1) - bc = 0$.
4. $\overline{S}_r(X_1)$ makes SNCs with $\overline{B}_2(X_1)$.
5. If C is a 2 curve on X_1, then C is not r small or r-1 big.

Proof. We may assume that the conclusions of Theorem 13.6 hold on X. Let $\{D_1, \dots, D_n\}$ be the curves D in X which intersect a r-1 big or r small 2 curve at a 2 point such that D is r big there. By assumption, $D_1, \dots, D_n$ are r big.

Let $\sigma_1 : W_1 \to X$ be the blow-up of D_1. By Theorem 13.7 and Lemma 8.8 $A_r(Z_1)$ holds. $\sigma_1^{-1}(D_1)$ contains no bad 2 points. If $q_1 \in \sigma_1^{-1}(D_1)$ is a 3 point with $\nu(q_1) = r - 1$, then $q_1 \in \sigma_1^{-1}(q)$ where q is a bad 2 point. In this case, 2. (b) of Theorem 13.7 holds at q_1.

Let $\{\overline{D}_2, \dots, \overline{D}_n\}$ be the strict transforms of $\{D_2, \dots, D_n\}$ on W_1. These curves are all r big, and are the curves D in Z_1 which intersect a r-1 big or r small curve at a 2 point such that D is r big there.

We can blow up successively the strict transforms of $\overline{D}_2, \dots, \overline{D}_n$ by a map $\lambda : W \to X$ to get a W such that $A_r(W)$ holds, the exceptional locus of λ contains no bad 2 points, and if q is an exceptional 3 point with $\nu(q) = r - 1$ then q must satisfy 2. (b) of Theorem 13.7.

Furthermore, if C is an r-1 big or r small 2 curve on W, and $p \in C$ is a bad 2 point, then there does not exist a curve $D \subset \overline{S}_r(W)$ such that D is r big at p.

By Theorem 13.7, after performing a sequence of quadratic transforms $X_1 \to Z$ over bad 2 points $p \in X$ such that if C is the 2 curve containing $p \in X$ then C is r-1 big or r small, we have

1'.: The conclusions of 1. - 2. of Theorem 13.6 hold.
2'.: Suppose that C is a 2 curve such that C is r-1 big or r small. If $p \in C$ is a 2 point, then p is good.
3'.: If p is a 3 point such that $\nu(p) = r - 1$, then either

    (a): p satisfies the assumptions of (145) and (146) of Theorem 12.4. If D_p is the 2 curve containing p, with local equations $y = z = 0$ at p, in the notation of Theorem 12.4, then F_q is resolved for all $p \ne q \in D_p$.

    or

    (b): p satisfies the assumptions of (147) and (148) of Theorem 12.4. If D_p is the 2 curve containing p, with local equations $y = z = 0$ at p, in the notation of Theorem 12.4, then F_q is resolved for all $p \ne q \in D_p$.

    or

    (c): p satisfies (151) or (152) of Theorem 13.6.

Let $\{p_1, \dots, p_n\}$ be the 3 points of X_1 which satisfy 3'. (a) or 3' (b). By Theorem 12.4, there exist sequences of permissible monoidal transforms, over sections of D_{p_i} for $1 \leq i \leq n$,

$$\lambda_{p_i} : Y_{p_i} \to \operatorname{spec}(\mathcal{O}_{X_1,p_i})$$

such that the conclusions of Theorem 12.4 hold.

Since D_{p_i} is resolved at points $p_i \neq q \in D_{p_i}$, the only obstruction to extending λ_{p_i} to a permissible sequence of monoidal transforms of sections over D_{p_i} in X_1 is if the corresponding sections over D_{p_i} in X_1 do not make SNCs with 2 curves. This difficulty can be resolved by performing quadratic transforms at the points where the section does not make SNCs with 2 curves, since these points are necessarily resolved.

We can thus extend the

$$Y_{p_i} \to \operatorname{spec}(\mathcal{O}_{X_1,p_i})$$

to a sequence of permissible monoidal transforms

$$\lambda : Y \to X_1$$

such that $Y \times_{X_1} \operatorname{spec}(\mathcal{O}_{X_1,p_i}) \cong Y_{p_i}$ for $1 \leq i \leq n$, $A_r(Y)$ holds, 1'. and 2'. hold on $Y - \lambda^{-1}(\{p_1, \dots, p_n\})$, and if $q \in Y - \lambda^{-1}(\{p_1, \dots, p_n\})$ is a 3 point such that $\nu(q) = r - 1$, then q satisfies (151) or (152) of Theorem 13.6. Let

$$\cdots \to Z_n \to \cdots \to Z_1 \to Y$$

be a sequence of permissible monoidal transforms such that $Z_n \to Z_{n-1}$ is the blow-up of a 2 curve C such that C is r-1 big or r small.

We will show that there exists $n < \infty$ such that Z_n does not contain a 2 curve C such that C is r-1 big or r small, and that Z_n satisfies the conclusions of the Theorem.

By Theorem 12.5, this holds above a neighborhood of $\lambda^{-1}(\{p_1, \dots, p_n\})$. We must verify this condition over $\overline{Y} = Y - \lambda^{-1}(\{p_1, \dots, p_n\})$.

Suppose that C is a 2 curve on $\overline{Y}$, such that C is r-1 big or r small and $p \in C$ is a 3 point. Then all 3 points q on $C \subset \overline{Y}$ have $\nu(q) = r - 1$, and satisfy (151) or (152).

Let $\pi : Y_1 \to \overline{Y}$ be the blow-up of C. The assumption that all 2 points of C are good and Lemma 8.6 imply that $q \in \pi^{-1}(C)$ a 2 point implies $\nu(q) \leq r$ and if $\nu(q) = r$ then either $\gamma(q) = r$ or $\tau(q) = 1$, q is a good 2 point, $C \subset \overline{S}_r(X)$ and if $\overline{C}$ is the 2 curve containing q, then $\overline{C}$ is a section over C such that $\overline{C} \subset \overline{S}_r(X)$. Lemma 8.6, the assumption that all 2 points are good, and Lemma 8.7 imply $A_r(Y_1)$ holds. Further, by Lemma 8.6, if $\overline{C} \subset \pi^{-1}(C)$ is a 2 curve such that $\overline{C}$ is r-1 big or r small, then $\overline{C}$ is a section over C. All 2 points of $\overline{C}$ are good points.

Suppose that $q \in C$ is a 3 point with permissible regular parameters (x, y, z) such that $y = z = 0$ are local equations of C and

$$\begin{aligned} u &= (x^a y^b z^c)^m \\ v &= P(x^a y^b z^c) + x^d y^e z^f F_q. \end{aligned}$$

$F_q \in (y, z)^{r-1}$.

Suppose that $q_1 \in \pi^{-1}(q)$, and q_1 has permissible parameters (x, y_1, z_1) such that

$$y = y_1, z = y_1 z_1$$

If (151) holds at q,

$$F_{q_1} = \frac{F_q}{y_1^{r-1}} = L_q(1, z_1) + y_1 \Omega + x \Lambda$$

implies $\nu(q_1) \leq r - 2$. If (152) holds at q we must have $\beta_j \geq j$ for all j, and

$$F_{q'} = \tau + z\Omega'$$

so that $\nu(q_1) = 0$.

Now suppose that $q_1 \in \pi^{-1}(q)$ and q_1 has permissible parameters (x, y_1, z_1) such that

$$y = y_1 z_1, z = z_1$$

If (151) holds at q,

$$F_{q_1} = \frac{F_q}{z_1^{r-1}} = L_q(y_1, 1) + z_1 \Omega + x \Lambda$$

implies $\nu(q_1) \leq r - 2$. If (152) holds at q, we must have $\beta_j \geq j$ for all j, and

$$F_{q_1} = \tau y_1^{r-1} + \sum_{j=1}^{r-1} a_j(x, z_1) x^{\alpha_j} z_1^{\beta_j - j} y_1^{r-1-j}$$

so that either $\nu(q_1) \leq r - 2$, or q_1 has the form of (152) also, but with $\frac{\beta_i}{i}$ decreased by 1.

By Lemma 13.1 and Lemma 13.2, after a finite number of blow-ups of 2 curves C such that C is r small or r-1 big, we reach $\tilde{Y} \to \overline{Y}$ such that $\tilde{Y}$ contains no 2 curves C such that C is r small or r-1 big. Since all 3 points q of $\tilde{Y}$ with $\nu(q) = r - 1$ must satisfy (151) or (152), which implies that $\alpha_j \geq j$ for all j or $\beta_j \geq j$ for all j, so that there exists a 2 curve through q which is r small or r-1 big, we must have $\nu(q) \leq r - 2$ if $q \in \tilde{Y}$ is a 3 point.

□

14. RESOLUTION 2

Throughout this section we will assume that $\Phi_X : X \to S$ is weakly prepared. We define a new condition on X

Definition 14.1. *Suppose that $r \geq 2$. We will say that $C_r(X)$ holds if:*

1. *If $p \in X$ is a 1 point then $\nu(p) \leq r$. If $\nu(p) = r$ then $\gamma(p) = r$.*
2. *If p is a 2 point then $\nu(p) \leq r$. If $\nu(p) = r$ then $\gamma(p) = r$. If $\nu(p) = r - 1$ then one of the following three cases must hold:*
   (a) *$\tau(p) > 0$ or*
   (b) *$\gamma(p) = r$ or*
   (c) *$r \geq 3$, $\nu(p) = r - 1$, $\tau(p) = 0$, $p \notin \overline{S}_r(X)$, there exists a unique curve $D \subset \overline{S}_{r-1}(X)$ containing a 1 point such that $p \in D$, and permissible parameters (x, y, z) at p such that $x = z = 0$ are local equations of D,*

$$
\begin{aligned}
u &= (x^a y^b)^m \\
v &= P(x^a y^b) + x^c y^d F_p \\
F_p &= \tau x^{r-1} + \sum_{j=1}^{r-1} \bar{a}_j(y, z) y^{d_j} z^{e_j} x^{r-1-j}
\end{aligned}
\tag{163}
$$

where τ is a unit, $\bar{a}_j$ are units (or 0). There exists i such that $\bar{a}_i \neq 0$, $e_i = i$, $0 < d_i < i$,

$$
\frac{d_i}{i} \leq \frac{d_j}{j}, \frac{e_i}{i} \leq \frac{e_j}{j}
$$

for all j, and

$$
\left\{ \frac{d_i}{i} \right\} + \left\{ \frac{e_i}{i} \right\} < 1.
$$

3. *If p is a 3 point then $\nu(p) \leq r - 2$.*
4. *$\overline{S}_r(X)$ makes SNCs with $\overline{B}_2(X)$.*

Remark 14.2. *If $C_r(X)$ holds then there does not exist a 2 curve C on X such that C is r small or r-1 big.*

In this section we will prove a condition stronger than $C_r(X)$ (Theorem 14.7).

Theorem 14.3. *Suppose that $r \geq 2$, $A_r(X)$ holds, $p \in X$ is a 2 point such that $\nu(p) = r$ and $2 \leq \tau(p) < r$, then either*

1. *There exists a sequence of quadratic transforms $\pi : Y \to X$ over p such that*
   (a) *$A_r(Y)$ holds.*
   (b) *If $q \in \pi^{-1}(p)$ is a 1 point then $\nu(q) \leq r$. $\nu(q) = r$ implies $\gamma(q) = r$.*
   (c) *If $q \in \pi^{-1}(p)$ is a 2 point then $\nu(q) \leq r - 1$.*
   (d) *If $q \in \pi^{-1}(p)$ is a 3 point, then $\nu(q) \leq r - 2$.*
   (e) *If $D \subset \pi^{-1}(p)$ is a 2 curve, then D is not r small or r-1 big.*
   or
2. *There exists a curve $C \subset \overline{S}_r(X)$ such that $p \in C$ and C is r big at p. There exists an affine neighborhood U of p such that the blow-up of $C \cap U$, $\pi : Y \to U$ is a permissible monoidal transform such that*
   (a) *$A_r(Y)$ holds.*

(b) *If $q \in \pi^{-1}(p)$ is a 2 point, then $\nu(q) \leq r - 1$.*
(c) *If $q \in \pi^{-1}(p)$ is the 3 point, then $\nu(q) \leq r - 2$.*
(d) *The 2 curve $D = \pi^{-1}(p)$ is not r small or r-1 big.*

In either case, if X satisfies the conclusions of Theorem 13.8, then Y satisfies the conclusions of Theorem 13.8.

Proof. p has permissible parameters (x, y, z) such that

$$
\begin{aligned}
u &= (x^a y^b)^m \\
v &= P(x^a y^b) + x^c y^d F_p \\
F_p &= \textstyle\sum_{i+j+k \geq r} a_{ijk} x^i y^j z^k
\end{aligned}
$$

Suppose that there does not exist a curve $C \subset \overline{S}_r(X)$ such that C is r big at p.

Let $\pi : X_1 \to X$ be the blow-up of p. We will first show that (a), (b) and (d) of 1. hold on X_1 and if $q \in \pi^{-1}(p)$ is a 2 point with $\nu(q) = r$ then $\tau(q) \geq \tau(p)$. This follows from Theorem 7.1, Theorem 7.3 and Lemma 7.9. All exceptional 2 curves D of π contain a 3 point q such that $\nu(q) \leq r - 2$. (e) thus holds by Lemmas 8.1 and 7.7.

By Lemma 8.1 there are at most finitely many 2 points $q \in \pi^{-1}(p)$ such that $\nu(q) = r$. Suppose that there exists a 2 point $q \in \pi^{-1}(p)$ and $\nu(q) = r$. After a permissible change of parameters at p, we have permissible parameters (x_1, y_1, z_1) at q such that $x = x_1, y = x_1 y_1, z = x_1 z_1$. $L_p = L_p(y, z)$ depends on both y and z.

Supppose there also exists a 2 point $q' \in \pi^{-1}(p)$ such that $\nu(q') = r$ and q' has permissible parameters (x', y', z') such that

$$ x = x'y', y = y', z = y'(z' + \alpha) $$

for some $\alpha \in k$. Then there exists a form $L(x, z - \alpha y)$ such that

$$
L_p(y, z) = \begin{cases} L(x, z - \alpha y) + \overline{c} x^{\overline{a}} y^{\overline{b}} & \text{if there exists } \overline{a}, \overline{b} \in \mathbf{N} \text{ such that} \\ & \overline{a} + \overline{b} = r, a(d + \overline{b}) - b(c + \overline{a}) = 0 \\ L(x, z - \alpha y) & \text{otherwise} \end{cases}
$$

Thus

$$ L_p = \overline{d}(z - \alpha y)^r + \overline{c} y^r $$

for some $\overline{d}, \overline{c} \in k$ with $\overline{d} \neq 0$, a contradiction to the assumption that $\tau(p) < r$. Let

$$ \cdots \to Y_n \to Y_{n-1} \to \cdots \to Y_1 \to X $$

be the sequence of quadratic transforms $\pi_n : Y_n \to Y_{n-1}$ constructed by blowing up all 2 points q' on Y_n which lie over p and have $\nu(q') = r$.

Suppose that this sequence has infinite length. Then there exists $q_n \in Y_n$ such that $\pi_n(q_n) = q_{n-1}$ and $\nu(q_n) = r$ for all n. There exists a series $\phi(x) = \sum \alpha_i x^i$ such that after replacing z with $z - \phi(x)$, q_n has permissible parameters (x_n, y_n, z_n) such that

$$ x = x_n, y = x_n^n y_n, z = x_n^n z_n $$

and

$$ F_{q_n} = L_q(y_n, z_n) + x_n \Omega_n. $$

$F_{q_n} = \frac{F_q}{x_n^{nr}}$ for all $n > 0$ implies $F_q \in (y, z)^r$.

$\hat{\mathcal{I}}_{\overline{S}_r(X),p} \subset (y, z)$ by Lemma 6.22. Since $\overline{S}_r(X)$ makes SNCs with $\overline{B}_2(X)$ at p, $y = z = 0$ are local equations at p of a curve $C \subset \overline{S}_r(X)$.

Now suppose that there exists a curve $C \subset \overline{S}_r(X)$ such that $p \in C$ and C is r big at p. There exists an affine neighborhood U of p such that $C \cap U$ makes SNCs with $\overline{B}_2(U)$. Let $\pi : Y \to U$ be the blow-up of $C \cap U$. There exist permissible parameters (x, y, z) at p such that $y = z = 0$ are local equations of C at p,

$$u = (x^a y^b)^m$$
$$F_p = \sum_{i+j \geq r} a_{ij}(x) y^i z^j.$$

At the 3 point $q \in \pi^{-1}(p)$, there are permissible parameters (x, y_1, z_1) such that

$$y = y_1 z_1, z = z_1$$

$$F_q = \frac{F_p}{z_1^r} = \sum_{i+j=r} a_{ij}(0) y_1^i + z_1 G + x\Omega$$

we have $\nu(q) \leq r - 2$, since $2 \leq \tau(p)$.

At a 2 point $q \in \pi^{-1}(p)$, after a permissible change of variables at p, there exist permissible parameters (x, y_1, z_1) at q such that

$$y = y_1, z = y_1 z_1$$

$$F_q = \frac{F_p}{y_1^r} = \sum_{i+j=r} a_{ij}(0) z_1^j + y_1 G + x\Omega.$$

$\nu(q) < r$ and $\gamma(q) < r$ since $\tau(p) < r$. Furthermore, if $D = \pi^{-1}(p)$, then $F_q \notin \hat{\mathcal{I}}_{D,q}^{r-1}$. There exists a possibly smaller affine neighborhood U of p such that $A_r(Y)$ holds by Lemma 8.8.

$\square$

Theorem 14.4. *Suppose that the conclusions of Theorem 13.8 hold on X with $r \geq 2$, $p \in X$ is a 2 point with permissible parameters (x, y, z) such that*

$$u = (x^a y^b)^m$$
$$v = P(x^a y^b) + x^c y^d F_p$$

and $\nu(p) = r - 1$, $\tau(p) = 0$, $L_p = f(x, y)$ depends on both x and y. Then there exists a sequence of quadratic transforms $\pi : Z \to X$ over p such that

1. *$q \in \pi^{-1}(p)$ a 1 point or a 2 point implies that $\nu(q) \leq r$. $\nu(q) = r$ implies $\gamma(q) = r$.*
2. *$q \in \pi^{-1}(p)$ a 2 point with $\nu(q) = r - 1$ implies that $\tau(q) > 0$ or $\gamma(q) = r$.*
3. *$q \in \pi^{-1}(p)$ a 3 point implies $\nu(q) \leq r - 2$.*
4. *The conclusions of Theorem 13.8 hold on Z*

Proof. Let

$$\pi : X_1 \to X \tag{164}$$

be the blow-up of p.

If $p_1 \in \pi^{-1}(p)$ is a 1 point then $\nu(p_1) \leq r$ and $\nu(p_1) = r$ implies $\gamma(p_1) = r$ by Theorem 7.1. If $p_1 \in \pi^{-1}(p)$ is a 2 point then we must have $\nu(p_1) \leq r - 2$, by our assumption on f. Suppose that $p_1 \in \pi^{-1}(p)$ is the 3 point. Then $\nu(p_1) \leq r-1$ and p_1 has permissible parameters (x_1, y_1, z_1) such that

$$x = x_1 z_1, y = y_1 z_1, z = z_1$$

Suppose that $\nu(p_1) = r - 1$. Then

$$L_{p_1} = f(x_1, y_1) + z_1 \Omega. \tag{165}$$

Let

$$F_p = \sum_{i+j+k \geq r-1} a_{ijk} x^i y^j z^k.$$

Suppose that we can construct an infinite sequence of quadratic transforms

$$\cdots \to X_n \to \cdots \to X_1 \to X$$

where $X_{n+1} \to X_n$ is the blow-up of a 3 point p_n lying over p_{n-1} with $\nu(p_n) = r - 1$. Then p_n has permissible parameters (x_n, y_n, z_n) such that

$$x = x_n z_n^n, y = y_n z_n^n, z = z_n$$

and

$$F_{p_n} = \frac{F_p}{z_n^{n(r-1)}} = \sum a_{ijk} x_n^i y_n^j z_n^{(k+n(i+j-r+1))}.$$

Thus $a_{ijk} = 0$ if $i + j < r - 1$, which implies that $F_p \in (x, y)^{r-1}$, a contradiction since the conclusions of Theorem 13.8 hold.

Thus by Theorem 7.1 and Lemma 7.9 there exists a finite sequence of quadratic transforms

$$\pi : X_m \to \cdots \to X_1 \to X$$

where $X_{n+1} \to X_n$ is the blow-up of a 3 point p_n lying over p_{n-1} with $\nu(p_n) = r - 1$, such that $A_r(X_m)$ holds, $\nu(q) \leq r - 2$ if $q \in \pi^{-1}(p)$ is a 3 point, and if $q \in \pi^{-1}(p)$ is a 1 point then $\nu(q) \leq r$, $\nu(q) = r$ implies $\gamma(q) = r$. Suppose that C is a 2 curve which is exceptional for π. Then C is not r-1 big or r small since C must contain a 3 point q' with $\nu(q') \leq r - 2$. Suppose that $q \in \pi^{-1}(p)$ is a 2 point and $\nu(q) \geq r - 1$. Then there exists a largest n such that q maps to a 3 point $p_n \in X_n$. The point q is then a 2 point on X_{n+1}. p_n has permissible parameters (x_1, y_1, z_1) such that

$$x = x_1 z_1^n, y = y_1 z_1^n, z = z_1$$

By assumption, $\nu(p_n) = r - 1$. Write

$$f = \sum_{i+j=r-1} a_{ij} x^i y^j.$$

We then have

$$u = (x_1^a y_1^b z_1^{n(a+b)})^m$$
$$v = P(x_1^a y_1^b z_1^{n(a+b)}) + x_1^c y_1^d z_1^{n(c+d+r-1)} F_{p_n}$$
$$F_{p_n} = \frac{F_p}{z_1^{n(r-1)}} = \sum_{i+j=r-1} a_{ij} x_1^i y_1^j + \sum_{i+j+k=r-1, k>0} b_{ijk} x_1^i y_1^j z_1^k + \Omega \tag{166}$$

with $\nu(\Omega) \geq r$.

Since q is a 2 point, $\hat{\mathcal{O}}_{X_{n+1},q}$ has regular parameters (x_2, y_2, z_2) of one of the following forms:

$$x_1 = x_2, y_1 = x_2(y_2 + \alpha), z_1 = x_2 z_2 \tag{167}$$

with $\alpha \neq 0$, or

$$x_1 = x_2, y_1 = x_2 y_2, z_1 = x_2(z_2 + \beta) \tag{168}$$

with $\beta \neq 0$, or

$$x_1 = x_2 y_2, y_1 = y_2, z_1 = y_2(z_2 + \beta) \tag{169}$$

with $\beta \neq 0$.

First suppose that (168) holds. (169) is symmetrical, and the analysis of that case is the same. Set

$$x_2 = \bar{x}_2(z_2 + \beta)^{-\frac{n}{n+1}}.$$

$$u = (\bar{x}_2^{(n+1)(a+b)} y_2^b)^m = (\bar{x}_2^{\bar{a}} y_2^{\bar{b}})^{\overline{m}}$$
$$v = P_q(\bar{x}_2^{\bar{a}} y_2^{\bar{b}}) + \bar{x}_2^{(n+1)(c+d+r-1)} y_2^d F_q$$
$$F_q = [\sum_{i+j=r-1} a_{ij} y_2^j$$
$$+ \sum_{k>0, i+j+k=r-1} b_{ijk} y_2^j (z_2 + \beta)^k + \bar{x}_2 \Omega^1] - \sum c_i \bar{x}_2^{\bar{a}_i} y_2^{\bar{b}_i}$$

where $(\bar{a}, \bar{b}) = 1$,

$$(n+1)(a+b)(\bar{b}_i + d) - b((n+1)(c+d+r-1) + \bar{a}_i) = 0.$$

If some $b_{ijk} \neq 0$ in (166), we have $\nu(q) \leq r - 1$ and $\nu(q) = r - 1$ implies that $\tau(q) > 0$. So suppose that all $b_{ijk} = 0$ in (166). If $\nu(q) \geq r - 1$, then we must have

$$\sum a_{ij} y_2^j = a_{i_0 j_0} y_2^{j_0} + a_{0,r-1} y_2^{r-1}$$

where $a_{0,r-1}$ could be zero, $0 < i_0$, $a_{i_0,j_0} \neq 0$ (since $f(x, y)$ depends on x and y) and

$$(n+1)(c+d+r-1)b - (n+1)(a+b)(j_0 + d) = 0.$$

Thus

$$(c+r-1-j_0)b - a(d+j_0) = 0 \tag{170}$$

We then have

$$x^c y^d f = a_{i_0 j_0} x^{c+r-1-j_0} y^{d+j_0} + a_{0,r-1} x^c y^{d+r-1}$$

which is normalized, so that $a_{i_0,j_0} = 0$ by (170). This contradiction shows that we must have that $\nu(q) \leq r - 2$ in this case.

Suppose that (167) holds. Substitute (167) into (166). Set

$$x_2 = \bar{x}_2(y_2 + \alpha)^{-\frac{b}{(n+1)(a+b)}}$$

$$u = (\bar{x}_2^{(n+1)} z_2^n)^{m(a+b)}$$
$$v = P(\bar{x}_2^{(n+1)(a+b)} z_2^{n(a+b)}) + \bar{x}_2^{(n+1)(c+d+r-1)} z_2^{n(c+d+r-1)} G$$

where

$$G = (y_2 + \alpha)^\lambda \left(\sum_{i+j=r-1} a_{ij}(y_2 + \alpha)^j + \sum_{k>0, i+j+k=r-1} b_{ijk}(y_2 + \alpha)^j z_2^k + \bar{x}_2 \Omega_2 \right),$$

with

$$\lambda = -\frac{b}{a+b}(c+d+r-1)+d.$$

The only term which can be removed from the first sum

$$(y_2 + \alpha)^\lambda \left(\sum_{i+j=r-1} a_{ij}(y_2 + \alpha)^j) \right)$$

of G in obtaining F_q is the constant term. Thus $\gamma(q) \le r$.   □

Theorem 14.5. *Suppose that $r \ge 2$ and the conclusions of Theorem 13.8 hold on X, so that if C is a 2 curve, then C is not r small or r-1 big. Suppose that $p \in X$ is a 1 or a 2 point and D is a generic curve through p on a component of E_X. Then there exists a sequence of quadratic transforms centered over a finite number of points on the strict transform of D, but not in the fiber over p, $\pi : X_1 \to X$, such that the following conditions hold.*

1. *There exists a neighborhood U of $D - p$ such that $C_r(\pi^{-1}(U))$ holds. The case 2. (c) of C_r does not occur in $\pi^{-1}(U)$.*
2. *Let D' be the strict transform of D on X_1. Suppose that $q \in D' - p$, and (x,y,z) are permissible parameters at q such that $x = z = 0$ are local equations of D' at q. If q is a 1 point then $\nu(F_q(0,y,0)) = 1$. If q is a 2 point then $\nu(q) = 0$.*
3. *The conclusions of Theorem 13.8 hold on X_1.*

Proof. Suppose that $q \in D$ is a 1 point. Then we can find permissible parameters (x,y,z) at q such that $x = y = 0$ are local equations of D at q. The multiplicity

$$\phi(q) = \nu(F_q(0,0,z))$$

is independent of such permissible parameters at q. Furthermore, the set

$$\{q \in D \cap (X - \overline{B}_2(X)) \mid \phi(q) \ge 2\}$$

is Zariski closed in $D \cap (X - \overline{B}_2(X))$. By Lemma 6.30, $F_q \notin \hat{I}_{D,q}$ if $q \in D$. At most 1 points q on D, $\phi(q) = 1$. Thus there are at most a finite number of points $q \in D - p$ such that the conclusions of the Theorem do not hold at q.

1) Suppose that $q \in D - p$ and $\nu(q) = r$. Then q is a generic point on a curve C of $\overline{S}_r(X)$. q is a 1 point.

1a) Suppose that C is r big. Then there exist permissible parameters (x,y,z) at q such that

$$\begin{aligned} u &= x^a \\ v &= P(x) + x^b F_q \\ F_q &= \sum_{i+k \ge r, j \ge 0} a_{ijk} x^i y^j z^k \end{aligned}$$

where $x = z = 0$ are local equations of C at q, $x = y = 0$ are local equations of D at q. $\gamma(q) = r$ implies $a_{00r} \ne 0$.

Let $\pi : Y \to X$ be the blow-up of q. Then $\nu(q') = 0$ if q' is the point on the intersection of the strict transform of D and $\pi^{-1}(q)$. Points of $\pi^{-1}(q)$ satisfy the condition of C_r by Theorem 7.3 and Lemma 7.9.

1b) Suppose that C is r small. By Lemma 6.24,

$$F_q = \sum_{i+j \geq r} a_{ij}(y)x^i z^j + \tau(y)x^{r-1}$$

where $x = z = 0$ are local equations of C at q, $x = y = 0$ are local equations of D at q, (with $\nu(\tau) \geq 1$). Since q is a generic point of C, $\nu(\tau) = 1$, and after a permissible change of parameters, we have $\tau = y$. $\gamma(q) = r$ implies $a_{0r}(y)$ is a unit. Let $\pi : Y \to X$ be the blow-up of q. Then $\nu(q') = 0$ if q' is the point on the intersection of the strict transform of D and $\pi^{-1}(q)$. Points of $\pi^{-1}(q)$ satisfy the condition of C_r by Theorem 7.3 and Lemma 7.9.

2) Suppose that $q \in D - p$, $\nu(q) = r - 1$ and the conclusions of the Theorem do not hold at q.

2a) Suppose that q is a 1 point and $r \geq 3$. Then q is a general point on a curve C in $\overline{S}_{r-1}(X)$. There are permissible parameters (x, y, z) at q such that $x = z = 0$ are local equations of D at q.

Let $\pi_1 : X_1 \to X$ be the blow-up of p. Theorem 7.1 implies $\nu(q') \leq r - 1$ for all $q' \in \pi_1^{-1}(q)$ and $q' \in \pi_1^{-1}(q)$ a 2 point with $\nu(q') = r - 1$ implies $\tau(q') > 0$.

At the 2 point $q_1 \in \pi_1^{-1}(q)$ on the strict transform of D, there are permissible parameters (x_1, y_1, z_1) such that

$$x = x_1 y_1, y = y_1, z = y_1 z_1$$

Suppose that $\nu(q_1) = r - 1$. We must have $\tau(q_1) > 0$. Let $\pi_1 : X_2 \to X_1$ be the blow-up of q_1. By Theorem 7.3, if $q' \in \pi^{-1}(q_1)$, then if q' is a 1 point $\nu(q') \leq r - 1$. If q' is a 2 point, $\nu(q') \leq r - 1$, $\nu(q') = r - 1$ implies $\tau(q') > 0$. q' a 3 point implies $\nu(q') \leq r - 2$. Let $q_2 \in \pi_1^{-1}(q)$ be the 2 point on the strict transform of D. There are permissible parameters (x_2, y_2, z_2) at q_2 such that

$$x_1 = x_2 y_2, y_1 = y_2, z_1 = y_2 z_2.$$

If $\nu(q_2) = r - 1$, then $\tau(q_2) > 0$.

Suppose that we can construct an infinite sequence of quadratic transforms

$$\cdots \to X_n \to X_{n-1} \to \cdots \to X_1 \to X$$

centered at the point q_n on the strict transform of D over q on X_n, where q_n are blown up as long as $\nu(q_n) = r - 1$.

By Theorem 7.3, all points q' on X_n lieing over p satisfy $\nu(q') \leq r - 1$, $\nu(q') \leq r - 2$ If q' is a 3 point and if q' is a 2 point with $\nu(q') = r - 1$. Then $\tau(q') > 0$.

Suppose that $\nu(q_n) = r - 1$ for all n. Then q_n has permissible parameters (x_1, y_1, z_1) such that

$$x = x_1 y_1^n, y = y_1, z = z_1 y_1^n$$

$$F_q = \sum_{i+j+k \geq r-1} a_{ijk} x^i y^j z^k$$

$$F_{q_n} = \frac{F_q}{y_1^{n(r-1)}} = \sum_{i+j+k \geq r-1} a_{ijk} x_1^i y_1^{j+n(i+k-(r-1))} z_1^k$$

implies $a_{ijk} = 0$ if $i + k < r - 1$, so that $F_q \in (x, z)^{r-1}$. This is a contradiction since $F_q \notin \hat{I}_{D,q}$.

Thus after a finite sequence of quadratic transforms, $\pi : Z \to X$ the strict transform of D intersects $\pi^{-1}(q)$ at a 2 point q_1 with $\nu(q_1) < r - 1$, so we are in case 3) below.

2b) Suppose that q is a 2 point. Suppose that C is the 2 curve through q. By Lemma 6.25, our assumption that C is not r-1 big, and since q is a generic point of C, we have $\tau(q) > 0$. There exist permissible parameters (x, y, z) at q such that

$$\begin{aligned}
u &= (x^a y^b)^m \\
v &= P(x^a y^b) + x^c y^d F_q \\
F_q &= \sum_{i+j \geq r-1, k \geq 0} a_{ijk} x^i y^j z^k + z x^{i_0} y^{j_0}
\end{aligned} \tag{171}$$

with $i_0 + j_0 = r - 2$. $x = z = 0$ are local equations of D at q and $\tau(q) > 0$.

Let $\pi : X_1 \to X$ be the blow-up of q. Then $\nu(q_1) \leq r - 1$ at all points $q_1 \in \pi^{-1}(q)$, $\nu(q_1) \leq r - 2$ if $q_1 \in \pi^{-1}(q)$ is a 3 point and all 2 points q_1 of $\pi^{-1}(q)$ with $\nu(q_1) = r - 1$ satisfy $\tau(q_1) > 0$ by Theorem 7.3.

The strict transform of D intersects $\pi^{-1}(q)$ at a 2 point q' such that

$$x = x_1 y_1, y = y_1, z = y_1 z_1$$

$$F_{q_1} = \frac{F_q}{y_1^{r-1}} = \sum_{i+j \geq r-1, k \geq 0} a_{ijk} x_1^i y_1^{(i+j+k)-(r-1)} z_1^k + z_1 x_1^{i_0}$$

$x_1 = z_1 = 0$ are local equations of the strict transform of D at q'. If $\nu(q') \leq r-2$ we are in case 3). Otherwise, q_1 is a 2 point with $\nu(q_1) = r - 1$ and $\tau(q_1) > 0$ (so that $i_0 = r - 2$). Let

$$\cdots \to X_n \to X_{n-1} \to \cdots \to X_1 \to X$$

be the sequence of quadratic transforms centered at the point q_n on the strict transform of D over q on X_n where points q_n are blown up as long as $\nu(q_n) = r - 1$. By Theorem 7.3, all points q' on X_n lieing over p satisfy $\nu(q') \leq r - 1$, $\nu(q') \leq r - 2$ if q' is a 3 point, and if q' is a 2 point with $\nu(q') = r - 1$, then $\tau(q') > 0$.

Suppose that $\nu(q_n) = r-1$ for all n. q_n has permissible parameters (x_1, y_1, z_1) such that

$$x = x_1 y_1^n, y = y_1, z = z_1 y_1^n$$

$$F_{q_n} = \frac{F_q}{y_1^{n(r-1)}} = \sum_{i+j \geq r-1, k \geq 0} a_{ijk} x_i^i y_1^{j+n(i+k-(r-1))} z_i^k + z_1 x_1^{r-2}$$

Thus $i+k-(r-1) \geq 0$ whenever $a_{ijk} \neq 0$ and $F_p \in (x, z)^{r-1}$, so that $F_p \in \hat{I}_{D,p}$, which is a contradiction.

After a finite sequence of quadratic transforms, $\pi' : X' \to X$, the strict transform of D thus intersects $(\pi')^{-1}(q)$ at a 2 point q' with $\nu(q') \leq r-2$, so the result follows from Case 3).

3) Suppose that $q \in D - p$, $\nu(q) \leq r-2$, and the conclusions of the Theorem do not hold at q. q is a 1 point or a 2 point and D makes SNCs with the 2 curve through C. The result then follows from a similar but slightly simpler argument to that of Case 2, by Theorems 7.1 and 7.3. $\qquad\square$

Theorem 14.6. *Suppose that X satisfies the conclusions of Theorem 13.8 with $r \geq 2$. Then there exists a sequence of permissible monoidal transforms $X_1 \to X$ such that $C_r(X_1)$ holds.*

Proof. Let T be the finite set of 2 points p on X such that (162) holds at p, and $p \notin D$ for any r big curve D which contains a 1 point.

By 1. of Theorem 13.7, there exists a sequence of quadratic transforms $\pi_0 : X_0 \to X$ centered over points $p \in T$ such that

1. If $p \in X_0$ is a 1 point then $\nu(p) \leq r$, $\nu(p) = r$ implies $\gamma(p) = r$.
2. If $p \in X_0$ is a 2 point then $\nu(p) \leq r$. $\nu(p) = r$ implies $\tau(p) \geq 2$, or (162) holds at p and there exists an r big curve $D \subset \bar{S}_r(X_0)$ containing p.
3. If $p \in X_0$ is a 3 point, then $\nu(p) \leq r-1$. $\nu(p) = r-1$ implies p satisfies the assumptions of (145) and (146) of Theorem 12.4. If D_p is the 2 curve containing p with local equations $y = z = 0$ at p in the notation of Theorem 12.4, then F_q is resolved for all $p \neq q \in D_p$.
4. $A_r(X_0)$ holds.
5. If C is a 2 curve on X_0, then C is not r small. If C is r-1 big, then $\nu(p) = r-1$ for all $p \in C$.

Let T_1 be the 3 points of X_0 which satisfy (145) and (146) of Theorem 12.4.

For $p \in T_1$, let $\lambda_p : Y_p \to \operatorname{spec}(\mathcal{O}_{X_0,p})$ be the sequence of monoidal transforms centered over sections of D_p such that the conclusions of Theorem 12.4 hold. By Theorem 12.5 (or Theorem 12.6 if $r = 2$), there exists a sequence of monoidal transforms $V_p \to Y_p$ centered at 2 curves C such that C is r-1 big so that V_p satisfies the conclusions of 1. - 3. of Theorem 12.5 (or Theorem 12.6 if $r = 2$).

Since D_p is resolved at all points $p \neq q \in D_p$, the only obstruction to extending λ_p to a permissible sequence of monoidal transforms of sections over D_p in X_0 is if the corresponding sections over D_p do not make SNCs with 2 curves. This difficulty can be removed by performing quadratic transforms at the (resolved) points where the section does not make SNCs with the 2 curves.

By 5. above and Lemmas 8.6, 8.7 and 13.2, we can thus construct a sequence of permissible monoidal transforms $\pi_0' : X_0' \to X_0$ such that

$$X_0' \times_{X_0} \operatorname{spec}(\mathcal{O}_{X_0,p}) \cong V_p$$

for $p \in T_1$,
    and X_0' satisfies:

1. $p \in X_0'$ a 1 point implies $\nu(p) \leq r$. $\nu(p) = r$ implies $\gamma(p) = r$.
2. $p \in X_0'$ a 2 point implies $\nu(p) \leq r$. $\nu(p) = r$ implies $\tau(p) \geq 2$ or (162) holds at p, and there exists a r big curve $D \subset \overline{S}_r(X_0)$ containing p.
3. $p \in X_0'$ a 3 point implies $\nu(p) \leq r - 2$.
4. $\overline{S}_r(X_0')$ makes SNCSs with $\overline{B}_2(X_0')$.
5. If C is a 2 curve on X_0, then C is not r small or r-1 big.

Let $\gamma_1, \ldots, \gamma_n$ be the r big curves in $\overline{S}_r(X_0')$. Each γ_i necessarily contains a 1 point.

Let $\pi : X_1 \to X_0'$ be the sequence of monoidal transforms (in any order) centered at the (strict transforms of) $\gamma_1, \ldots, \gamma_n$.

By Lemma 8.8, 2. of Theorem 14.3, and 2. of Theorem 13.7,

1. If $p \in X_1$ is a 1 point then $\nu(p) \leq r$. $\nu(p) = r$ implies $\gamma(p) = r$.
2. If $p \in X_1$ is a 2 point then $\nu(p) \leq r$. $\nu(p) = r$ implies $\tau(p) \geq 2$. If $\nu(p) = r$ and $\tau(p) < r$, then p does not lie on a r big curve E in $\overline{S}_r(X_1)$.
3. $p \in X_1$ a 3 point implies $\nu(p) \leq r - 1$. $\nu(p) = r - 1$ implies p satisfies the assumptions of (147) and (148) of Theorem 12.4. If D_p is the 2 curve containing p with local equations $y = z = 0$ at p in the notation of Theorem 12.4, then F_q is resolved for all $p \neq q \in D_p$.
4. $A_r(X_1)$ holds.
5. If C is a 2 curve on X_0, then C is not r small or r-1 big.
6. There are only finitely many 2 points $p \in X_1$ such that $\nu(p) = r$.

6. is a consequence of 5. Let T_2 be the set of 3 points on X_1 satisfying (147) and (148) of Theorem 12.4.

By Theorem 14.3, there exists a sequence of quadratic transforms $\pi_2 : X_2 \to X_1$ centered over the 2 points p of X_1 with $\nu(p) = r$ and $2 \leq \tau(p) < r$ such that

1. If $p \in X_2$ is a 1 or 2 point then $\nu(p) \leq r$. If $\nu(p) = r$ then $\gamma(p) = r$.
2. If $p \in X_2$ is a 3 point then $\nu(p) \leq r - 1$. If $\nu(p) = r - 1$, then $p \in T_2$.
3. $A_r(X_2)$ holds.
4. If $D \subset X_2$ is a 2 curve, then D is not r small or r-1 big.

For $p \in T_2$, let

$$\lambda^p : Y_p \to \operatorname{spec}(\mathcal{O}_{X_2,p}) \tag{172}$$

be a sequence of permissible monoidal transforms over sections of D_p such that the conclusions of Theorem 12.4 hold.

Since D_p is resolved at all points $p \neq q \in D_p$, the only obstruction to extending λ^p to a permissible sequence of monoidal transforms of sections over D_p in X_2 is if the corresponding sections over D_p do not make SNCs with 2 curves. This difficulty can be removed by performing quadratic transforms at the points where the section does not make SNCs with the 2 curves.

By Theorems 7.1 and 7.3, we can thus construct a permissible sequence of monoidal transforms $\pi_3 : X_3 \to X_2$ such that

1. If $p \in T_2$, then $X_3 \times_{X_2} \operatorname{spec}(\mathcal{O}_{X_2,p}) \cong Y_p$.

2. If $q \in X_3 - \pi_3^{-1}(T_2)$, and q is a 1 or 2 point then $\nu(q) \leq r$. If $\nu(q) = r$, then $\gamma(q) = r$.
3. If $q \in X_3 - \pi_3^{-1}(T_2)$ and q is a 3 point then $\nu(q) \leq r - 2$.
4. $A_r(X_3)$ holds.
5. If $D \subset X_3 - \pi_3^{-1}(T_2)$ is a 2 curve, then D is not r small or r-1 big.

By 5. and Theorem 12.5 (or Theorem 12.6 if $r = 2$), we can perform a sequence of permissible monoidal transforms $\sigma : \overline{Z}_2 \to X_3$ centered at r-1 big 2 curves C to get that

1. If $p \in \overline{Z}_2$ is a 1 or 2 point, then $\nu(p) \leq r$. $\nu(p) = r$ implies $\gamma(p) = r$.
2. If $p \in \overline{Z}_2$ is a 3 point, then $\nu(p) \leq r - 2$.
3. There are no 2 curves C in $\overline{Z}_2$ which are r small or r-1 big.
4. $\overline{S}_r(\overline{Z}_2)$ makes SNCs with $\overline{B}_2(Z_2)$.

Since 3. holds, there are only finitely many 2 points $\{q_1, \ldots, q_m\}$ on $\overline{Z}_2$ such that $\nu(q_i) = r - 1$ and $\tau(q_i) = 0$.

By Theorem 14.4, we can perform a sequence of quadratic transforms $\sigma_1 : W_1 \to \overline{Z}_2$ over the finitely many 2 points q_i in $\overline{Z}_2$ such that $\nu(q_i) = r - 1$, $\tau(q_i) = 0$ and L_q depends on both x and y (where (x, y, z) are permissible parameters at q_i) so that

1. $\nu(q) \leq r$ and $\nu(q) = r$ implies $\gamma(q) = r$ at 1 and 2 points of W_1.
2. $\nu(q) \leq r - 2$ at 3 points of W_1
3. If $q \in W_1$ is a 2 point with $\nu(q) = r - 1$ and $\tau(q) = 0$, then either $\gamma(q) = r$ or there exist permissible parameters (x, y, z) at q such that L_q depends only on x.
4. There are no 2 curves C on W_1 which are r small or r-1 big.
5. $\overline{S}_r(W_1)$ makes SNCs with $\overline{B}_2(W_1)$.

Over the (finitely many) points $\{a_1, \ldots, a_n\}$ of W_1 which are 2 points with $\nu(p) = r - 1$, $\gamma(p) > r$, $\tau(p) = 0$, and there exist permissible parameters (x, y, z) at a_i such that L_{a_i} depends only on x, by Theorem 12.1, there exist sequences of permissible monoidal transforms

$$Y_{a_i} \to \operatorname{spec}(\mathcal{O}_{W_1, a_i})$$

where $Y_{a_i} \to \operatorname{spec}(\mathcal{O}_{W_1, a_i})$ is a sequence of blow-ups of sections over a general curve through a_i, and satisfies the conclusions of Theorem 12.1.

By Theorem 14.5 there exists a sequence of permissible monoidal transforms $\tilde{\pi} : W_2 \to W_1$ such that

$$W_2 \times_{W_1} \operatorname{spec}(\mathcal{O}_{W_1, a_i}) \cong Y_{a_i}$$

for all i, and for $q \in \tilde{\pi}^{-1}(W_2 - \{a_1, \ldots, a_n\})$,

1. $\nu(q) \leq r$, $\nu(q) = r$ implies $\gamma(q) \leq r$ if q is a 1 or 2 point.
2. q a 2 point and $\nu(q) = r - 1$ implies $\tau(q) > 0$ or $\gamma(q) = r$.
3. $\nu(q) \leq r - 2$ if q is a 3 point.
4. There are no 2 curves C in $\tilde{\pi}^{-1}(W_2 - \{a_1, \ldots, a_n\})$ which are r small or r-1 big.
5. $A_r(W_2)$ holds.

Since all 2 curves $C \subset W_2$ which are r-1 big must map to some a_i by 4., there exists a sequence of permissible monoidal transforms $\pi_3 : W_3 \to W_2$ centered at 2 curves C which are r-1 big such that

$$W_3 - (\tilde{\pi} \circ \pi_3)^{-1}(\{a, \dots, a_n\}) \cong W_2 - \tilde{\pi}^{-1}(\{a_1, \dots, a_n\})$$

and $W_3 \times_{W_1} \operatorname{spec}(\mathcal{O}_{W_1,a_i})$ satisfies the conclusions of V_{a_i} of Theorem 12.2 (or of V_{a_i} of Theorem 12.3).

Then the sequence of quadratic transforms $W_{a_i} \to V_{a_i}$ of Theorem 12.2 (or of $W_{a_i} \to V_{a_i}$ of Theorem 12.3) extend to $\pi_4 : W_4 \to W_3$ such that

$$W_4 - (\tilde{\pi} \circ \pi_3 \circ \pi_4)^{-1}(\{a_1, \dots, a_n\}) \cong W_2 - \tilde{\pi}^{-1}(\{a_1, \dots, a_n\})$$

and $W_4 \times_{W_1} \operatorname{spec}(\mathcal{O}_{W_1,a_i}) \cong W_{a_i}$ for $1 \leq i \leq n$ in the notation of Theorem 12.2 (or of Theorem 12.3).

Now assume that $r \geq 3$. Let $\{D\}$ be the strict transform on W_4 of the curves $\{\overline{D}\}$ in $\overline{S}_r(W_1)$ which contain some a_i. Each $\overline{D}$ contains a 1 point and $\overline{D}$ is r small since $\overline{D}$ makes SNCs with $\overline{B}_2(W_1)$ and $\nu(a_i) = r - 1$. By Theorem 12.2 and Lemma 8.10, and since by Lemma 6.26 there does not exist a 2 point $q \in D$ such that $\nu(q) = r - 1$ and $\tau(q) > 0$, there exists a finite sequence of quadratic transforms $W_5 \to W_4$ centered at points disjoint from any fiber over some a_i such that if $W_6 \to W_5$ is a sequence of monoidal transforms centered at the strict transforms of the D then $C_r(W_6)$ holds.

Now suppose that $r = 2$. Let $\{D\}$ be the strict transforms on W_4 of the curves $\{\overline{D}\}$ in $\overline{S}_r(W_1)$ which contain some a_i. Each $\overline{D}$ contains a 1 point and $\overline{D}$ is not r big. By Theorem 12.3, Lemmas 8.10 and 8.11 there exists a sequence of quadratic transforms $W_5 \to W_4$ centered at points disjoint from any fiber over some a_i such that if $W_6 \to W_5$ is a sequence of monoidal transforms centered at the strict transforms of the D, and then followed by a sequence of monoidal transforms $W_7 \to W_6$ centered at the strict transforms of 2 curves C on W_6 which are sections over one of the D blown up in $W_6 \to W_5$ and such that C is 1 big, then $C_r(W_6)$ holds. $\qquad\square$

Theorem 14.7. *Suppose that $C_r(X)$ holds with $r \geq 2$. Then there exists a sequence of quadratic transforms $\pi : X_1 \to X$ such that $C_r(X_1)$ holds and if C is a 2 curve on X_1 such that C contains a 2 point p with $\nu(p) = r$ and p lies on a curve D in $\overline{S}_r(X_1)$, then for $p \neq q \in C - B_3(X)$, $\nu(q) \leq r - 1$ and $\nu(q) = r - 1$ implies $\gamma(q) = r - 1$.*

Proof. Let $\pi : X_1 \to X$ be the product of quadratic transforms centered at all 2 points $q \in X$ such that $\nu(q) = r$ and q is on a curve $D \subset \overline{S}_r(X)$.

Suppose that $q \in X$ is a 2 point on a 2 curve C such that $q \in D$, for a curve $D \subset \overline{S}_r(X)$. There exist permissible parameters (x, y, z) at q such that

$$
\begin{aligned}
u &= (x^a y^b)^m \\
v &= P(x^a y^b) + x^c y^d F_q
\end{aligned}
$$

where $x = z = 0$ are local equations of D at q. Suppose that $\nu(q) = r$ so that $\gamma(q) = r$. By Lemma 8.5, after a permissible change of variables,

$$F_q = \tau z^r + \sum_{i=2}^{r} a_i(x, y) z^{r-i}$$

with τ a unit, $\nu(a_i) \geq i$. At a 1 point $q' \in \pi^{-1}(q)$ we have $\nu(q') \leq r$, $\nu(q') = r$ implies $\gamma(q') = r$.

The 2 points $q' \in \pi^{-1}(q)$ have permissible parameters (x_1, y_1, z_1) such that

$$x = x_1 y_1, y = y_1, z = y_1(z_1 + \alpha)$$

or

$$x = x_1, y = x_1 y_1, z = x_1(z_1 + \alpha).$$

In either case

$$F_{q'} = \tau(z_1 + \alpha)^r + \text{ terms of order } \leq r - 2 \text{ in } z_1$$

implies $\gamma(q') \leq r$, $\gamma(q') \leq r - 1$ if $\alpha \neq 0$. Thus each exceptional curve $\overline{C}$ of π contains at most one 2 point q' such that $\nu(q') = r$. At the 3 point $q' \in \pi^{-1}(q)$,

$$x = x_1 z_1, y = y_1 z_1, z = z_1$$

and $\nu(q') = 0$.

Thus by Lemma 7.9, $C_r(X_1)$ holds. Let C' be the strict transform of C, D' the strict transform of D on X_1. C' and D' are disjoint. C' intersects $\pi_1^{-1}(q)$ at the 3 point q' with $\nu(q') = 0$. By Lemma 7.9 there is at most one curve E in $\overline{S}_r(X_1)$ such that $E \subset \pi^{-1}(q)$, and E intersects each 2 curve in at most one point. If E intersects an exceptional 2 curve $\overline{C}$ in a point q' such that $\nu(q') = r$, then for $q' \neq q'' \in \overline{C}$, $\gamma(q'') \leq r - 1$ (by the above analysis). Thus each exceptional 2 curve $\overline{C}$ for π_1 satisfies the conditions of the conclusions of the Theorem.

The strict transform C' of a 2 curve C on X_1 contains no 2 points q with $\nu(q) = r$ which are contained in a curve D in $\overline{S}_r(X_1)$.

□

15. Resolution 3

Throughout this section we will assume that $\Phi_X : X \to S$ is weakly prepared.

Lemma 15.1. *Suppose that $C \subset X$ is a 2 curve. Suppose that t is a natural number or ∞. Then the set*

$$\{q \in C \mid q \text{ is a 2 point and } \gamma(q) \geq t\}$$

is Zariski closed in $C - B_3(X)$.

Proof. Suppose that $p \in C$ is a 2 point. There exist permissible parameters (x, y, z) at p such that (x, y, z) are uniformizing parameters in an étale neighborhood U of p in X. At p,

$$\begin{aligned} u &= (x^a y^b)^m \\ v &= P(x^a y^b) + x^c y^d F(x, y, z). \end{aligned}$$

Set

$$w = \frac{v - P_\lambda(x^a y^b)}{x^c y^d}$$

with $\lambda > c + d$. $w \in \Gamma(U, \mathcal{O}_X)$. If $q \in C \cap U$, there are permissible parameters $(x, y, z_q = z - \alpha)$ at q for some $\alpha \in k$. There exist $a_i(q) \in k$ such that

$$F_q = w - \sum_{i=0}^{\infty} a_i(q) \frac{(x^a y^b)^i}{x^c y^d}.$$

$$\{q \in C \cap U \mid \nu(F_q(0, 0, z_q)) \geq t\} = \begin{cases} \{q \in C \cap U \mid \frac{\partial^i w}{\partial z^i}(0, 0, \alpha) = 0, 0 \leq i < t\} \\ \quad \text{if } ad - bc \neq 0 \\ \{q \in C \cap U \mid \frac{\partial^i w}{\partial z^i}(0, 0, \alpha) = 0, 0 < i < t\} \\ \quad \text{if } ad - bc = 0 \end{cases}$$

is Zariski closed.

Since U is an étale cover of an affine neighborhood V of p,

$$\{q \in C \mid q \text{ is a 2 point and } \gamma(q) \geq t\} \cap V$$

is Zariski closed in $V \cap C$. $\qquad \square$

Lemma 15.2. *Suppose that C is a 2 curve and there exists $p \in C$ with permissible parameters (x_p, y_p, z_p) at p such that $x_p = y_p = 0$ are local equations of C at p and $\nu(F_p(0, 0, z_p)) < \infty$. If $q \in C$ then $\nu(F_q(0, 0, z_q)) < \infty$, where (x_q, y_q, z_q) are permissible parameters at q and $x_q = 0, y_q = 0$ are local equations for C at q.*

Proof. If $\nu(F_q(0, 0, z_q)) = \infty$, then $F_q \in \hat{\mathcal{I}}_{\overline{C}, q}$ so that $F_p \in \hat{\mathcal{I}}_{\overline{C}, p}$ for all $p \in C$ by Lemma 8.1. Thus $\nu(F_p(0, 0, z_p)) = \infty$ for all $p \in C$, a contradiction. $\qquad \square$

Theorem 15.3. *Suppose that $C_r(X)$ holds with $r \geq 2$ and the conclusions of Theorem 14.7 hold on X. Then there exists a sequence of permissible monoidal transforms $\pi : Y \to X$ centered at r big curves C in $\overline{S}_r$ such that $C_r(Y)$ and the conclusions of Theorem 14.7 hold on Y and if D is a curve in $\overline{S}_r(Y)$, then D is not r big.*

Proof. Suppose that the $C \subset \bar{S}_r(X)$ is r big. C must contain a 1 point. Let $\pi : Y \to X$ be the blow-up of C.

By Lemma 8.8, $C_r(Y)$ holds and the conclusions of Theorem 14.7 hold on Y. There is at most one curve $D \subset \bar{S}_r(Y) \cap \pi^{-1}(C)$. If this curve exists it must be a section over C.

Let $p \in C$ be a 1 point. As in (71) of the proof of Lemma 8.8, there exist permissible parameters (x, y, z) at p such that $\hat{I}_{C,p} = (x, z)$,

$$\begin{aligned} u &= x^a \\ F_p &= \tau z^r + \textstyle\sum_{i=2}^r a_i(x, y)z^{r-i} \end{aligned} \tag{173}$$

where τ is a unit, $x^i \mid a_i$ for $2 \le i \le r$.

As shown in the proof of Lemma 8.8, the only point $q \in \pi^{-1}(p)$ which could be in $\bar{S}_r(Y)$ is the 1 point with permissible parameters $x = x_1, z = x_1 z_1$.

$$\begin{aligned} u &= x_1^a \\ F_q &= \tau z_1^r + \textstyle\sum_{i=2}^r \frac{a_i(x_1, y)}{x_1^i} z_1^{r-i} \end{aligned} \tag{174}$$

In this case, (174) has the form of (173) with

$$\min\{\tfrac{j}{i} \text{ such that } x^j \mid a_i, x^{j+1} \nmid a_i \text{ for } 2 \le i \le r\}$$

decreased by 1.

By induction on

$$\min\{\tfrac{j}{i} \text{ such that } x^j \mid a_i, x^{j+1} \nmid a_i \text{ for } 2 \le i \le r\}$$

we can construct a sequence of permissible blow-ups of r big curves in $\bar{S}_r$ such that the conclusions of the Theorem hold. $\square$

Theorem 15.4. *Suppose that $C_r(X)$ holds with $r \ge 2$, the conclusions of Theorem 14.7 hold on X and if C is a curve in $\bar{S}_r(X)$, then C is not r big. Suppose that $p \in \bar{S}_r(X)$ is a 1 point, D is a general curve through p. For a 1 point $q \in D$, define*

$$\epsilon(D, q) = \nu(F_q(0, 0, z))$$

where (x, y, z) are permissible parameters at q so that $\hat{I}_{D,q} = (x, y)$ and

$$\begin{aligned} u &= x^a \\ v &= P(x) + x^c F_q. \end{aligned}$$

Then there exists a sequence of blow-ups of points on the strict transform of D, but not at p, $\lambda : Z \to X$ such that

1. *$C_r(Z)$ and the conclusions of Theorem 14.7 hold on Z.*
2. *Let $\tilde{D}$ be the strict transform of D on Z. Then $\epsilon(\tilde{D}, q) = 1$ for all 1 points $q \ne p$ on $\tilde{D}$, $\nu(q) = 0$ if $q \in \tilde{D}$ is a 2 point, and there are no 3 points on $\tilde{D}$.*
3. *Suppose that $\tilde{p}$ is a fundamental point of λ.*
    (a) *If $q \in \lambda^{-1}(\tilde{p})$ is a 1 point then $\nu(q) \le r - 1$.*

(b) *If $q \in \lambda^{-1}(\tilde{p})$ is a 2 point, then $\nu(q) \le r$. If $\nu(q) = r$ then $\gamma(q) = r$. If $\nu(q) = r - 1$, but $\tau(q) = 0$, then $\gamma(q) = r$ and q is on the strict transform of a curve in $\overline{S}_r(X)$.*

(c) *If $q \in \lambda^{-1}(\tilde{p})$ is a 3 point then $\nu(q) \le r - 2$.*

4. *There does not exist a curve $C \subset \overline{S}_r(Z)$ such that C is r big.*

Proof. The existence of Z and the validity of $C_r(Z)$ and 2. follow from Theorem 14.5.

Suppose that D contains a 1 point $q \ne p$ such that $\nu(q) = r$. Then D intersects a curve C in $\overline{S}_r(X)$ transversally at q, and q is a generic point of C. By Lemma 6.24 and Lemma 8.5, since C is not r big, $\gamma(q) = r$ and q is a generic point of C, there are thus permissible parameters (x, y, z) at q such that

$$u = x^a$$
$$F_q = \tau z^r + \sum_{i=2}^{r-1} a_i(x, y)x^{\alpha_i} z^{r-i} + x^{r-1}y$$

where $\alpha_i \ge i$, τ is a unit and $x \not| a_i$ for $2 \le i \le r-1$, $\hat{\mathcal{I}}_{C,q} = (x, z)$, $\hat{\mathcal{I}}_{D,q} = (x, y)$.

Let $\pi_1 : X_1 \to X$ be the blow-up of q.

Suppose that $q_1 \in \pi_1^{-1}(q)$ and there are permissible parameters (x_1, y_1, z_1) at q_1 such that

$$x = x_1, y = x_1(y_1 + \alpha), z = x_1(z_1 + \beta).$$

Then

$$u = x_1^a$$
$$\frac{F_q}{x_1^r} = \tau(z_1 + \beta)^r + \sum_{i=2}^{r-1} a_i x_1^{\alpha_i - i}(z_1 + \beta)^{r-i} + (y_1 + \alpha)$$

Thus, after normalizing to get $F_{q'}$, we have that $\gamma(q') \le r - 1$ if $\beta \ne 0$, and $\gamma(q') = 1$, if $\beta = 0$.

Suppose that $q_1 \in \pi_1^{-1}(q)$ and there are permissible parameters (x_1, y_1, z_1) at q_1 so that

$$x = x_1 y_1, y = y_1, z = y_1(z_1 + \alpha)$$

If $\alpha \ne 0$, then q' is a 2 point with $\gamma(q') \le r - 1$. If $\alpha = 0$, then q' is a 2 point on the strict transform of C, $\nu(q') \le r - 1$ and $\gamma(q') \le r$.

Suppose that $q_1 \in \pi_1^{-1}(q)$ and there are regular parameters (x_1, y_1, z_1) in $\hat{\mathcal{O}}_{X_1, q_1}$ so that

$$x = x_1 z_1, y = y_1 z_1, z = z_1$$

Then q' is a 2 point with $\nu(q') = 0$. q' is the point in $\pi_1^{-1}(q)$ on the strict transform of D. Thus 3. holds for $\lambda^{-1}(q)$.

If $q \in D$ is a 1 point with $\nu(q) \le r - 1$ or a 2 point with $\nu(q) \le r - 2$, 3. for $\lambda^{-1}(q)$ follows from Theorems 7.1 and 7.3 (or the proof of Theorem 14.5).

If $q \in D$ is a 2 point with $\nu(q) = r - 1$, then q is a generic point of a 2 curve $C \subset \overline{S}_{r-1}(X)$ (or such that $F_q \in \hat{\mathcal{I}}_{C,q}$ if $r = 2$). Since $C_r(X)$ holds, C is r-1 small. Since q is a generic point of C, we must have $\tau(q) > 0$ by Lemma 6.25. 1. - 3. for $\lambda^{-1}(q)$ then follow from Theorem 7.1 and Theorem 7.3.

The conclusions of Theorem 14.7 hold since this condition is stable under quadratic transforms.   □

Definition 15.5. *Suppose that C is a 2 curve of X. Then C satisfies (E) if For $q \in C$,*

1. *$\nu(q) = 0$ if q is a 3 point.*
2. *$\gamma(q) \leq 1$ at all but finitely many 2 points $q \in C$, where either*
   (a) *$\nu(q) = \gamma(q) = r$ or*
   (b) *$\nu(q) = r - 1$, $\gamma(q) = r$ and $\tau(q) = 0$.*

Theorem 15.6. *Suppose that $C_r(X)$ holds with $r \geq 2$, the conclusions of Theorem 14.7 hold on X and C is a 2 curve of X containing a 2 point p such that either $\nu(p) = \gamma(p) = r$, or $\nu(p) = r - 1$, $\gamma(p) = r$ and $\tau(q) = 0$. Then there exists a sequence of quadratic transforms $\pi : Y \to X$ (over points in C) such that the following properties hold. Let $\tilde{C}$ be the strict transform of C. Suppose that q is an exceptional point of π. Then*

1. *If q is a 1 point, then $\nu(q) \leq r - 1$.*
2. *If q is a 2 point, then $\nu(q) \leq r - 1$. If $\nu(q) = r - 1$ then $\tau(q) > 0$.*
3. *If q is a 3 point then $\nu(q) \leq r - 2$*

Furthermore, $\tilde{C}$ satisfies (E), $C_r(Y)$ holds and the conclusions of Theorem 14.7 hold on Y.

Proof. By our assumption on p, and Lemma 8.1, $F_{q'} \notin \hat{I}_{C,q'}$ at all points $q' \in C$. There thus cannot exist $q' \in C$ which satisfies (163), since then $F_{q'} \in \hat{I}_{C,q'}$.

Suppose that $q \in C$. Then there exist permissible parameters (x, y, z) at q such that $\hat{I}_{C,q} = (x, y)$ and $F_q \notin \hat{I}_{C,q}$.

First suppose that q is a 3 point and $\nu(q) > 0$. Suppose that $\pi_1 : X_1 \to X$ is the blow-up of q. Suppose that $q' \in \pi_1^{-1}(q)$. Since $\nu(q) \leq r - 2$, we have that $\nu(q') \leq r - 1$ if q' is a 1 point, $\nu(q') \leq r - 1$ if q' is a 2 point, $\nu(q') = r - 1$ implies $\tau(q') > 0$, and $\nu(q') \leq r - 2$ if q' is a 3 point by Theorem 7.1. The strict transform of C intersects $\pi_1^{-1}(q)$ in a 3 point.

Consider the infinite sequence of blow-ups of points $X_{n+1} \to X_n$, centered at the points q_n on the strict transform of C on X_n over q,

$$\cdots \to X_n \to \cdots \to X_1 \to X$$

q_n has permissible parameters (x_n, y_n, z_n) defined by

$$x = x_n z_n^n, y = y_n z_n^n, z = z_n$$

$$F_{q_n} = \frac{F_q}{z_n^{\sum_{i=1}^n s_i}}$$

where $s_i = \nu(q_i)$. Since $F_q \notin (x, y)$ we have that $\nu(q_n) = 0$ for all sufficiently large n.

Let

$$W = \{q \in C | q \text{ is a 2 point with } \gamma(q) > 1\}.$$

W is a finite set by Lemma 15.1, and since $\gamma(q) \leq 1$ at a generic point of C.

Suppose that $q \in W$ is a 2 point. Suppose that either $\nu(q) \leq r - 2$ or q is such that $\nu(q) = r - 1$ and $\tau(q) > 0$. Then arguing as in the case when q is a 3 point, and using Theorems 7.1 and 7.3 we can produce a sequence of blow-ups of points $\pi : X_m \to X$, centered at the points q_n on the strict transform of C on

X_n over q, such that $\nu(q_m) = 0$, and the conclusions of the Theorem hold in a neighborhood of $\pi^{-1}(q)$.   □

Theorem 15.7. *Suppose that $C_r(X)$ holds with $r \geq 2$ and the conclusions of Theorem 14.7 hold on X. Then there exists a permissible sequence of blow-ups $\pi : Y \to X$ such that for $p \in Y$*

1. $\nu(p) \leq r - 1$ *if p is a 1 point or a 2 point.*
2. *If p is a 2 point and $\nu(p) = r - 1$, then $\tau(p) > 0$ or $r \geq 3$ and (163) holds at p.*
3. $\nu(p) \leq r - 2$ *if p is a 3 point*

Proof. By Theorem 15.3, we can assume that if C is a curve in $\overline{S}_r(X)$ then C is not r big. Furthermore, since $C_r(X)$ holds, each curve in $\overline{S}_r(X)$ contains a 1 point. There are finitely many 1 points $\{p_1, \ldots, p_m\}$ in X such that each p_i is in $\overline{S}_r(X)$, and p_i is either an isolated point in $\overline{S}_r(X)$, or is a special point of a curve in $\overline{S}_r(X)$ (A special 1 point on a curve in $\overline{S}_r(X)$ is a point which is not generic in the sense that the conclusions of 1. (a) of Lemma 8.10 do not hold). Let D_{p_i} be a general curve through p_i for $1 \leq i \leq m$. By Theorem 15.4, after possibly performing a finite sequence of quadratic transforms at points $\neq p_i$ on the D_{p_i}, we may assume that $\epsilon(D_{p_i}, q) = 1$ for all 1 points $q \neq p_i$ on D_{p_i}, $\nu(q) = 0$ if $q \in D_{p_i}$ is a 2 point, and there are no 3 points on any D_{p_i}.

There are no exceptional 1 points in $\overline{S}_r(X)$ created by the sequence of blow-ups in Theorem 15.4.

For each p_i, let t_{p_i} be the number l computed in Theorem 11.5 (or Theorem 11.6 if $r = 2$) for p_i. By Theorem 11.4, for $1 \leq i \leq m$, there exist sequences of monoidal transforms

$$\lambda^{p_i} : X_1(i) \to \text{spec}(\mathcal{O}_{X,p_i})$$

where $X_1(i) \to \text{spec}(\mathcal{O}_{X,p_i})$ is a sequence of permissible monoidal transforms centered at sections over D_{p_i}, such that the conclusions of Theorem 11.4 hold on $X_1(i)$ (with $t \geq t_{p_i}$). Since D_{p_i} is resolved at all points $p_i \neq q \in D_{p_i}$, the only obstruction to extending λ^{p_i} to a permissible sequence of monoidal transforms of sections over D_{p_i} in X is if the corresponding sections over D_{p_i} in X do not make SNCs with the 2 curves. This difficulty can be resolved by performing quadratic transforms at the points where the section does not make SNCs with the 2 curves.

By Theorem 15.6, we can then perform a sequence of quadratic transforms centered at 2 points, so that if C is a 2 curve containing a point p such that either

$$\nu(p) = \gamma(p) = r$$

or

$$\nu(p) = r - 1, \gamma(p) = r \text{ and } \tau(p) = 0$$

then C satisfies (E). There are no exceptional 1 points in $\overline{S}_r(X)$ in this sequence of blow-ups.

We can thus extend the maps $X_1(i) \to \text{spec}(\mathcal{O}_{X,p_i})$ to a sequence of permissible monoidal transforms centered at sections over D_{p_i} and points, $\lambda :$ $Y \to X$. $C_r(Y - \lambda^{-1}(\{p_1, \ldots, p_m\}))$ holds and the conclusions of Theorem

14.7 hold on $Y - \lambda^{-1}(\{p_1, \ldots, p_m\})$, there are no special or isolated 1 points in $\overline{S}_r(Y) - \lambda^{-1}(\{p_1, \ldots, p_m\})$,

$$Y \times_X \text{spec}(\mathcal{O}_{X,p_i}) \cong X_1(i)$$

for $1 \le i \le m$ and if $q \in \overline{S}_r(Y) - \lambda^{-1}(\{p_1, \ldots, p_m\})$ is a 2 point with

$$\nu(q) = r, \gamma(q) = r,$$

or

$$\nu(q) = r - 1, \gamma(q) = r \text{ and } \tau(q) = 0$$

then the 2 curve containing q satisfies (E).

Let $\{q_1, \ldots, q_n\}$ be the 2 points of Y such that $\nu(q_i) = r$ and q_i is contained in a curve of $\overline{S}_r(Y)$. Let C_{q_i} be the 2 curve containing q_i for $1 \le i \le m$. Then $\gamma(q) \le 1$ if $q_i \ne q \in C_{q_i}$ is a 2 point, and $\gamma(q) = 0$ if $q \in C_{q_i}$ is a 3 point since C_{q_i} satisfies (E) and the conclusions of Theorem 14.7 hold on Y. For each q_i, let t_{q_i} be the number l computed in Theorem 11.5 (or Theorem 11.6 if $r = 2$) for p_i. For $1 \le i \le n$, let

$$\lambda^{q_i} : Y_1(i) \to \text{spec}(\mathcal{O}_{Y,q_i})$$

be a permissible sequence of monoidal transforms centered at sections over C_{q_i} such that the conclusions of Theorem 11.4 hold on $Y_1(i)$ (with $t \ge t_{q_i}$).

Since C_{q_i} is resolved at all points $q_i \ne q \in C_{q_i}$, the only obstruction to extending λ^{q_i} to a permissible sequence of monoidal transforms of sections over C_{p_i} in Y is if the corresponding sections over C_{p_i} in Y do not make SNCs with the 2 curves. This difficulty can be resolved by performing quadratic transforms at the points where the section does not make SNCs with the 2 curves.

We can thus extend the $Y_1(i) \to \text{spec}(\mathcal{O}_{Y,q_i})$ to a sequence of permissible monoidal transforms centered at points and sections over C_{q_i}, $\phi : Z \to Y$, so that

$$Z \times_Y \text{spec}(\mathcal{O}_{Y,q_i}) \cong Y_1(i)$$

for $1 \le i \le n$, and if

$$Z_0 = Z - \phi^{-1}(\{q_1, \ldots, q_n\}) - (\lambda \circ \phi)^{-1}(\{p_1, \ldots, p_m\})$$

then $C_r(Z_0)$ holds and Z_0 satisfies the conclusions of Theorem 14.7. Z_0 contains no special or isolated 1 points. If $D \subset \overline{S}_r(Z)$ and D is r big then $D \cap Z_0 = \emptyset$. If $q \in Z_0$ is a 2 point which does not satisfy the conclusions of the Theorem, then $\gamma(q) = r$ and $\nu(q) = r - 1$ or $\nu(q) = r$. If $C \subset Z$ is the 2 curve containing q, then C satisfies (E). If $\nu(q) = r$, then q is not contained in a curve $D \subset \overline{S}_r(Z)$.

By Lemma 8.10 and Theorem 11.5 if $r \ge 3$ (or Lemma 8.10, Lemma 8.11 and Theorem 11.6 if $r = 2$) there exists a sequence of permissible monoidal transforms $\psi : W \to Z$ consisting of a sequence of blow-ups of r big curves $D \subset \overline{S}_r$, followed by a sequence of blow-ups of r small curves $D \subset \overline{S}_r$, and finally followed by a sequence of quadratic transforms if $r \ge 3$ (or quadratic transforms and monoidal transforms centered at 2 curves C such that C is 1 big and C is a section over a 2 small curve blown up in constructing ϕ if $r = 2$) such that $C_r(W)$ holds, $\overline{S}_r(W)$ is a finite union of 2 points, and the conclusions of the Theorem hold everywhere in W, except possibly at a finite number of 2 points p. If C is a 2 curve on W containing a 2 point p where the theorem fails to hold, then (E) holds on C. In particular, there are no 2 curves in $\overline{S}_r(W)$.

Suppose that $C \subset W$ is a 2 curve containing a 2 point such that the conclusions of the theorem do not hold. Then C satisfies (E).

Let $\{q_1, \ldots, q_s\}$ be the two points on C such that

1. $\nu(q_i) = r$, $\gamma(q_i) = r$ or
2. $\nu(q_i) = r - 1$, $\gamma(q_i) = r$ and $\tau(q_i) = 0$.

$\gamma(q) \leq 1$ if $q \in C - \{q_1, \ldots, q_s\}$ is a 2 point, and $\nu(q) = 0$ if $q \in C$ is a 3 point. No q_i is contained in a curve in $\overline{S}_r(W)$, since $\overline{S}_r(W)$ is finite.

We will now show that there exists an affine neighborhood U of $\{q_1, \ldots, q_s\}$ and uniformizing parameters $\tilde{x}, y, \tilde{z}$ on U such that $\tilde{x} = y = 0$ are local equations of C on U, and all points of $C \cap U$ are 2 points.

Let A_1 and A_2 be the components of E_W such that C is a connected component of $A_1 \cap A_2$. There exist very ample divisors H_1, H_2, H_3, H_4 on W such that $A_1 \sim H_1 - H_2$, $A_2 \sim H_3 - H_4$ and $q_i \notin H_j$ for $1 \leq i \leq s$, $1 \leq j \leq 4$.

Let $U = W - (H_1 \cup H_2 \cup H_3 \cup H_4)$. $U = \operatorname{spec}(A)$ is affine and there exist $\tilde{x}, y \in A$ such that $\tilde{x} = 0$ is an equation for $A_1 \cap U$ in U, $y = 0$ is an equation for $A_2 \cap U$ in U. After possibly replacing U with a smaller affine neighborhood of $\{q_1, \ldots, q_s\}$, we may assume that $U \cap C = U \cap A_1 \cap A_2$ and $E_W \cap U = (A_1 \cup A_2) \cap U$.

There exists a morphism $\pi : C \to \mathbf{P}^1$ such that π is étale over $\pi(q_i)$, $1 \leq i \leq s$, and $\pi(q_i) \neq \infty$ for any i (We can take π to be a generic projection). Let z be a coordinate on $\mathbf{P}^1 - \{\infty\}$. After replacing U with a possibly smaller affine neighborhood of $\{q_1, \ldots, q_s\}$ we have an inclusion $\pi^* : k[z] \to A$, so that $U \to \operatorname{spec}(k[\tilde{x}, y, z])$ is étale.

There exists a component E of D_S such that $\Phi_W(A_1) \subset E$ and $\Phi_W(A_2) \subset E$ (since $\Phi_W : W \to S$ is weakly prepared). There exists an affine neighborhood $\overline{V}$ of $\{\Phi_W(q_1), \ldots, \Phi_W(q_s)\}$ in S and $u \in \Gamma(\overline{V}, \mathcal{O}_S)$ such that $u = 0$ is a local equation of D_S. Then $u = 0$ is a local equation of E_W in $\Phi_W^{-1}(\overline{V}) \cap U$. Thus if we replace U with $\Phi_W^{-1}(\overline{V}) \cap U$, u extends to a system of permissible parameters at $\Phi_W(p)$ for all $p \in C \cap \Phi_W^{-1}(\overline{V}) \cap U$.

There exist $a, b \in \mathbf{N}$ such that $u = \tilde{x}^a y^b \overline{\gamma}$ where $\overline{\gamma} \in A$ is a unit in A. Let $\gamma = \overline{\gamma}^{\frac{1}{a}}$, $B = A[\gamma]$, $V = \operatorname{spec}(B)$. Then $h : V \to U$ is étale.

Let $x = \gamma \tilde{x}$. $k[x, y, z] \to B$ defines a morphism $g : V \to \mathbf{A}^3$. $q \in g^{-1}(x = 0)$ if and only if $x \in m_q$ which holds if and only if $\tilde{x} \in m_q$. Thus $g^{-1}(x = 0) = h^{-1}(\tilde{x} = 0)$. g is étale at all points of $g^{-1}(x = 0)$. Since this is an open condition (c.f. Prop 4.5 SGA1) there exists a Zariski closed subset Z_1 of V which is disjoint from $h^{-1}(\tilde{x} = 0)$ such that $g \mid V - Z_1$ is étale. Let U_1 be an affine neighborhood of $\{q_1, \ldots, q_s\}$ in U which is disjoint from $h(Z_1)$. Let $V_1 = h^{-1}(U_1)$.

After replacing U with U_1 and V with V_1, we have that $V \to U$ is an étale cover and (x, y, z) are uniformizing parameters on V.

There exist $v_i \in \mathcal{O}_{S, \Phi_W(q_i)}$ such that (u, v_i) are permissible parameters at $\Phi_W(q_i)$ and $u = 0$ is a local equation of E_W at q_i for $1 \leq i \leq s$. For each q_i there exist z_i such that (x, y, z_i) are permissible parameters at q_i for (u, v_i) for $1 \leq i \leq s$ which satisfy the conclusions of Lemma 8.5.

The morphism $\pi_1 : Y_{q_1} \to \text{spec}(\hat{\mathcal{O}}_{W,q_1})$ of Theorem 11.2 (or Theorem 11.7 if $\nu(q_i) = r - 1$) extends to a sequence of permissible monoidal transforms $\pi_1 : \tilde{Y}_1 \to V$ centered at sections over C.

$$\tilde{Y}_1 \times_{\text{spec}(\mathcal{O}_{W,q_2})} \text{spec}(\hat{\mathcal{O}}_{W,q_2}) \to \text{spec}(\hat{\mathcal{O}}_{W,q_2})$$

extends to a sequence of permissible monoidal transforms $Y_{q_2} \to \text{spec}(\hat{\mathcal{O}}_{W,q_2})$ of the form of the conclusions of Theorem 11.2 (or Theorem 11.7).

$Y_{q_2} \to \text{spec}(\hat{\mathcal{O}}_{W,q_2})$ extends to a sequence of permissible monoidal transforms

$$\tilde{Y}_2 \overset{\pi_2}{\to} \tilde{Y}_1 - \pi_1^{-1}(h^{-1}(q_1)) \to V - \{h^{-1}(q_1)\}.$$

Preceeding inductively, we extend

$$\tilde{Y}_{s-1} \times_{\text{spec}(\mathcal{O}_{W,q_s})} \text{spec}(\hat{\mathcal{O}}_{W,q_s}) \to \text{spec}(\hat{\mathcal{O}}_{W,q_s})$$

to a sequence of permissible monoidal transforms $Y_{q_s} \to \text{spec}(\hat{\mathcal{O}}_{W,q_s})$ of the form of the conclusions of Theorem 11.2 (or Theorem 11.7).

$Y_{q_s} \to \text{spec}(\hat{\mathcal{O}}_{W,q_s})$ extends to a sequence of permissible monoidal transforms

$$\tilde{Y}_s \overset{\pi_s}{\to} \tilde{Y}_{s-1} - (\pi_1 \circ \cdots \circ \pi_{s-1})^{-1}(h^{-1}(q_{s-1})) \to V - \{q_1, \ldots, q_{s-1}\}.$$

For $1 \leq i \leq s$, let t_{q_i} be the value of l in the statement of Theorem 11.5 (or Theorem 11.7) for the point q_i.

Let $\omega_i \in A$, $1 \leq i \leq s$ be such that

$$\omega_i \equiv \gamma \mod m_{q_i}^{t_{q_i}} \hat{\mathcal{O}}_{W,q_i}.$$

By the Chinese Remainder Theorem, there exists $\omega \in A$ such that after possibly replacing U with a smaller affine neighborhood of $\{q_1, \ldots, q_s\}$, we have that $(\omega\tilde{x}, y, \tilde{z})$ are uniformizing parameters on U and

$$\omega\tilde{x} \equiv x \mod m_{q_i}^{t_{q_i}} \hat{\mathcal{O}}_{W,q_i}$$

for $1 \leq i \leq s$.

We can thus replace $\tilde{x}$ with $\omega\tilde{x}$ in (112) of Theorem 11.4 for $1 \leq i \leq s$. With this choice of $\tilde{x}$, The map $\overline{Y}_{q_1} \to \text{spec}(\mathcal{O}_{W,q_1})$ of Theorem 11.4 (or Theorem 11.7) then satisfies the assumptions of Theorem 11.5, and extends to a permissible sequence of monoidal transforms centered at sections over C

$$\lambda_1 : \hat{Y}_1 \to U.$$

The map $\overline{Y}_{q_2} \to \text{spec}(\mathcal{O}_{W,q_2})$ of Theorem 11.4 (or Theorem 11.7) satisfies the assumptions of Theorem 11.5 (or Theorem 11.7) and extends to a permissible sequence of monoidal transforms centered at sections over C,

$$\hat{Y}_2 \overset{\lambda_2}{\to} \hat{Y}_1 - \lambda_1^{-1}(q_1) \to U - \{q_1\}.$$

Preceeding inductively, the map $\overline{Y}_{q_s} \to \text{spec}(\mathcal{O}_{W,q_s})$ of Theorem 11.4 (or Theorem 11.7) satisfies the assumptions of Theorem 11.5, and extends to a permissible sequence of monoidal transforms centered at sections over C

$$\hat{Y}_s \overset{\lambda_s}{\to} \hat{Y}_{s-1} - (\lambda_1 \circ \cdots \circ \lambda_{s-1})^{-1}(q_{s-1}) \to U - \{q_1, \ldots, q_{s-1}\}.$$

Since all points of $C - U$ are resolved, there exists a sequence of permissible monoidal transforms $\tilde{\lambda}_1 : Y_1' \to W$ consisting of quadratic transforms centered at points over $C - U$ and permissible monoidal transforms centered at sections of C such that all points of $\tilde{\lambda}_1^{-1}(C - U)$ are resolved, and $Y_1' \times_W U \cong \hat{Y}_1$.

By Theorems 11.5 and 11.7 (or Theorems 11.6 and Theorem 11.8 if $r = 2$), there exists a sequence of permissible monoidal transforms $\tilde{Z}_1 \to Y_1'$ (with induced maps $\psi_1 : \tilde{Z}_1 \to W$) centered over points and curves which map to q_1 such that all points of $\psi_1^{-1}(q_1)$ satisfy the conclusions of the Theorem.

By Theorem 7.1 and 7.3,

$$\hat{Y}_2 \to \hat{Y}_1 - \lambda_1^{-1}(q_1)$$

then extends to a sequence of permissible monoidal transforms $\tilde{\lambda}_2 : Y_2' \to \tilde{Z}_1$ consisting of quadratic transforms centered at points over $C - (U - \{q_1\})$ and permissible monoidal transforms centered at sections over C such that all points of $(\tilde{\lambda}_1 \circ \tilde{\lambda}_2)^{-1}(C - U)$ are resolved, all points of $(\tilde{\lambda}_1 \circ \tilde{\lambda}_2)^{-1}(q_1)$ satisfy the conclusions of the Theorem and

$$Y_2' \times_W (U - \{q_1\}) \cong \hat{Y}_2.$$

By Theorems 11.5 and 11.7 (or Theorems 11.6 and 11.8 if $r = 2$), there exists a sequence of permissible monoidal transforms $\tilde{Z}_2 \to Y_2'$ with induced maps $\psi_2 : \tilde{Z}_2 \to W$ centered at points and curves that map to q_2 such that all points of $\psi_2^{-1}(q_2)$ satisfy the conclusions of the Theorem.

By induction on s, we can then construct a sequence of permissible monoidal transforms $\psi_s : \tilde{Z}_s \to W$ centered at points and curves supported over C such that all points of $\psi_s^{-1}(C)$ satisfy the conclusions of the Theorem, and all points of $\psi_s^{-1}(C - \{q_1, \ldots, q_s\})$ are resolved.

By induction on the number of 2 curves $C \subset W$ which contain a 2 point which does not satisfy the conclusions of the Theorem, we can construct a sequence of permissible monoidal transforms $\overline{W} \to W$ such that $\overline{W}$ satisfies the conclusions of the Theorem. $\square$

16. RESOLUTION 4

Throughout this section we will assume that $\Phi_X : X \to S$ is weakly prepared.

Theorem 16.1. *Suppose that $r \geq 1$ and for $p \in X$,*

1. $\nu(p) \leq r$ *if p is a 1 point or a 2 point.*
2. *If p is a 2 point and $\nu(p) = r$, then $\tau(p) > 0$ or $r \geq 2$, $\tau(p) = 0$ and there exists a unique curve $D \subset \overline{S}_r(X)$ (containing a 1 point) such that $p \in D$, and permissible parameters (x, y, z) at p such that $x = z = 0$ are local equations of D.*

$$
\begin{aligned}
u &= (x^a y^b)^m \\
v &= P(x^a y^b) + x^c y^d F_p \\
F_p &= \tau x^r + \sum_{j=1}^{r} \bar{a}_j(y, z) y^{d_j} z^{e_j} x^{r-j}
\end{aligned}
\tag{175}
$$

where τ is a unit, $\bar{a}_j$ are units (or 0), there exists i such that $\bar{a}_i \neq 0$, $e_i = i$, $0 < d_i < i$,

$$
\frac{d_i}{i} \leq \frac{d_j}{j}, \frac{e_i}{i} \leq \frac{e_j}{j}
$$

for all j and

$$
\left\{\frac{d_i}{i}\right\} + \left\{\frac{e_i}{i}\right\} < 1.
$$

3. $\nu(p) \leq r - 1$ *if p is a 3 point*

Then there exists a sequence of permissible monoidal transforms $\pi : X_1 \to X$ such that $\overline{A}_r(X_1)$ holds. That is,

1. $\nu(p) \leq r$ *if $p \in X$ is a 1 point or a 2 point.*
2. *If $p \in X$ is a 1 point and $\nu(p) = r$, then $\gamma(p) = r$.*
3. *If $p \in X$ is a 2 point and $\nu(p) = r$, then $\tau(p) > 0$.*
4. $\nu(p) \leq r - 1$ *if $p \in X$ is a 3 point*

Proof. If p is a 1 point such that $\nu(p) = 1$, then $\gamma(p) = 1$. Thus $\overline{A}_r(X)$ holds if $r = 1$. For the rest of the proof we will assume that $r \geq 2$.

Let

$$
W(X) = \{p \in 1 \text{ points of } X \mid \nu(p) = r \text{ and } \gamma(p) > r\}.
$$

$W(X)$ is Zariski closed in the open subset of 1 points of X. Let $\overline{W}(X)$ be the Zariski closure of $W(X)$ in X.

Suppose that $p \in \overline{W}(X)$ is a point where $\overline{W}(X)$ does not make SNCs with $\overline{B}_2(X)$. Then p can not satisfy (175). Let $\pi : X_1 \to X$ be the quadratic transform with center p. By Theorems 7.1 and 7.3, all points of $\pi^{-1}(p)$ satisfy the assumptions of Theorem 16.1, and there are no points of $\pi^{-1}(p)$ which satisfy (175). If p is a 1 point, $x \mid L_p$ implies $\nu(q) \leq r - 1$ if $q \in \pi^{-1}(p)$ is a 1 point, so $\pi^{-1}(p)$ contains no curves of $\overline{W}(X_1)$. By Theorems 7.1 and 7.3, $\pi^{-1}(p)$ contains no curves of $\overline{W}(X_1)$ if p is a 2 or 3 point.

Thus there exists a sequence of quadratic transforms $\pi : X_1 \to X$ such that $\overline{W}(X_1)$ is a disjoint union of nonsingular curves and isolated points, X_1 satisfies the assumptions of Theorem 16.1, and $\overline{W}(X_1)$ makes SNCs with $\overline{B}_2(X_1)$. By Theorems 7.1 and 7.3 and Lemma 7.9, we can further assume that $\overline{S}_r(X_1)$ makes

SNCs with $\overline{B}_2(X_1)$, except possibly at some 3 points of X_1, and if $C \subset \overline{S}_r(X_1)$ is a curve which contains a 2 point satisfying (175), then C contains no 3 points. We can then without loss of generality assume that $X = X_1$.

Suppose that $C \subset \overline{W}(X)$ is a curve. C makes SNCs with the locus of 2 curves. We either have that C is r big or r small.

For a curve C, or isolated point p in $\overline{W}(X)$, We will show that we can construct a sequence of monoidal transforms $\pi : Y \to X$, centered at points and curves over C (or over p), such that the assumptions of the theorem hold on Y, and 2. of the conclusions of the theorem hold at points over C (over p).

We can then iterate this process to obtain $Z \to X$ such that the assumptions of the theorem hold on Z, and if $p \in Z$ is a 1 point with $\nu(p) = r$, then $\gamma(p) = r$.

Suppose that C is r small Since C is r small, (175) cannot hold at any $p \in C$. By Lemma 8.9, we can construct a sequence of monoidal transforms $\pi : Y \to X$, centered at points on C and the strict transform of C, such that the assumptions of the theorem hold on Y, and the conclusions of the theorem hold at points of $\pi^{-1}(C)$.

Suppose that C is r big

Let $\pi : X_1 \to X$ be the blow-up of C. We will show that the assumptions of the theorem and 2. of then conclusions of the theorem hold at points above C.

Suppose that $p \in C$ is a 2 point with $\tau(p) > 0$ or a 1 point. Then all points of $\pi^{-1}(p)$ satisfy the conclusions of the Theorem by Lemma 8.8.

Suppose that $p \in C$ is a 2 point such that (175) holds. Then $x = z = 0$ are local equations of C at p.

Suppose that $q \in \pi^{-1}(p)$ is a 2 point. q has permissible parameters (x_1, y, z_1) such that $x = x_1, z = x_1(z_1 + \alpha)$.

$$
\begin{aligned}
u &= (x_1^a y^b)^m \\
v &= P(x_1^a y^b) + x_1^{c+r} y^d \frac{F_p}{x_1^r}.
\end{aligned}
$$

$$F_p = \tau x^r + y\Omega$$

implies

$$\frac{F_p}{x_1^r} = \tau + y\frac{\Omega}{x_1^r}$$

$ad - b(c + r) \neq 0$, since F_p is normalized, which implies that $\nu(q) = 0$.

Suppose that $q \in \pi^{-1}(p)$ is the 3 point. q has permissible parameters (x_1, y, z_1) such that

$$x = x_1 z_1, z = z_1.$$

$$
\begin{aligned}
u &= (x_1^a y^b z_1^a)^m \\
v &= P(x_1^a y^b z_1^a) + x_1^c y^d z_1^{c+r} F_q
\end{aligned}
$$

where

$$F_q = \frac{F_p}{z_1^r} = \tau x_1^r + \sum_{j=1}^r \overline{a}_j(y, z_1) y^{d_j} x_1^{r-j} z_1^{e_j - j}$$

By assumption $d_i + r - i + e_i - i < r$. Thus $\nu(q) \leq r - 1$.

Suppose that p is an isolated point in $\overline{W}(X)$

There are permissible parameters (x, y, z) at p such that

$$
\begin{aligned}
u &= x^a \\
v &= P(x) + x^c F_p \\
L_p &= x^t \Omega(x, y, z)
\end{aligned}
$$

with $0 < t < r$, $x \nmid \Omega$.

Let $\pi_1 : X_1 \to X$ be the blow-up of p. If $q \in \pi_1^{-1}(p)$ is a 2 point then $\nu(q) \leq r$ and $\nu(q) = r$ implies $\tau(q) > 0$ by Theorem 7.1. If $q \in \pi_1^{-1}(p)$ is a 1 point then $\nu(q) \leq r - t < r$

We are now reduced to assuming that $\overline{W}(X) = \emptyset$, so that $\gamma(p) = r$ if $p \in X$ is a 1 point with $\nu(p) = r$.

Now suppose that $p \in X$ satisfies (175) so that the curve D in $\overline{S}_r(X)$ that p lies on satisfies $\gamma(q) = r$ if $q \in D$ is a 1 point.

By our initial reduction, we may assume that D is nonsingular, and makes SNCs with $\overline{B}_2(X)$. Since $F_p \in \hat{\mathcal{I}}^r_{C,p}$, C is r big.

Let $\pi : X_1 \to X$ be the blow-up of D. If $p \in D$ is a 2 point with $\tau(p) > 0$ or a 1 point, then all points of $\pi^{-1}(p)$ satisfy the conclusions of the Theorem by Lemma 8.8. The case when p satisfies (175) is exactly as in the case when $C \subset \overline{W}(X)$ is r big. $\qquad\square$

17. Proof of the Main Theorem

Theorem 17.1. *Suppose that $\Phi_X : X \to S$ is weakly prepared, $r \geq 2$ and $\overline{A}_r(X)$ holds. Then there exists a permissible sequence of monoidal transforms $Y \to X$ such that $\overline{A}_{r-1}(Y)$ holds.*

Proof. The Theorem follows from successive application of Lemma 10.3 and Theorems 13.8, 14.6, 14.7, 15.7 and 16.1 □

Theorem 17.2. *Suppose that $\Phi_X : X \to S$ is weakly prepared. Then there exists a sequence of permissible monoidal transforms $Y \to X$ such that $\Phi_Y : Y \to S$ is prepared.*

Proof. For $r \gg 0$ $\overline{A}_r(X)$ holds by Zariski's Subspace Theorem (Theorem 10.6 [3]). The theorem then follows from successive application of Theorem 17.1, and the fact that $\overline{A}_1(X)$ holds if and only if $\Phi_X : X \to S$ is prepared. □

Theorem 17.3. *Suppose that $\Phi : X \to S$ is a dominant morphism from a 3 fold to a surface and $D_S \subset S$ is a reduced 1 cycle such that $E_X = \Phi^{-1}(D_S)_{red}$ contains $sing(X)$ and $sing(\Phi)$. Then there exist sequences of monoidal transforms with nonsingular centers $\pi_1 : S_1 \to S$ and $\pi_2 : X_1 \to X$ such that $\Phi_{X_1} : X_1 \to S_1$ is prepared with respect to $D_{S_1} = \pi_2^{-1}(D_S)_{red}$.*

Proof. This follows from Lemma 6.2 and Theorem 17.2. □

Throughout this section we will suppose that $\Phi : X \to S$ is a dominant morphism from a nonsingular 3 fold to a nonsingular surface, D_S is a reduced SNC divisor on S, $E_X = \Phi^{-1}(D_S)_{\mathrm{red}}$ is a SNC divisor on X, and $\mathrm{sing}(\Phi) \subset E_X$.

If $p \in E_X$ we will say that p is a 1, 2 or 3 point depending on if p is contained in 1, 2 or 3 components of E_X. $q \in D_S$ will be called a 1 or 2 point depending on if q is contained in 1 or 2 components of D_S.

Regular parameters (u, v) in $\mathcal{O}_{S,q}$ with $q \in D_S$ are permissible if:

1. $u = 0$ is a local equation of D_S if q is a 1 point or
2. $uv = 0$ is a local equation of D_S if q is a 2 point.

Definition 18.1. *We will say that Φ is Strongly Prepared at $p \in X$ (with respect to D_S) if one on the following forms hold.*

1. *Φ is prepared at p (as defined in Definition 6.5) or*
2. *There exist permissible parameters (u, v) at q and regular parameters (x, y, z) in $\hat{\mathcal{O}}_{X,p}$ such that one of the following hold:*
   (a) *p is a 2 point and*
   $$u = x^a, v = y^b.$$
   (b) *p is a 3 point and*
   $$u = x^a, v = y^b z^c$$
   (with $a, b, c > 0$).
   (c) *p is a 3 point and*
   $$u = x^a y^b, v = y^c z^d$$
   (with $a, b, c, d > 0$).

Suppose that $p \in X$ is strongly prepared and (u, v) are permissible parameters at $\Phi(p)$. Regular parameters (x, y, z) in $\hat{\mathcal{O}}_{X,p}$ are called $*$-permissible parameters at p for (u, v) if one of the forms of Definition 18.1 holds in $\hat{\mathcal{O}}_{X,p}$. We will also say that (u, v) are strongly prepared at p. If a form 1. holds at p, $*$-permissible parameters are permissible as defined in Definition 6.4.

Throughout this section we will assume that $\Phi : X \to S$ is strongly prepared.

Lemma 18.2. *Suppose that $\mathcal{O}_{X,p} \to R$ is finite étale, and there exists $\overline{x}, \overline{y}, \overline{z} \in R$ such that $(\overline{x}, \overline{y}, \overline{z})$ are regular parameters in R_q for all primes $q \subset R$ such that $q \cap \mathcal{O}_{X,p} = m_p$. Then there exists an étale neighborhood U of p such that $(\overline{x}, \overline{y}, \overline{z})$ are uniformizing parameters on U.*

Proof. There exists an affine neighborhood $V_1 = \mathrm{spec}(A)$ of $p \in X$ and a finite étale extension B of A such that $B \otimes_A A_{m_p} \cong R$. Set $U_1 = \mathrm{spec}(B)$. Let $\pi : U_1 \to V_1$ be the natural map. There exists an open neighborhood U_2 of $\pi^{-1}(p)$ such that $(\overline{x}, \overline{y}, \overline{z})$ are uniformizing parameters on U_2. Let $Z = U_1 - U_2$. Set $W = \pi(Z)$, $U_3 = U_1 - \pi^{-1}(W)$. $U_3 \to V_2 = V_1 - W$ is finite étale. Thus there exists an étale neighborhood U of p where $(\overline{x}, \overline{y}, \overline{z})$ are uniformizing parameters. $\square$

Lemma 18.3. *Suppose that permissible parameters (u, v) for $\Phi(p) \in D_S$ are strongly prepared at $p \in E_X$. Then there exist $*$-permissible parameters (x, y, z) at p such that (x, y, z) are uniformizing parameters on an étale neighborhood of p, and one of the following forms hold:*

1. *p is a 1 point, $u = 0$ is a local equation of E_X and*

$$u = x^a$$
$$v = P(x) + x^b y$$

   where $P(x)$ is a polynomial of degree $\leq b$.

2. *p is a 2 point, $u = 0$ is a local equation of E_X and*

$$u = (x^a y^b)^m$$
$$v = P(x^a y^b) + x^c y^d$$

   where $(a, b) = 1$, $ad - bc \neq 0$, $P(t)$ is a polynomial of degree $\leq \left[\max\{\frac{c}{a}, \frac{d}{b}\} \right]$.

3. *p is a 2 point, $u = 0$ is a local equation of E_X and*

$$u = (x^a y^b)^m$$
$$v = P(x^a y^b) + x^c y^d z$$

   where $(a, b) = 1$, $P(t)$ is a polynomial of degree $\leq \left[\max\{\frac{c}{a}, \frac{d}{b}\} \right]$.

4. *p is a 3 point, $u = 0$ is a local equation of E_X and*

$$u = (x^a y^b z^c)^m$$
$$v = P(x^a y^b z^c) + x^d y^e z^f$$

   where $(a, b, c) = 1$, $P(t)$ is a polynomial of degree $\leq \left[\max\{\frac{d}{a}, \frac{e}{b}, \frac{f}{c}\} \right]$.

5. *p is a 2 point, $uv = 0$ is a local equation of E_X and*

$$u = x^a, v = y^b.$$

6. *p is a 3 point, $uv = 0$ is a local equation of E_X and*

$$u = x^a y^b, v = z^c$$

   (with $a, b, c > 0$).

7. *p is a 3 point, $uv = 0$ is a local equation of E_X and*

$$u = x^a y^b, v = y^c z^d$$

   (with $a, b, c, d > 0$).

Proof. Suppose there exist regular parameters (x, y, z) in $\hat{\mathcal{O}}_{X,p}$ such that

$$u = x^a$$
$$v = P(x) + x^b y.$$

There exist $\alpha \in \hat{\mathcal{O}}_{X,p}$ and $\overline{x} \in \mathcal{O}_{X,p}$ such that $x = \alpha \overline{x}$, and $\alpha^a \in \mathcal{O}_{X,p}$. Set $R = \mathcal{O}_{X,p}[\alpha]$. Let L be the quotient field of R. R is finite étale over $\mathcal{O}_{X,p}$.

$$v - P_b(\alpha \overline{x}) = \left(\alpha^b y + \frac{(P(\alpha \overline{x}) - P_b(\alpha \overline{x}))}{\overline{x}^b} \right) \overline{x}^b$$

implies

$$\overline{y} = \frac{v - P_b(\alpha \overline{x})}{\overline{x}^b} \in (\hat{R}_q) \cap L = R_q$$

(by Lemma 2.1 [14]) for all maximal ideals $q \subset R$. Thus $\overline{y} \in \cap R_q = R$. Choose $\overline{z} \in \mathcal{O}_{X,p}$ such that

$$z \equiv \overline{z} \bmod m_p^2 \hat{\mathcal{O}}_{X,p}.$$

Then $m_p R = (\overline{x}, \overline{y}, \overline{z})$. By Lemma 18.2 there exists an étale neighborhood U of p such that $(\overline{x}, \overline{y}, \overline{z})$ are uniformizing parameters on U.

Suppose there exist regular parameters (x, y, z) in $\hat{\mathcal{O}}_{X,p}$ such that

$$
\begin{aligned}
u &= (x^a y^b)^m \\
v &= P(x^a y^b) + x^c y^d
\end{aligned}
$$

There exists $\alpha_1, \alpha_2 \in \hat{\mathcal{O}}_{X,p}$ and $\overline{x}, \overline{y} \in \mathcal{O}_{X,p}$ such that $x = \alpha_1 \overline{x}, y = \alpha_2 \overline{y}$. Set

$$e = \left\lceil \max \left\{ \frac{c}{a}, \frac{d}{b} \right\} \right\rceil.$$

$$u = (\alpha_1^a \alpha_2^b)^m (\overline{x}^a \overline{y}^b)^m.$$

Set $\gamma = \alpha_1^a \alpha_2^b$. Let K be the quotient field of $\mathcal{O}_{X,p}$.

$$\gamma^m = \frac{u}{(\overline{x}^a \overline{y}^b)^m} \in \hat{\mathcal{O}}_{X,p} \cap K = \mathcal{O}_{X,p}.$$

Set $R = \mathcal{O}_{X,p}[\gamma]$. R is finite étale over $\mathcal{O}_{X,p}$. Let L be the quotient field of R. Set

$$w = \alpha_1^c \alpha_2^d + \frac{P(\alpha_1^a \alpha_2^b \overline{x}^a \overline{y}^b) - P_e(\alpha_1^a \alpha_2^b \overline{x}^a \overline{y}^b)}{\overline{x}^c \overline{y}^d} = \frac{v - P_e(\alpha_1^a \alpha_2^b \overline{x}^a \overline{y}^b)}{\overline{x}^c \overline{y}^d} \in (\hat{R}_q) \cap L = R_q$$

for all maximal ideals q of R. Thus $w \in \cap R_q = R$. Set $f = ad - bc$. Set

$$\tilde{x} = (\gamma^d \omega^{-b})^{\frac{1}{f}} \overline{x},$$

$$\tilde{y} = (\gamma^{-c} \omega^a)^{\frac{1}{f}} \overline{y}.$$

$$
\begin{aligned}
u &= (\tilde{x}^a \tilde{y}^b)^m \\
v &= P_e(\tilde{x}^a \tilde{y}^b) + \tilde{x}^c \tilde{y}^d.
\end{aligned}
$$

Choose $\tilde{z} \in \mathcal{O}_{X,p}$ such that $z \equiv \tilde{z} \bmod m_p^2 \hat{\mathcal{O}}_{X,p}$. Then

$$\tilde{x}, \tilde{y}, \tilde{z} \in R_1 = R[(\gamma^d \omega^{-b})^{\frac{1}{f}}, (\gamma^{-c} \omega^a)^{\frac{1}{f}}]$$

are regular parameters at all maximal ideals of R_1. By Lemma 18.2, there exists an étale neighborhood U of p such that $\tilde{x}, \tilde{y}, \tilde{z}$ are uniformizing parameters on U.

Suppose there exist regular parameters (x, y, z) in $\hat{\mathcal{O}}_{X,p}$ such that

$$
\begin{aligned}
u &= (x^a y^b)^m \\
v &= P(x^a y^b) + x^c y^d z.
\end{aligned}
$$

There exist $\alpha_1, \alpha_2 \in \hat{\mathcal{O}}_{X,p}$ and $\overline{x}, \overline{y} \in \mathcal{O}_{X,p}$ such that $x = \alpha_1 \overline{x}, y = \alpha_2 \overline{y}$. Set $e = \left\lceil \max \left\{ \frac{c}{a}, \frac{d}{b} \right\} \right\rceil$. Let K be the quotient field of $\mathcal{O}_{X,p}$.

$$u = (\alpha_1^a \alpha_2^b)^m (\overline{x}^a \overline{y}^b)^m.$$

Set $\gamma = \alpha_1^a \alpha_2^b$.

$$\gamma^m = \frac{u}{(\overline{x}^a \overline{y}^b)^m} \in \hat{\mathcal{O}}_{X,p} \cap K = \mathcal{O}_{X,p}.$$

Set $R = \mathcal{O}_{X,p}[\gamma]$. R is finite étale over $\mathcal{O}_{X,p}$. Let L be the quotient field of R. Set

$$\overline{z} = \frac{v - P_e(\alpha_1^a \alpha_2^b \overline{x}^a \overline{y}^b)}{\overline{x}^c \overline{y}^d} \in (\hat{R}_q) \cap L = R_q$$

for all maximal ideals q of R. Thus $\overline{z} \in \cap R_q = R$. Set $\tilde{x} = \alpha_1 \alpha_2^{\frac{b}{a}} \overline{x}$. $\tilde{x}, \overline{y}, \overline{z} \in$ $R_1 = R[\alpha_1 \alpha_2^{\frac{b}{a}}]$ and $(\tilde{x}, \overline{y}, \overline{z}) = m_p R$. By Lemma 18.2 there exists an étale neighborhood U of p such that $(\tilde{x}, \overline{y}, \overline{z})$ are uniformizing parameters on U.

Suppose there exist regular parameters (x, y, z) in $\hat{\mathcal{O}}_{X,p}$ such that

$$\begin{aligned} u &= (x^a y^b z^c)^m \\ v &= P(x^a y^b z^c) + x^d y^e z^f. \end{aligned}$$

There exist $\alpha_1, \alpha_2, \alpha_3 \in \hat{\mathcal{O}}_{X,p}$ and $\overline{x}, \overline{y}, \overline{z} \in \mathcal{O}_{X,p}$ such that $x = \alpha_1 \overline{x}$, $y = \alpha_2 \overline{y}$, $z = \alpha_3 \overline{z}$. Set $g = \left[\max \left\{ \frac{d}{a}, \frac{e}{b}, \frac{f}{c} \right\} \right]$.

$$u = (\alpha_1^a \alpha_2^b \alpha_3^c)^m (\overline{x}^a \overline{y}^b \overline{z}^c)^m.$$

Set $\gamma = \alpha_1^a \alpha_2^b \alpha_3^c$. Let K be the quotient field of $\mathcal{O}_{X,p}$.

$$\gamma^m = \frac{u}{(\overline{x}^a \overline{y}^b \overline{z}^c)^m} \in \hat{\mathcal{O}}_{X,p} \cap K = \mathcal{O}_{X,p}$$

Set $R = \mathcal{O}_{X,p}[\gamma]$. R is finite étale over $\mathcal{O}_{X,p}$. Let L be the quotient field of R. Set

$$\omega = \frac{v - P_g(\alpha_1^a \alpha_2^b \alpha_3^c \overline{x}^a \overline{y}^b \overline{z}^c)}{\overline{x}^d \overline{y}^d \overline{z}^f} \in (\hat{R}_q) \cap L = R_q$$

for all maximal ideals q of R. After possibly permuting x, y, z, we can assume that $h = ae - bd \neq 0$. Set

$$\tilde{x} = (\gamma^e \omega^{-b})^{\frac{1}{h}} \overline{x}, \tilde{y} = (\gamma^{-d} \omega^a)^{\frac{1}{h}} \overline{y}.$$

Set $R_1 = R[(\gamma^e \omega^{-b})^{\frac{1}{h}}, (\gamma^{-d} \omega^a)^{\frac{1}{h}}]$. $\tilde{x}, \tilde{y}, \overline{z} \in R_1$.

$$\begin{aligned} u &= (\tilde{x}^a \tilde{y}^b \overline{z}^c)^m \\ v &= P_g(\tilde{x}^a \tilde{y}^b \overline{z}^c) + \tilde{x}^d \tilde{y}^e \overline{z}^f \end{aligned}$$

$(\tilde{x}, \tilde{y}, \overline{z}) = m_p R_1$. By Lemma 18.2 there exists an étale neighborhood U of p such that $(\tilde{x}, \tilde{y}, \overline{z})$ are uniformizing parameters on U.

The arguments for the remaining cases 5., 6. and 7. are easier. $\square$

Remark 18.4. *Suppose that $p \in X$ is a prepared 3 point, so that*

$$\begin{aligned} u &= (x^a y^b z^c)^m \\ v &= P(x^a y^b z^c) + x^d y^e z^f, \end{aligned}$$

$u = 0$ is a local equation of E_X and

$$\text{rank} \begin{pmatrix} a & b & c \\ d & e & f \end{pmatrix} = 2.$$

Then at most one of $ae - bd, af - cd, bf - ce$ is zero.

Proof. By assumption, a, b, c are all nonnegative. Suppose that two of these forms are zero. After permuting x, y, z, we may assume that $ae - bd = 0$ and $af - cd = 0$. Then $e = \frac{bd}{a}$, $f = \frac{cd}{a}$ and $bf - ce = \frac{bcd}{a} - \frac{cbd}{a} = 0$, a contradiction. $\square$

Definition 18.5. *Suppose that* $\Phi : X \to S$ *is strongly prepared with respect to* D_S. *Suppose that* $p \in E_X$ *We will say that* p *is a good point for* Φ *if there exist permissible parameters* (u, v) *at* $\Phi(p)$ *and* *-permissible parameters* (x, y, z) *at* p *for* (u, v) *such that one of the following forms hold:*
p *is a 3 point,* $u = 0$ *is a local equation of* E_X *at* p *and*

$$\begin{aligned} u &= x^a y^b z^c \\ v &= x^d y^e z^f \end{aligned} \tag{176}$$

with

$$\text{rank} \begin{pmatrix} a & b & c \\ d & e & f \end{pmatrix} = 2$$

p *is a 3 point,* $uv = 0$ *is a local equation of* E_X *at* p,

$$\begin{aligned} u &= x^a y^b \\ v &= z^c \end{aligned} \tag{177}$$

p *is a 3 point,* $uv = 0$ *is a local equation of* E_X *at* p

$$\begin{aligned} u &= x^a y^b \\ v &= y^c z^d \end{aligned} \tag{178}$$

with $a, b, c, d > 0$.
p *is a 2 point,* $u = 0$ *is a local equation of* E_X *at* p,

$$\begin{aligned} u &= x^a y^b \\ v &= x^c y^d \end{aligned} \tag{179}$$

with $ad - bc \neq 0$
p *is a 2 point,* $u = 0$ *is a local equation of* E_X *at* p *and there exists* $\alpha \in k$ *such that*

$$\begin{aligned} u &= (x^a y^b)^m \\ v &= \alpha(x^a y^b)^t + (x^a y^b)^t z \end{aligned} \tag{180}$$

with $(a, b) = 1$.
p *is a 2 point,* $u = 0$ *is a local equation of* E_X *at* p

$$\begin{aligned} u &= x^a y^b \\ v &= x^c y^d z \end{aligned} \tag{181}$$

with $ad - bc \neq 0$
p *is a 2 point,* $uv = 0$ *is a local equation of* E_X *at* p

$$\begin{aligned} u &= x^a \\ v &= y^b \end{aligned} \tag{182}$$

p *is a 1 point,* $u = 0$ *is a local equation of* E_X *at* p *and there exists* $\alpha \in k$ *such that*

$$\begin{aligned} u &= x^a \\ v &= \alpha x^c + x^c y \end{aligned} \tag{183}$$

$p \in X$ will be called a bad point if p is not a good point.

Remark 18.6. *Suppose that $p \in X$ is a good point of one of the forms (176), (177), (178), (179), (180), (181), (182) or (183). Then (as in Lemma 18.3) there exist $*$-permissible parameters (x, y, z) at p such that (x, y, z) are uniformizing parameters on an étale neighborhood of p, and one of the forms (176), (177), (178), (179), (180), (181), (182) or (183) hold.*

Suppose that $p \in X$ is a 1 point and (u, v) are permissible parameters at $\Phi(p)$, (x, y, z) are $*$-permissible parameters at p for (u, v) such that

$$\begin{aligned} u &= x^a \\ v &= P(x) + x^c y. \end{aligned}$$

with deg $(P) \le c$. Set $d = \text{ord } (P) \in \mathbf{N} \cup \{\infty\}$.

Suppose that (u_1, v_1) are also permissible parameters at $\Phi(p)$ and (x_1, y_1, z_1) are $*$-permissible parameters at p for (u_1, v_1) such that

$$\begin{aligned} u_1 &= x_1^{a_1} \\ v_1 &= P_1(x_1) + x_1^{c_1} y_1. \end{aligned}$$

with deg $(P_1) \le c_1$. Set $d_1 = \text{ord } (P_1) \in \mathbf{N} \cup \{\infty\}$.

We will compare a, c, d and a_1, c_1, d_1.

If $(u, v) = (u_1, v_1)$ then there exists an a-th root of unity $\omega \in k$ such that $x = \omega x_1$, so that $c = c_1$, and

$$P_1(x_1) = P(\omega x_1).$$

Thus $a = a_1$, $c = c_1$, $d = d_1$.

Suppose that (u, v) and (u_1, v_1) are related by a change of parameters of the type of Case 1.1 of the proof of Lemma 6.7. This case can only occur if $\Phi(p)$ is a 2 point. We have $v_1 = u$ and $u_1 = v$. Then $d = \text{ord}(P) \le c$. The analysis of Case 1.1 in Lemma 6.7 shows that there are $*$-permissible parameters $(\overline{x}, \overline{y}, \overline{z})$ for (u_1, v_1) such that

$$\begin{aligned} u_1 &= v = \overline{x}^d \\ v_1 &= u = \overline{P}(\overline{x}) + \overline{x}^{a+c-d} \overline{y} \end{aligned}$$

where $\text{ord}(\overline{P}) = a$. Thus $a_1 = d$, $c_1 = a + c - d$ and $d_1 = a$.

Suppose that (u, v) and (u_1, v_1) are related by a change of parameters of the type of Case 1.2 of the proof of Lemma 6.7. We have $u_1 = \alpha u$ and $v_1 = v$ where $\alpha(u, v)$ is a unit series. The analysis of Case 1.2 in Lemma 6.7 shows that there are $*$-permissible parameters $(\overline{x}, \overline{y}, \overline{z})$ for (u_1, v) such that

$$\begin{aligned} u_1 &= \overline{x}^a \\ v &= \overline{P}(\overline{x}) + \overline{x}^c \overline{y} \end{aligned}$$

where $\text{ord}(\overline{P}) = d$. Thus $a_1 = a$, $c_1 = c$ and $d_1 = d$.

Suppose that (u, v) and (u_1, v_1) are related by a change of parameters of the type of Case 1.3 of the proof of Lemma 6.7. We have $u_1 = u$ and $v_1 = \alpha u + \beta v$ where $\alpha(u, v), \beta(u, v)$ are series, β is a unit series. If $\Phi(p)$ is a 2 point then

$\alpha = 0$. The analysis of Case 1.3 in Lemma 6.7 shows that there are *-permissible parameters $(\bar{x}, \bar{y}, \bar{z})$ for (u_1, v_1) such that

$$\begin{aligned} u_1 &= x^a \\ v_1 &= \overline{P}(x) + x^c \bar{y} \end{aligned}$$

such that

$$\begin{aligned} \overline{P}(x) &= \sum \alpha_{ij} x^{a(i+1)} P(x)^j + \sum \beta_{ij} x^{a_i} P(x)^{j+1} \\ &\equiv \beta_{00} P(x) + \sum \alpha_{i0} x^{a(i+1)} \mod x^{d+1}. \end{aligned}$$

$a_1 = a$, $c_1 = c$, $d_1 \leq d$ if $a \nmid d$. If $\alpha = 0$, we have $a_1 = a$, $c_1 = c$, $d_1 = d$.

Suppose that E is a component of E_X, $p \in E$, $f \in \hat{\mathcal{O}}_{X,p}$, $x = 0$ is a local equation of E at p. Then define

$$\nu_E(f) = \max \{n \text{ such that } x^n \mid f\}.$$

Definition 18.7. *Suppose that $p \in X$ is a 1 point, and E is the component of E_X containing p. Suppose that (u, v) are permissible parameters at $\Phi(p)$ such that $u = 0$ is a local equation of E at p. If (x, y, z) are *-permissible parameters at p for (u, v), then there is an expression*

$$\begin{aligned} u &= x^a \\ v &= P(x) + x^c y. \end{aligned}$$

For fixed (u, v), a, c and $\nu_E(v)$ are independent of the choice of permissible parameters (x, y, z) for (u, v). Define

$$A(\Phi, p) = \min (c - \nu_E(v))$$

where the minimum is over permissible parameters (u, v) at $\Phi(p)$ such that $u = 0$ is a local equation of E at p.
If $A(\Phi, p) > 0$, define

$$C(\Phi, p) = \min (c - \nu_E(v), \nu_E(v) + a)$$

where the minimum (in the lexicographic order) is over permissible parameters (u, v) at $\Phi(p)$ such that $u = 0$ is a local equation of E at p.

Suppose that E is a component of E_X, $p \in E$ is a 1 point. Suppose that (u, v) are permissible parameters for $\Phi(p) = q$ such that $u = 0$ is a local equation of E at p, (x, y, z) are *-permissible parameters for (u, v) at p. There is an expression

$$\begin{aligned} u &= x^a \\ v &= P(x) + x^c y. \end{aligned} \tag{184}$$

$c > 0$ is equivalent to $\Phi(E) = q$. $c = 0$ is equivalent to $\Phi(E)$ is a component of D_S with local equation $u = 0$ at q.

Suppose that $\Phi(E) = q$ is a 1 point on S. By the discussion before Definition 18.7, $A(\Phi, p) = c - \nu_E(v)$ if and only if $a \nmid \text{ord}(P)$ or $c = \text{ord}(P)$. If $P(x) = \sum a_i x^i$, we can make a permissible change of parameters at q, replacing v with $v - \sum a_{ia} u^i$ to achieve $A(\Phi, p) = c - \nu_E(v)$.

Suppose that $\Phi(E) = q$ is a 2 point on S. By the discussion before Definition 18.7, $A(\Phi, p) = c - \nu_E(v)$.

Suppose that $\Phi(E)$ is a component D of D_S. This is equivalent to $c = 0$ in (184). Then

$$0 = A(\Phi, p) = c - \nu_E(v).$$

In all these cases, if $A(\Phi, p) = c - \nu_E(v) > 0$, then we have

$$C(\Phi, p) = (c - \nu_E(v), a + \nu_E(v)).$$

and there exists an open neighborhood U of p such that $A(\Phi, p') = A(\Phi, p)$ for all $p' \in E \cap U$ and $C(\Phi, p') = C(\Phi, p)$ if $A(\Phi, p) > 0$. Then $A(\Phi, p') = A(\Phi, p)$ and $C(\Phi, p') = C(\Phi, p)$ at all 1 points $p' \in E$. We can then define

$$A(\Phi, E) = A(\Phi, p)$$

and

$$C(\Phi, E) = C(\Phi, p)$$

for $p \in E$ a 1 point.

Lemma 18.8. *Suppose that $p \in X$ is a 2 point and E_1, E_2 are the components of E_X containing p. Then there exist permissible parameters (u, v) at $q = \Phi(p)$ and $*$-permissible parameters (x, y, z) for (u, v) at p such that, if p satisfies (18) of Definition 6.5,*

$$\begin{aligned} u &= (x^a y^b)^k \\ v &= P(x^a y^b) + x^c y^d \end{aligned}$$

or if p satisfies (19) of Definition 6.5,

$$\begin{aligned} u &= (x^a y^b)^k \\ v &= P(x^a y^b) + x^c y^d z \end{aligned}$$

where $x = 0$ is a local equation of E_1, $y = 0$ is a local equation of E_2, then

$$A(\Phi, E_1) = c - \nu_{E_1}(v), A(\Phi, E_2) = d - \nu_{E_2}(v).$$

If $A(\Phi, E_1) > 0$ then

$$C(\Phi, E_1) = (c - \nu_{E_1}(v), \nu_{E_1}(v) + ak).$$

If $A(\Phi, E_2) > 0$ then

$$C(\Phi, E_2) = (d - \nu_{E_2}(v), \nu_{E_2}(v) + bk).$$

Suppose that $p \in X$ is a 3 point, p satisifes (20) of Definiton 6.5, and E_1, E_2, E_3 are the components of E_X containing p. Then there exist permissible parameters (u, v) at $q = \Phi(p)$ and $*$-permissible parameters (x, y, z) for (u, v) at p such that

$$\begin{aligned} u &= (x^a y^b z^c)^m \\ v &= P(x^a y^b z^c) + x^d y^e z^f \end{aligned}$$

where $x = 0$ is a local equation of E_1, $y = 0$ is a local equation of E_2, $z = 0$ is a local equation of E_3, and

$$A(\phi, E_1) = d - \nu_{E_1}(v), A(\Phi, E_2) = e - \nu_{E_2}(v), A(\Phi, E_3) = f - \nu_{E_3}(v).$$

If $A(\Phi, E_1) > 0$, then

$$C(\Phi, E_1) = (d - \nu_{E_1}(v), \nu_{E_1}(v) + am),$$

If $A(\Phi, E_2) > 0$, then

$$C(\Phi, E_2) = (e - \nu_{E_2}(v), \nu_{E_2}(v) + bm),$$

If $A(\Phi_{E_3}) > 0$, then

$$C(\Phi, E_3) = (f - \nu_{E_3}(v), \nu_{E_3}(v) + cm),$$

Proof. Suppose that $p \in X$ is a 2 point satisfying (18), (u, v) are permissible parameters at q and (x, y, z) are uniformizing parameters for (u, v) at p such that

$$\begin{aligned} u &= (x^a y^b)^k \\ v &= P(x^a y^b) + x^c y^d \end{aligned}$$

and (x, y, z) are uniformizing parameters on an étale neighborhood of p. Let $P(t) = \sum a_i t^i$. If q is a 1 point, then we can replace v with $v - \sum_i a_{ik} u^i$, so that $k \nmid \text{ord}(P)$.

If $c = 0$, then $0 = A(\Phi, E_1) = c - \nu_{E_1}(v)$, and if $d = 0$, then $0 = A(\Phi, E_2) = d - \nu_{E_2}(v)$.

Suppose that $c > 0$. Then $\Phi(E_1) = q$. If p' is a 1 point on E_1 near p then there exist *-permissible parameters $(\bar{x}, \bar{y}, \bar{z})$ at p' such that

$$\begin{aligned} u &= \bar{x}^{ak} \\ v &= P_{p'}(\bar{x}) + \bar{x}^c \bar{y} \end{aligned}$$

where $P_{p'}(\bar{x}) = P(\bar{x}^a) + \alpha \bar{x}^c$ for some nonzero $\alpha \in k$.

If q is a 1 point we have $ak \nmid \text{ord}(P_{p'})$ or $c = \text{ord}(P_{p'})$. By the discussion before Definition 18.7, we have that $A(\Phi, E_1) = c - \nu_{E_1}(v)$, and if $A(\Phi, E_1) > 0$, then

$$C(\Phi, E_1) = (c - \nu_{E_1}(v), \nu_{E_1}(v) + ak).$$

A similar argument shows that $A(\Phi, E_2) = d - \nu_{E_2}(v)$ if $d > 0$, and if $A(\Phi, E_2) > 0$, then

$$C(\Phi, E_2) = (d - \nu_{E_2}(v), \nu_{E_2}(v) + bk).$$

If p satisfies (19) or (20) then the proof is similar.   □

Remark 18.9. *If p is a 1 point then $A(\Phi, p) = 0$ if and only if p is a good point.*

Set

$$A(\Phi) = \max \{A(\Phi, E) \mid E \text{ is a component of } E_X\}.$$

If $A(\Phi) > 0$, define

$$C(\Phi) = \max \{C(\Phi, E) \mid E \text{ is a component of } E_X\}.$$

Lemma 18.10. *Suppose that $p \in X$ is a 1 point, (u, v) are permissible parameters at $\Phi(p)$ such that $u = 0$ is a local equation of E_X at p, (x, y, z) are *-permissible parameters at p for (u, v) such that*

$$\begin{aligned} u &= x^a \\ v &= P(x) + x^c y \end{aligned}$$

with $\deg(P) \leq c$. Set $d = \text{ord}(P) \in \mathbf{N} \cup \{\infty\}$.
   1. *Suppose that $\Phi(p)$ is a 1 point. Then p is a bad point if $d < c$ and $a \nmid d$.*
   2. *Suppose that $\Phi(p)$ is a 2 point. Then p is a bad point if $d < c$.*

Proof. Suppose that (u_1, v_1) are permissible parameters at $q = \Phi(p)$ such that $u_1 = 0$ is a local equation of E_X at p, and (u_1, v_1) realize p as a good point. We will find a contradiction.

If q is a 1 point then there exist series $\overline{\alpha}, \overline{\beta}, \overline{\gamma}$ in u, v such that

$$u_1 = \overline{\alpha}u$$
$$v_1 = \overline{\beta}u + \overline{\gamma}v.$$

Thus (u_1, v_1) is obtained by transformations of the form of Case 1.2 and Case 1.3 of Lemma 6.7. The conclusions of the Lemma now follow from the analysis preceeding Definition 18.7.

Suppose that q is a 2 point. Then there exist unit series $\overline{\alpha}, \overline{\beta}$ in u, v such that

$$u_1 = \overline{\alpha}u$$
$$v_1 = \overline{\beta}v$$

or

$$u_1 = \overline{\alpha}v$$
$$v_1 = \overline{\beta}u.$$

In the first case we have, with the notation preceeding Definition 18.7, that $a_1 = a$, $c_1 = c$ and $d_1 = d$ so that $d_1 < c_1$. In the second case we have $a_1 = d$, $c_1 = a + c - d$ and $d_1 = a$ so that $d_1 < c_1$. p is thus a bad point. $\square$

Theorem 18.11. *Suppose that $\Phi : X \to S$ is strongly prepared. Then the locus of bad points in X is a Zariski closed set of pure codimension 1, consisting of a union of components of E_X.*

Proof. We will first show that the good points of X are a Zariski open set in E_X.

Suppose that $p \in X$ is a good 3 point. Then there exists an open neighborhood U of p, uniformizing parameters (x, y, z) in an étale cover of U such that $u = 0$ is a local equation of E_X in U and

$$u = x^a y^b z^c$$
$$v = x^d y^e z^f.$$

If $q \in U$ is a 2 point, then we have (after possibly permuting x, y, z) that (x, y, z_1) are regular parameters at q where $z = z_1 + \alpha$ (with $\alpha \neq 0$). Set $x = x_1(z_1 + \alpha)^{-\frac{c}{a}}$. Then (x_1, y, z_1) are permissible parameters at q, and

$$u = x_1^a y^b$$
$$v = x_1^d y^e (z_1 + \alpha)^{f - \frac{cd}{a}}.$$

If $ae - bd \neq 0$, we can make a permissible change of variables $(\tilde{x}, \tilde{y}, \tilde{z})$ at q to get

$$u = \tilde{x}^a \tilde{y}^b$$
$$v = \tilde{x}^d \tilde{y}^e.$$

If $ae - bd = 0$ then $f - \frac{cd}{a} \neq 0$, so that we can make a permissible change of parameters to get

$$u = (x_1^{a_1} y^{b_1})^k$$
$$v = \beta(x_1^{a_1} y^{b_1})^t + (x_1^{a_1} y^{b_1})^t z_1.$$

If $q \in U$ is a 1 point, then we have (after possibly permuting x, y, z) that (x, y_1, z_1) are regular parameters at q where $y = y_1 + \alpha$, $z = z_1 + \beta$ (with $\alpha, \beta \neq 0$). Set $x = x_1(y_1 + \alpha)^{-\frac{b}{a}}(z_1 + \beta)^{-\frac{c}{a}}$. Then (x_1, y_1, z_1) are permissible parameters at q, and

$$
\begin{aligned}
u &= x_1^a \\
v &= \gamma x_1^d + x_1^d(\gamma_1 y_1 + \gamma_2 z_1 + \cdots)
\end{aligned}
$$

where $\gamma, \gamma_1, \gamma_2 \in k$, $\gamma \neq 0$ and either $\gamma_1 \neq 0$ or $\gamma_2 \neq 0$ since we cannot have both $e - \frac{db}{a} = 0$ and $f - \frac{dc}{a} = 0$. Thus all points in U are good points.

Suppose that $p \in X$ is a good 2 point and (180) holds at p. Then there exists an open neighborhood U of p, uniformizing parameters (x, y, z) in an étale cover of U such that $u = 0$ is a local equation of E_X in U and

$$
\begin{aligned}
u &= (x^a y^b)^k \\
v &= \overline{\beta}(x^a y^b)^t + (x^a y^b)^t z
\end{aligned}
$$

If $q \in U$ is a 2 point, then we have that (x, y, z_1) are permissible parameters at q where $z = z_1 + \alpha$ and q is a good point.

If $q \in U$ is a 1 point, then we have (after possibly permuting x, y) that (x, y_1, z_1) are regular parameters at q where $y = y_1 + \alpha$, $z = z_1 + \beta$ (with $\alpha \neq 0$). Set $x = x_1(y_1 + \alpha)^{-\frac{b}{a}}$. Then (x_1, y_1, z_1) are permissible parameters at q, and

$$
\begin{aligned}
u &= x_1^{ak} \\
v &= (\overline{\beta} + \beta)x_1^{at} + x_1^{at} z_1
\end{aligned}
$$

Thus all points in U are good points.

Suppose that $p \in X$ is a good 2 point, and (179) holds at p. Then there exists an open neighborhood U of p, uniformizing parameters (x, y, z) in an étale cover of U such that $u = 0$ is a local equation of E_X in U and

$$
\begin{aligned}
u &= x^a y^b \\
v &= x^c y^d
\end{aligned}
$$

where $ad - bc \neq 0$. If $q \in U$ is a 2 point, then we have that (x, y, z_1) are permissible parameters at q where $z = z_1 + \alpha$, and q is a good point.

If $q \in U$ is a 1 point, then we have (after possibly permuting x, y) that (x, y_1, z_1) are regular parameters at q where $y = y_1 + \alpha$, $z = z_1 + \beta$ (with $\alpha \neq 0$). Set $x = x_1(y_1 + \alpha)^{-\frac{b}{a}}$. Set $\gamma = \alpha^{d - \frac{bc}{a}}$, $\tilde{y}_1 = (y_1 + \alpha)^{d - \frac{bc}{a}} - \gamma$. Then $(x_1, \tilde{y}_1, z_1)$ are permissible parameters at q, and

$$
\begin{aligned}
u &= x_1^a \\
v &= \gamma x_1^c + x_1^c \tilde{y}_1
\end{aligned}
$$

Thus all points in U are good points.

If p is a good point satisfying (177), (178), (179), (181), (182) or (183), a similar argument shows that there is a Zariski open neighborhood U of p of good points.

We will now show that the bad points of X have pure codimension 1 in X. It suffices to show that any bad point lies on a surface of bad points.

First suppose that p is a bad 3 point. Then there exists an open neighborhood U of p, uniformizing parameters (x, y, z) in an étale cover of U such that $u = 0$ is a local equation of E_X in U and

$$
\begin{aligned}
u &= (x^a y^b z^c)^k \\
v &= P(x^a y^b z^c) + x^d y^e z^f
\end{aligned}
$$

where (after possibly permuting x, y, z) we have

$$
\max\{\frac{d}{a}, \frac{e}{b}, \frac{f}{c}\} = \frac{f}{c}
$$

Thus $\mathrm{ord}(P) < \frac{f}{c}$, since $\mathrm{ord}\,(P) \geq \frac{f}{c}$ implies that $x^d y^e z^f | P(x^a y^b z^c)$, and p is thus a good point.

If $\Phi(p)$ is a 1 point, we can make a permissible change of parameters so that we have that $k \nmid \mathrm{ord}\,(P)$.

Let $q \in U$ be a 1 point on the surface $z = 0$. $c, f > 0$ imply $z = 0$ is a local equation of a component of E_X which maps to $\Phi(p)$. There are regular parameters (x_1, y_1, z) at q where $x = x_1 + \alpha$, $y = y_1 + \beta$ with $\alpha, \beta \neq 0$. There are permissible parameters (x_1, y_1, z_1) at q where

$$
z = (x_1 + \alpha)^{-\frac{a}{c}} (y_1 + \beta)^{-\frac{b}{c}} z_1
$$

$$
\begin{aligned}
u &= z_1^{ck} \\
v &= P(z_1^c) + (x_1 + \alpha)^{d - \frac{fa}{c}} (y_1 + \beta)^{e - \frac{fb}{c}} z_1^f \\
&= P(z_1^c) + \alpha^{d - \frac{fa}{c}} \beta^{e - \frac{fb}{c}} z_1^f + z_1^f (\gamma_1 x_1 + \gamma_2 y_1 + \cdots)
\end{aligned}
$$

where $\gamma_1, \gamma_2 \in k$ and γ_1 or $\gamma_2 \neq 0$.

Suppose that $\Phi(p)$ is a 1 point. Then $\Phi(q) = \Phi(p)$ is a 1 point. q is a bad point by Lemma 18.10, since $ck \nmid c\,\mathrm{ord}(P)$ and $c\,\mathrm{ord}(P) < f$

Suppose that $\Phi(p)$ is a 2 point. Then $\Phi(q) = \Phi(p)$ is a 2 point. q is a bad point by Lemma 18.10 since $c\,\mathrm{ord}\,(P) < f$.

Suppose that p is a bad 2 point satisfying (19). There exists an open neighborhood U of p and uniformizing parameters (x, y, z) on an étale cover of U such that

$$
\begin{aligned}
u &= (x^a y^b)^k \\
v &= P(x^a y^b) + x^c y^d z.
\end{aligned}
$$

We can (after possibly permuting x, y) assume that $ad - bc \geq 0$. Since p is a bad point, $\mathrm{ord}\,(P) < \frac{d}{b}$. If $\Phi(p)$ is a 1 point we can make a permissible change of parameters so that $k \nmid \mathrm{ord}\,(P)$. Let $q \in U$ be a 1 point on the surface $y = 0$. $b, d > 0$ implies $y = 0$ is a local equation of a component of E_X which maps to $\Phi(p)$. There are regular parameters (x_1, y_1, z) at q where $x = x_1 + \alpha$, $z = z_1 + \beta$ (with $\alpha \neq 0$). There are permissible parameters (x_1, y_1, z_1) at q where

$$
y = (x_1 + \alpha)^{-\frac{a}{b}} y_1
$$

$$
\begin{aligned}
u &= y_1^{bk} \\
v &= P(y_1^b) + (x_1 + \alpha)^{c - \frac{da}{b}} (z_1 + \beta) y_1^d \\
&= P(y_1^b) + \alpha^{c - \frac{da}{b}} \beta y_1^d + \tilde{z}_1 y_1^d
\end{aligned}
$$

$\Phi(q) = \Phi(p)$ so that $\Phi(q)$ is a 1 point if and only if $\Phi(p)$ is a 1 point. Since b ord $(P) < d$ and $bk \nmid b$ ord (P) if $\Phi(q)$ is a 1 point, q is a bad point by Lemma 18.10.

A similar argument shows that there is a surface of bad points passing through a bad point satisfying (18) or (17).   □

Lemma 18.12. *Suppose that* $\Phi : X \to S$ *is strongly prepared,* $q \in D_S$ *and* $p \in \Phi^{-1}(q)$ *is such that one of the forms 1. - 7. of Lemma 18.3 hold at p. Then* $m_q \mathcal{O}_{X,p}$ *is not invertible if and only if one of the following holds:*
p is a 1 point

$$\begin{aligned} u &= x^k \\ v &= x^c y \end{aligned} \tag{185}$$

with $c < k$.
p is a 2 point

$$\begin{aligned} u &= (x^a y^b)^k \\ v &= P(x^a y^b) + x^c y^d \end{aligned} \tag{186}$$

with $a, b > 0$, $(a, b) = 1$, $ad - bc \neq 0$,

$$min\{\frac{c}{a}, \frac{d}{b}\} < ord \ (P) < max\{\frac{c}{a}, \frac{d}{b}\},$$

$$min\{\frac{c}{a}, \frac{d}{b}\} < k.$$

p is a 2 point

$$\begin{aligned} u &= (x^a y^b)^k \\ v &= x^c y^d \end{aligned} \tag{187}$$

with $a, b > 0$, $(a, b) = 1$, $ad - bc \neq 0$,

$$min\{\frac{c}{a}, \frac{d}{b}\} < k < max\{\frac{c}{a}, \frac{d}{b}\}.$$

p is a 2 point

$$\begin{aligned} u &= (x^a y^b)^k \\ v &= P(x^a y^b) + x^c y^d z \end{aligned} \tag{188}$$

with $a, b > 0$, $(a, b) = 1$, $ad - bc \neq 0$,

$$min\{\frac{c}{a}, \frac{d}{b}\} < ord \ (P) < max\{\frac{c}{a}, \frac{d}{b}\},$$

$$min\{\frac{c}{a}, \frac{d}{b}\} < k.$$

p is a 2 point

$$\begin{aligned} u &= (x^a y^b)^k \\ v &= x^c y^d z \end{aligned} \tag{189}$$

with $a, b > 0$, $(a, b) = 1$, $ad - bc \neq 0$,

$$min\{\frac{c}{a}, \frac{d}{b}\} < k.$$

p is a 2 point

$$\begin{aligned} u &= (x^a y^b)^k \\ v &= (x^a y^b)^t z \end{aligned}$$ (190)

with $a, b > 0$, $(a, b) = 1$, $t < k$.

p is a 2 point

$$\begin{aligned} u &= x^a \\ v &= y^b \end{aligned}$$ (191)

p is a 3 point

$$\begin{aligned} u &= (x^a y^b z^c)^k \\ v &= P(x^a y^b z^c) + x^d y^e z^f \end{aligned}$$ (192)

with $a, b, c > 0$, $(a, b, c) = 1$,

$$min\{\frac{d}{a}, \frac{e}{b}, \frac{f}{c}\} < ord\ (P) < max\{\frac{d}{a}, \frac{e}{b}, \frac{f}{c}\},$$

$$k > min\{\frac{d}{a}, \frac{e}{b}, \frac{f}{c}\}.$$

p is a 3 point

$$\begin{aligned} u &= (x^a y^b z^c)^k \\ v &= x^d y^e z^f \end{aligned}$$ (193)

with $a, b, c > 0$, $(a, b, c) = 1$,

$$min\{\frac{d}{a}, \frac{e}{b}, \frac{f}{c}\} < k < max\{\frac{d}{a}, \frac{e}{b}, \frac{f}{c}\}.$$

p is a 3 point

$$\begin{aligned} u &= x^a y^b \\ v &= z^c \end{aligned}$$ (194)

with $a, b, c > 0$.

p is a 3 point

$$\begin{aligned} u &= x^a y^b \\ v &= y^c z^d \end{aligned}$$ (195)

with $a, b, c, d > 0$.

Proof. Suppose that p is a 1 point. Then (185) follows easily.

Suppose that p is a 2 point with

$$\begin{aligned} u &= (x^a y^b)^k \\ v &= P(x^a y^b) + x^c y^d, \end{aligned}$$

$P \neq 0$ and $e = ord\ (P) < max\{\frac{c}{a}, \frac{d}{b}\}$. Set $\lambda_2 = max\{\frac{c}{a}, \frac{d}{b}\}$, $\lambda_1 = min\{\frac{c}{a}, \frac{d}{b}\}$.
$u \mid v$ if and only if $e \geq k$ and $\lambda_1 \geq k$. $v \mid u$ if and only if $e \leq \lambda_1$ and $e \leq k$.
Thus $(u, v)\mathcal{O}_{X,p}$ is not invertible if and only if $\lambda_1 < k$ and $\lambda_1 < e$.

Suppose that p is a 2 point with

$$\begin{aligned} u &= (x^a y^b)^k \\ v &= x^c y^d \end{aligned}$$

Set $\lambda_1 = \min\{\frac{c}{a}, \frac{d}{b}\}$, $\lambda_2 = \max\{\frac{c}{a}, \frac{d}{b}\}$. $u \mid v$ if and only if $k \leq \lambda_1$, $v \mid u$ if and only if $k \geq \lambda_2$. So $(u, v)\mathcal{O}_{X,p}$ is not invertible if and only if $\lambda_1 < k < \lambda_2$.

Suppose that p is a 2 point with

$$
\begin{aligned}
u &= (x^a y^b)^k \\
v &= P(x^a y^b) + x^c y^d z
\end{aligned}
$$

with $ad - bc \neq 0$, $e = \text{ord}\,(P) < \max\{\frac{c}{a}, \frac{d}{b}\}$. Set $\lambda_1 = \min\{\frac{c}{a}, \frac{d}{b}\}$, $\lambda_2 = \max\{\frac{c}{a}, \frac{d}{b}\}$. $u \mid v$ if and only if $e \geq k$ and $k \leq \lambda_1$. $v \mid u$ if and only if $e \leq k$ and $e \leq \lambda_1$. So $(u, v)\mathcal{O}_{X,p}$ is not invertible if and only if $\lambda_1 < k$ and $\lambda_1 < e$.

Suppose that p is a 2 point with

$$
\begin{aligned}
u &= (x^a y^b)^k \\
v &= x^c y^d z
\end{aligned}
$$

and $ad - bc \neq 0$. (u, v) is invertible at p if and only if $c \geq ka$ and $d \geq bk$. Thus (u, v) is not invertible at p if and only if $k > \min\{\frac{c}{a}, \frac{d}{b}\}$, and we get (189).

Suppose that p is a 2 point with

$$
\begin{aligned}
u &= (x^a y^b)^k \\
v &= P(x^a y^b) + (x^a y^b)^t z
\end{aligned}
$$

with $P \neq 0$ and $e = \text{ord}\,(P) \leq t$. We will show that (u, v) is invertible at p. If $k \leq e$ then $u|v$. Suppose that $k > e$. There are new permissible parameters $(\overline{x}, \overline{y}, \overline{z})$ such that

$$
\begin{aligned}
v &= (\overline{x}^a \overline{y}^b)^e \\
u &= \overline{P}(\overline{x}^a \overline{y}^b) + (\overline{x}^a \overline{y}^b)^{t-e+k}\overline{z}
\end{aligned}
$$

with $\text{ord}\,(\overline{P}) = k$. Thus $v|u$.

Suppose that p is a 2 point with

$$
\begin{aligned}
u &= (x^a y^b)^k \\
v &= (x^a y^b)^t z
\end{aligned}
$$

Then (u, v) not invertible at p if and only if $t < k$, and we get (190).

Suppose that p is a 3 point with

$$
\begin{aligned}
u &= (x^a y^b z^c)^k \\
v &= P(x^a y^b z^c) + x^d y^e z^f
\end{aligned}
$$

with $P \neq 0$ and $\text{ord}\,(P) < \max\{\frac{d}{a}, \frac{e}{b}, \frac{f}{c}\}$. Set

$$
\lambda_2 = \max\{\frac{d}{a}, \frac{e}{b}, \frac{f}{c}\}, \quad \lambda_1 = \min\{\frac{d}{a}, \frac{e}{b}, \frac{f}{c}\}
$$

$u|v$ is equivalent to $\text{ord}\,(P) \geq k$, $k \leq \lambda_1$. $v|u$ is equivalent to $\text{ord}\,(P) \leq k$ and $\text{ord}\,(P) \leq \lambda_1$. That is, (u, v) is not invertible at p if and only if $\text{ord}\,(P) > \lambda_1$ and $k > \lambda_1$. We thus get (192).

Suppose that p is a 3 point with

$$
\begin{aligned}
u &= (x^a y^b z^c)^k \\
v &= x^d y^e z^f
\end{aligned}
$$

Set

$$
\lambda_2 = \max\{\frac{d}{a}, \frac{e}{b}, \frac{f}{c}\}, \quad \lambda_1 = \min\{\frac{d}{a}, \frac{e}{b}, \frac{f}{c}\}
$$

$u|v$ is equivalent to $k \leq \lambda_1$. $v|u$ is equivalent to $k \geq \lambda_2$.

Thus (u, v) is not invertible at p if and only if $\lambda_1 < k < \lambda_2$, and we get (193). $\qquad\square$

Lemma 18.13. *Suppose that $\Phi : X \to S$ is strongly prepared. Let S_1 be the blow-up of S at a point $q \in D_S$. Let U be the largest open set of X such that the rational map $X \to S_1$ is a morphism $\Phi_1 : U \to S_1$. Then Φ_1 is strongly prepared.*

Proof. This follows from the analysis of Lemma 18.12. $\qquad\square$

Theorem 18.14. *Suppose that $\Phi : X \to S$ is strongly prepared, $p \in X$ is a 1 point and the rational map Φ_1 from X to the blow-up S_1 of $q = \Phi(p)$ is a morphism in a neighborhood of p. Then $A(\Phi_1, p) \leq A(\Phi, p)$. If $A(\Phi_1, p) = A(\Phi, p) > 0$, then $C(\Phi_1, p) < C(\Phi, p)$.*

Proof. At p we have permissible parameters such that

$$
\begin{aligned}
u &= x^k \\
v &= P(x) + x^c y
\end{aligned}
$$

and $C(\Phi, p) = (c - \nu_E(v), \nu_E(v) + k)$.

First suppose that $P \neq 0$ and $e = \text{ord}\,(P) \leq c$. If $e > k$ then we have permissible parameters u_1, v_1 at $q_1 = \Phi_1(p)$ such that

$$
u = u_1, v = u_1 v_1.
$$

Then

$$
\begin{aligned}
u_1 &= x^k \\
v_1 &= \frac{P(x)}{x^k} + x^{c-k} y
\end{aligned}
$$

$A(\Phi_1, p) \leq (c - k - (e - k)) = A(\Phi, p)$ and if $A(\Phi_1, p) = A(\Phi, p)$ then $C(\Phi_1, p) \leq (c - k - (e - k), e - k + k) = (c - e, e) < (c - e, e + k) = C(\Phi, p)$.

If $e = k$ then there exists $0 \neq \alpha \in k$ such that $P(x) = \alpha x^k + \cdots$. There exist permissible parameters (u_1, v_1) at $q_1 = \Phi_1(p)$ such that

$$
u = u_1, v = u_1(v_1 + \alpha).
$$

$$
\begin{aligned}
u_1 &= x^k \\
v_1 &= \frac{P(x)}{x^k} - \alpha + x^{c-k} y.
\end{aligned}
$$

Thus $A(\Phi_1, p) < (c - k) - (e - k) = A(\Phi, p)$.

If $e < k$ then we have permissible parameters u_1, v_1 at $q_1 = \Phi_1(p)$ such that

$$
u = u_1 v_1, v = v_1.
$$

We have permissible parameters $(\overline{x}, \overline{y}, z)$ at p such that

$$
\begin{aligned}
v &= \overline{x}^e \\
u &= \overline{P}(\overline{x}) + \overline{x}^{k+c-e} \overline{y}
\end{aligned}
$$

where $\text{ord}\,(\overline{P}) = k$.

Then

$$
\begin{aligned}
v_1 &= \overline{x}^e \\
u_1 &= \frac{\overline{P}(\overline{x})}{\overline{x}^e} + \overline{x}^{k+c-2e} \overline{y}
\end{aligned}
$$

$$A(\Phi_1, p) \leq (k + c - 2e - (k - e)) = A(\Phi, p)$$

and if $A(\Phi_1, p) = A(\Phi, p)$ then $C(\Phi_1, p) \leq (k + c - 2e - (k - e), (k - e) + e) < C(\Phi, p)$.

Now suppose that $P(x) = 0$. Then

$$u = x^k$$
$$v = x^c y$$

with $c \geq k$. There exist permissible parameters (u_1, v_1) at $q_1 = \Phi_1(p)$ such that

$$u = u_1, v = u_1 v_1$$

and

$$A(\Phi_1, p) = A(\Phi, p) = 0.$$

$\square$

Theorem 18.15. *Suppose that* $\Phi : X \to S$ *is strongly prepared and* $q \in S$. *Then the locus of points Z in X where Φ does not factor through the blow-up of q is a pure codimension 2 subscheme. Z makes SNCs with $\overline{B}_2(X)$ except possibly at 3 points of the form (192).*

Suppose that C is a component of this locus which makes SNCs with $\overline{B}_2(X)$, and $\pi : X_1 \to X$ is the blow-up of C, $E_1 = \pi^{-1}(C)_{red}$, $\Phi_1 = \Phi \circ \pi$. Then Φ_1 is strongly prepared and either $A(\Phi_1, E_1) = 0$ or

$$A(\Phi_1, E_1) < A(\Phi)$$

Proof. **Suppose that** $p \in X$ **is a 3 point such that** $m_q \mathcal{O}_{X,p}$ **is not invertible and (192) holds at** p. We may assume that there exists an open neighborhood U of p such that (x, y, z) are uniformizing parameters on an étale cover of U.

After possibly interchanging x, y, z, we can assume that

$$\frac{d}{a} = \min\{\frac{d}{a}, \frac{e}{b}, \frac{f}{c}\}$$

and

$$\frac{f}{c} = \max\{\frac{d}{a}, \frac{e}{b}, \frac{f}{c}\}.$$

We will now determine the locus of points in U where $m_q \mathcal{O}_U$ is not invertible. First suppose that q' is a 2 point on the curve $x = z = 0$. q' has regular parameters (x, y_1, z) where $y = y_1 + \alpha$. Thus q' has permissible parameters (x_1, y_1, z) where x_1 is defined by

$$x = x_1(y_1 + \alpha)^{-\frac{b}{a}}$$

Set $\lambda = (a, c)$, $a_1 = \frac{a}{\lambda}$, $c_1 = \frac{c}{\lambda}$.

$$u = (x_1^{a_1} z^{c_1})^{k\lambda}$$
$$v = P((x_1^{a_1} z^{c_1})^\lambda) + x_1^d z^f (y_1 + \alpha)^{e - \frac{db}{a}}.$$

We have $a_1 f - c_1 d > 0$ and $\lambda \text{ord}(P) < \frac{f}{c_1}$. We can make a permissible change of variables to get

$$u = (\overline{x}_1^{a_1} \overline{z}^{c_1})^{k\lambda}$$
$$v = P((\overline{x}_1^{a_1} \overline{z}^{c_1})^\lambda) + \overline{x}_1^d \overline{z}^f.$$

$k > \frac{d}{a}$ implies $\lambda k > \frac{d}{a_1}$ and ord $(P) > \frac{d}{a}$ implies $\lambda \text{ord}\,(P) > \frac{d}{a_1}$. q thus has the form of (186), and we see that (u,v) is not invertible on the curve with local equations $x = z = 0$.

Now suppose that q' is a 2 point on the curve $y = z = 0$. q' has regular parameters (x_1, y, z) where $x = x_1 + \alpha$. Thus q' has permissible parameters (x_1, y_1, z) where y_1 is defined by

$$y = y_1(x_1 + \alpha)^{-\frac{a}{b}}$$

Set $\lambda = (b, c)$, $b_1 = \frac{b}{\lambda}$, $c_1 = \frac{c}{\lambda}$.

$$\begin{aligned} u &= (y_1^{b_1} z^{c_1})^{k\lambda} \\ v &= P((y_1^{b_1} z^{c_1})^{\lambda}) + (x_1 + \alpha)^{d - \frac{ea}{b}} y_1^e z^f \end{aligned}$$

First suppose that $bf - ce \neq 0$. Then $bf - ce > 0$, and $b_1 f - c_1 e > 0$. Since $\lambda \text{ord}\,(P) < \frac{f}{c_1}$, we have by (186) that (u, v) is not invertible at q' if and only if $\lambda \text{ord}\,(P) > \frac{e}{b_1}$ and $\lambda k > \frac{e}{b_1}$ so that (u,v) is not invertible at 2 points q' on $y = z = 0$ if and only if ord $(P) > \frac{e}{b}$ and $k > \frac{e}{b}$.

Now suppose that $bf - ce = 0$, so that $b_1 f - c_1 e = 0$. Since $\lambda \text{ord}\,(P) < \frac{f}{c_1}$, (190) cannot hold, and we then have that (u, v) is invertible at 2 points q' on $y = z = 0$.

Now suppose that q' is a 2 point on the curve $x = y = 0$. q' has regular parameters (x, y, z_1) where $z = z_1 + \alpha$. Thus q' has permissible parameters (x_1, y, z_1) where x_1 is defined by

$$x = x_1(z_1 + \alpha)^{-\frac{c}{a}}$$

Set $\lambda = (a, b)$, $a_1 = \frac{a}{\lambda}$, $b_1 = \frac{b}{\lambda}$.

$$\begin{aligned} u &= (x_1^{a_1} y^{b_1})^{k\lambda} \\ v &= P((x_1^{a_1} y^{b_1})^{\lambda}) + x_1^d y^e (z_1 + \alpha)^{f - \frac{dc}{a}} \end{aligned}$$

First suppose that $ae - bd \neq 0$ and ord $(P) < \frac{e}{b}$. Then $a_1 e - b_1 d > 0$ and $\lambda \text{ord}\,(P) < \frac{e}{b_1}$. By assumption $\lambda k > \frac{d}{a_1}$ and $\lambda \text{ord}\,(P) > \frac{d}{a_1}$. By (186), (u, v) is not invertible at 2 points q' on $x = y = 0$.

Now suppose that $ae - bd \neq 0$ and ord $(P) \geq \frac{e}{b}$. Then $a_1 e - b_1 d > 0$ and $\lambda \text{ord}\,(P) \geq \frac{e}{b_1}$, so that we can choose permissible coordinates at q' so that

$$\begin{aligned} u &= (x_1^{a_1} y_1^{b_1})^{k\lambda} \\ v &= x_1^d y_1^e \end{aligned}$$

By assumption $\lambda k > \frac{d}{a_1}$, so that by (187), (u, v) is not invertible at 2 points q' on $x = y = 0$ if and only if $k < \frac{e}{b}$.

Suppose that $ae - bd = 0$ and ord $(P) < \frac{e}{b}$. Then $a_1 e - b_1 d = 0$ and $\lambda \text{ord}\,(P) < \frac{e}{b_1}$. Since (190) can then not hold at q', we have that (u, v) are invertible at 2 points q' on $x = y = 0$.

Now suppose that $ae - bd = 0$ and ord $(P) \geq \frac{e}{b}$. Then $a_1 e - b_1 d = 0$ and $\lambda \text{ord}\,(P) \geq \frac{e}{b_1}$, so that we can choose permissible coordinates at q' so that

$$\begin{aligned} u &= (x_1^{a_1} y_1^{b_1})^{k\lambda} \\ v &= (\beta + \alpha^{f - \frac{dc}{a}})(x_1^{a_1} y_1^{b_1})^t + (x_1^{a_1} y_1^{b_1})^t z_1 \end{aligned}$$

where $t = \frac{e}{b}\lambda$, $\beta \in k$ is the degree t coefficient of P. For q' in a possibly smaller neighborhood of p_1, (190) can then not hold at q', so that (u, v) are invertible at 2 points q' on $x = y = 0$.

Suppose that q' is a 1 point in U on $z = 0$. q' has regular parameters (x_1, y_1, z) where $x = x_1 + \alpha$, $y = y_1 + \beta$ with $\alpha, \beta \neq 0$. Thus q' has permissible parameters (x_1, y_1, z_1) where z_1 is defined by

$$z = (x_1 + \alpha)^{-\frac{a}{c}}(y_1 + \beta)^{-\frac{b}{c}}z_1$$

$$u = z_1^{ck}$$
$$v = P(z_1^c) + (x_1 + \alpha)^{d-\frac{af}{c}}(y_1 + \beta)^{e-\frac{bf}{c}}z_1^f.$$

Since by assumption $c \text{ ord}(P) < f$, q' cannot be in the form of (185), so that (u, v) is invertible at 1 points on $z = 0$.

Suppose that q' is a 1 point in U on $y = 0$. q' has regular parameters (x_1, y, z_1) where $x = x_1 + \alpha$, $z = z_1 + \beta$ with $\alpha, \beta \neq 0$. Thus q' has permissible parameters (x_1, y_1, z_1) where y_1 is defined by

$$y = (x_1 + \alpha)^{-\frac{a}{b}}(z_1 + \beta)^{-\frac{c}{b}}y_1$$

$$u = y_1^{bk}$$
$$v = P(y_1^b) + (x_1 + \alpha)^{d-\frac{ae}{b}}(z_1 + \beta)^{f-\frac{ec}{b}}y_1^e$$

If $b \text{ ord }(P) < e$ or $b \text{ ord }(P) > e$ then q' cannot have the form of (185), so that (u, v) is invertible at all 1 points on $y = 0$. If $k \leq \frac{e}{b}$, then (u, v) is invertible at all 1 points on $y = 0$.

Suppose that $b \text{ ord }(P) = e$ and $k > \frac{e}{b}$. Then we can write $P(t) = \gamma t^{\frac{e}{b}} + \cdots$ where $\gamma \neq 0$. We have (u, v) is invertible at q' on $y = 0$ unless

$$\alpha^{d-\frac{ae}{b}}\beta^{f-\frac{ec}{b}} + \gamma = 0$$

which holds only if (α, β) are on the algebraic curve

$$\beta^{bf-ce} = (-\gamma)^b\alpha^{ae-bd}.$$

In this case (u, v) is not invertible on the curve with local equations

$$y = 0, z^{bf-ec} + (-\gamma)^b x^{ae-bd} = 0.$$

If q' is a 1 point in U on $x = 0$, then there are permissible parameters (x_1, y_1, z_1) at q' such that

$$u = x_1^{ak}$$
$$v = P(x_1^a) + x_1^d(y_1 + \alpha)^{e-\frac{db}{a}}(z_1 + \beta)^{f-\frac{dc}{a}}$$

with $\alpha, \beta \neq 0$. Thus (u, v) is invertible at 1 points on $x = 0$ in U since $a \text{ ord }(P) > d$.

If $\pi : X_1 \to X$ is the blow-up of a 2 curve through p, then $\Phi_1 = \Phi \circ \pi$ is strongly prepared above p. $\pi^{-1}(p)$ is a 2 curve, so there are no 1 points in $\pi^{-1}(p)$.

Suppose that $p \in X$ is a 3 point such that $m_q\mathcal{O}_{X,p}$ is not invertible, and (193) holds at p. We may assume that there exists an open neighborhood U of p such that (x, y, z) are uniformizing parameters on an étale cover of U.

After possibly interchanging x, y, z, we can assume that

$$\frac{d}{a} = \min\{\frac{d}{a}, \frac{e}{b}, \frac{f}{c}\}$$

and

$$\frac{f}{c} = \max\{\frac{d}{a}, \frac{e}{b}, \frac{f}{c}\}$$

We will determine the locus of points in U where $m_q \mathcal{O}_U$ is not invertible. First suppose that q' is a 2 point on the curve $x = z = 0$. q' has regular parameters (x, y_1, z) where $y = y_1 + \alpha$. Thus q' has permissible parameters (x_1, y_1, z_1) where x_1 is defined by

$$x = x_1(y_1 + \alpha)^{-\frac{b}{a}}$$

Set $\lambda = (a, c)$, $a_1 = \frac{a}{\lambda}$, $c_1 = \frac{c}{\lambda}$.

$$u = (x_1^{a_1} z^{c_1})^{k\lambda}$$
$$v = x_1^d z_1^f (y_1 + \alpha)^{e - \frac{db}{a}}$$

We have $a_1 f - c_1 d > 0$. $\frac{d}{a} < k < \frac{f}{c}$ implies $\frac{d}{a_1} < k\lambda < \frac{f}{c_1}$. q' thus has the form of (187), and we see that (u, v) is not invertible on the curve with local equations $x = z = 0$.

Suppose that q' is a 2 point on the curve $y = z = 0$. q' has permissible parameters (x_1, y_1, z) where $x = x_1 + \alpha$, y_1 is defined by $y = y_1(x_1 + \alpha)^{-\frac{a}{b}}$. Set $\lambda = (b, c)$, $b_1 = \frac{b}{\lambda}$, $c_1 = \frac{c}{\lambda}$.

$$u = (y_1^{b_1} z^{c_1})^{k\lambda}$$
$$v = (x_1 + \alpha)^{d - \frac{ea}{b}} y_1^e z^f.$$

First suppose that $bf - ce \neq 0$. Then $bf - ce > 0$ and $b_1 f - c_1 e > 0$. Since $k\lambda < \frac{f}{c_1}$, we have by (187) that (u, v) is not invertible at 2 points q' on $y = z = 0$ if and only if $\frac{e}{b_1} < k\lambda$, which holds if and only if $\frac{e}{b} < k$.

Now suppose that $bf - ce = 0$, so that $b_1 f - c_1 e = 0$. Then (u, v) is invertible at 2 points q' on $y = z = 0$.

Now suppose that q' is a 2 point on the curve $x = y = 0$. q' has regular parameters (x, y, z_1) where $z = z_1 + \alpha$. q' has permissible parameters (x_1, y, z_1) where x_1 is defined by $x = x_1(z_1 + \alpha)^{-\frac{c}{a}}$. Set $\lambda = (a, b)$, $a_1 = \frac{a}{\lambda}$, $b_1 = \frac{b}{\lambda}$.

$$u = (x_1^{a_1} y^{b_1})^{k\lambda}$$
$$v = x_1^d y^e (z_1 + \alpha)^{f - \frac{dc}{a}}$$

First suppose that $ae - bd \neq 0$. Then $a_1 e - b_1 d > 0$. By assumption $\frac{d}{a_1} < k\lambda$. We have by (187) that (u, v) is not invertible at 2 points q' on $x = y = 0$ if and only if $k\lambda < \frac{e}{b_1}$ which holds if and only if $k < \frac{e}{b}$.

Now suppose that $ae - bd = 0$. Then $a_1 e - b_1 d = 0$ and (u, v) is invertible at 2 points on the curve $x = y = 0$.

Suppose that q' is a 1 point in U on $z = 0$. q' has regular parameters (x_1, y_1, z) where $x = x_1 + \alpha$, $y = y_1 + \beta$ (with $\alpha, \beta \neq 0$). Thus q' has permissible parameters

(x_1, y_1, z_1) where z_1 is defined by $z = (x_1 + \alpha)^{-\frac{a}{c}}(y_1 + \beta)^{-\frac{b}{c}} z_1$.

$$
\begin{aligned}
u &= z_1^{ck} \\
v &= (x_1 + \alpha)^{d - \frac{af}{c}}(y_1 + \beta)^{e - \frac{bf}{c}} z_1^f
\end{aligned}
$$

Thus (u, v) is invertible at all 1 points of $z = 0$.

Similarly, (u, v) is invertible at all 1 points of $x = 0$ and $y = 0$.

If $\pi : X_1 \to X$ is the blow-up of a 2 curve C through p, then Φ_{X_1} is strongly prepared above p. $\pi^{-1}(p)$ is a 2 curve, so there are no 1 points in $\pi^{-1}(p)$.

Suppose that $p \in X$ is a 2 point such that $m_q \mathcal{O}_{X,p}$ is not invertible, and (188) holds at p. We may assume that there exists an open neighborhood U of p such that (x, y, z) are uniformizing parameters on an étale cover of U, and the conclusions of Lemma 18.8 hold for p. After possibly interchanging x and y we may assume that $ad - bc > 0$. We will determine the locus of points in U where $m_q \mathcal{O}_U$ is not invertible. First suppose that q' is a 2 point on the curve $x = y = 0$. q' has regular parameters (x, y, z_1) where $z = z_1 + \alpha$. Thus q' has permissible parameters $(\overline{x}, \overline{y}, \overline{z})$ such that

$$
\begin{aligned}
u &= (\overline{x}^a \overline{y}^b)^k \\
v &= P(\overline{x}^a \overline{y}^b) + \overline{x}^c \overline{y}^d
\end{aligned}
$$

Since $\frac{c}{a} < e = \text{ord}\,(P) < \frac{d}{b}$, and $k > \frac{c}{a}$, we are in the form of (186). Thus (u, v) is not invertible along the curve $x = y = 0$.

Suppose that q' is a 1 point near q. q' has permissible parameters (x_1, y_1, z_1) where either

$$
x = x_1(y_1 + \alpha)^{-\frac{b}{a}}, y = y_1 + \alpha, z = z_1 + \beta \tag{196}
$$

with $\alpha \neq 0$ or

$$
x = x_1 + \alpha, y = y_1(x_1 + \alpha)^{-\frac{a}{b}}, z = z_1 + \beta \tag{197}
$$

with $\alpha \neq 0$. If q' has permissible parameters satisfying (196), then since $\text{ord}\,(P) > \frac{c}{a}$,

$$
\begin{aligned}
u &= x_1^{ak} \\
v &= P(x_1^a) + x_1^c(y_1 + \alpha)^{d - \frac{bc}{a}}(z_1 + \beta) \\
&= \beta \alpha^{d - \frac{bc}{a}} x_1^c + x_1^c \overline{z}
\end{aligned} \tag{198}
$$

(u, v) is not invertible at q' if and only if q' satisfies (185). Since $c < ak$ by assumption, this holds if and only if $\beta = 0$.

If q' is a 1 point near p on $x = 0$ (so that (198) holds) then $A(\Phi, q') = 0$.

If q' has permissible parameters satisfying (197), then

$$
\begin{aligned}
u &= y_1^{bk} \\
v &= P(y_1^b) + (x_1 + \alpha)^{c - \frac{da}{b}} y_1^d(z_1 + \beta)
\end{aligned} \tag{199}
$$

(u, v) is invertible at q', since $b \,\text{ord}\, P < d$ by assumption, so that q' cannot satisfy (185).

If q' is a 1 point near p on $y = 0$ (so that (199) holds) then

$$
A(\Phi, q') = d - b \,\text{ord}\,(P).
$$

We will now consider the invariant A on the blow-up of $V(x,y)$ or $V(x,z)$ over p.

Let $\pi_1 : X_1 \to X$ be the blow-up of $C = V(x,y)$. $\Phi_1 = \Phi \circ \pi_1$ is strongly prepared above p. If $q \in \pi_1^{-1}(p)$ is a 1 point, then q has regular parameters (x, y_1, z) defined by

$$x = x_1, y = x_1(y_1 + \alpha)$$

with $\alpha \neq 0$. There are permissible parameters $(\overline{x}_1, y, z)$ at q where $\overline{x}_1$ is defined by

$$x_1 = \overline{x}_1(y_1 + \alpha)^{-\frac{b}{a+b}}$$

Thus

$$
\begin{aligned}
u &= \overline{x}_1^{(a+b)k} \\
v &= P(\overline{x}_1^{a+b}) + \overline{x}_1^{c+d}(y_1 + \alpha)^{d - \frac{(c+d)b}{a+b}} z
\end{aligned}
$$

If $(a+b)\mathrm{ord}(P) \geq c + d$, then $A(\Phi_1, q) = 0$. Assume that $(a+b)\mathrm{ord\ p} < c + d$. Since $\mathrm{ord}\ (P) > \frac{c}{a}$, we have that

$$c + d - (a+b)\,\mathrm{ord}(P) = (d - b\,\mathrm{ord}(P)) + (c - a\,\mathrm{ord}(P)) < d - b\,\mathrm{ord}(P)$$

Thus

$$A(\Phi_1, q) \leq c + d - (a+b)\,\mathrm{ord}(P) < d - b\,\mathrm{ord}(P) \leq A(\Phi).$$

If $\pi_1 : X_1 \to X$ is the blow-up of $C = V(x, z)$, then Φ_{X_1} is strongly prepared above p, and there are no 1 points in $\pi_1^{-1}(p)$.

Suppose that $p \in X$ is a 2 point such that $m_q\mathcal{O}_{X,p}$ is not invertible, and (189) holds at p. We may assume that there exists an open neighborhood U of p such that (x, y, z) are uniformizing parameters on an étale cover of U. After possibly interchanging x and y, we may assume that $ad - bc > 0$. We will determine the locus of points in U where $m_q\mathcal{O}_U$ is not invertible. First suppose that q' is a 2 point on the curve $x = y = 0$. q' has regular parameters (x, y, z_1) where $z = z_1 + \alpha$. Thus q' has permissible parameters $(\overline{x}, \overline{y}, \overline{z})$ such that

$$
\begin{aligned}
u &= (\overline{x}^a \overline{y}^b)^k \\
v &= \overline{x}^c \overline{y}^d
\end{aligned}
$$

Since $\frac{c}{a} < k$, we are in the form of (187), and (u, v) is not invertible along the curve $x = y = 0$ if and only if $k < \frac{d}{b}$.

Suppose that q' is a 1 point near p. q' has permissible parameters (x_1, y_1, z_1) where either

$$x = x_1(y_1 + \alpha)^{-\frac{b}{a}}, y = y_1 + \alpha, z = z_1 + \beta \qquad (200)$$

with $\alpha \neq 0$ or

$$x = x_1 + \alpha, y = y_1(x_1 + \alpha)^{-\frac{a}{b}}, z = z_1 + \beta \qquad (201)$$

with $\alpha \neq 0$. If q' has permissible parameters satisfying (200), then

$$
\begin{aligned}
u &= x_1^{ak} \\
v &= x_1^c(y_1 + \alpha)^{d - \frac{bc}{a}}(z_1 + \beta) \\
&= \beta\alpha^{d - \frac{bc}{a}} x_1^c + x_1^c \overline{z}
\end{aligned}
$$

(u, v) is not invertible at q' if and only if q' satisfies (185). Since $c < ak$ by assumption, this holds if and only if $\beta = 0$.

If q' is a 1 point near p on $x = 0$ (so that (200) holds) then $A(\Phi, q') = 0$.

If q' has permissible parameters satisfying (201), then

$$
\begin{aligned}
u &= y_1^{bk} \\
v &= (x_1 + \alpha)^{c - \frac{da}{b}} y_1^d (z_1 + \beta)
\end{aligned}
$$

$$
\begin{aligned}
u &= y_1^{bk} \\
v &= \alpha^{c - \frac{da}{b}} \beta y_1^d + y_1^d z_1
\end{aligned}
$$

Thus (u, v) is invertible at q' if $\beta \neq 0$, and if $\beta = 0$, then (u, v) is invertible at q' if and only if $d \geq kb$.

If q' is a 1 point near p on $y = 0$ (so that (201) holds) then $A(\Phi, q') = 0$.

We will now consider the invariant A on the blow-up of a curve $V(x, y)$, $V(y, z)$ or $V(x, z)$ where (u, v) is not invertible on the curve.

Let $\pi_1 : X_1 \to X$ be the blow-up of $C = V(x, y)$. $\Phi_1 = \Phi \circ \pi_1$ is strongly prepared over p. If $q \in \pi^{-1}(p)$ is a 1 point, then q has regular parameters (x, y_1, z) defined by

$$
x = x_1, y = x_1(y_1 + \alpha)
$$

with $\alpha \neq 0$. There are permissible parameters (x_1, y, z) at q where x_1 is defined by

$$
x_1 = \overline{x}_1 (y_1 + \alpha)^{-\frac{b}{a+b}}
$$

Thus

$$
\begin{aligned}
u &= \overline{x}_1^{(a+b)k} \\
v &= \overline{x}_1^{c+d} (y_1 + \alpha)^{d - \frac{(c+d)b}{a+b}} z
\end{aligned}
$$

and $A(\Phi_1, q) = 0$.

If $\pi_1 : X_1 \to X$ is the blow-up of $C = V(x, z)$ or $V(y, z)$, then $\Phi_1 = \Phi \circ \pi_1$ is strongly prepared over p, and there are no 1 points in $\pi^{-1}(p)$.

Suppose that $p \in X$ is a 2 point such that $m_q \mathcal{O}_{X,p}$ is not invertible, and (190) holds at p. We may assume that there exists an open neighborhood U of p such that (x, y, z) are uniformizing parameters on an étale cover of U. We will determine the locus of points in U where $m_q \mathcal{O}_U$ is not invertible. First suppose that q' is a 2 point on the curve $x = y = 0$. q' has permissible parameters (x, y, z_1) where $z = z_1 + \alpha$.

$$
\begin{aligned}
u &= (x^a y^b)^k \\
v &= \alpha (x^a y^b)^t + (x^a y^b)^t z_1
\end{aligned}
$$

Thus (u, v) is invertible along the curve $x = y = 0$, if $\alpha \neq 0$.

Suppose that q' is a 1 point near p. q' has permissible parameters (x_1, y_1, z_1) where either

$$
x = x_1 (y_1 + \alpha)^{-\frac{b}{a}}, y = y_1 + \alpha, z = z_1 + \beta \tag{202}
$$

with $\alpha \neq 0$ or

$$
x = x_1 + \alpha, y = y_1 (x_1 + \alpha)^{-\frac{a}{b}}, z = z_1 + \beta \tag{203}
$$

with $\alpha \neq 0$. If q' has permissible parameters satisfying (202), then

$$
\begin{aligned}
u &= x_1^{ak} \\
v &= x_1^{at}(z_1 + \beta)
\end{aligned}
$$

Thus (u, v) is invertible at q' if $\beta \neq 0$, and q' is not invertible along $V(x, z)$ since $t < k$ by assumption.

If q' has permissible parameters satisfying (203), then

$$
\begin{aligned}
u &= y_1^{bk} \\
v &= y_1^{bt}(z_1 + \beta)
\end{aligned}
$$

Thus (u, v) is invertible at q' if $\beta \neq 0$, and q' is not invertible along $V(y, z)$ since $t < k$ by assumption.

If $\pi_1 : X_1 \to X$ is the blow-up of $V(x, z)$ or $V(y, z)$, then $\Phi_1 = \Phi \circ \pi_1$ is strongly prepared over p and there are no 1 points in $\pi^{-1}(p)$.

Suppose that $p \in X$ is a 2 point such that $m_q \mathcal{O}_{X,p}$ is not invertible, and (186) holds at p. After possibly interchanging x and y, we may assume that $ad - bc > 0$. We may assume that there exists an open neighborhood U of p such that (x, y, z) are uniformizing parameters on an étale cover of U and the conclusions of Lemma 18.8 hold for p. We will determine the locus of points in U where $m_q \mathcal{O}_U$ is not invertible. If q' is a 2 point on the curve $x = y = 0$, then q' has the form of (186), so that Thus (u, v) is not invertible along the curve $x = y = 0$.

Suppose that q' is a 1 point near q'. q' has permissible parameters (x_1, y_1, z_1) where either

$$
x = x_1(y_1 + \alpha)^{-\frac{b}{a}}, y = y_1 + \alpha, z = z_1 + \beta \tag{204}
$$

with $\alpha \neq 0$ or

$$
x = x_1 + \alpha, y = y_1(x_1 + \alpha)^{-\frac{a}{b}}, z = z_1 + \beta \tag{205}
$$

with $\alpha \neq 0$. If q' has permissible parameters satisfying (204), then

$$
\begin{aligned}
u &= x_1^{ak} \\
v &= P(x_1^a) + x_1^c(y_1 + \alpha)^{d - \frac{bc}{a}} \\
&= \alpha^{d - \frac{bc}{a}} x_1^c + x_1^c \overline{y}
\end{aligned}
$$

for some permissible parameters $(x_1, \overline{y}, z)$, since a ord $(P) > c$. Thus (u, v) is invertible at q', since we have $\alpha \neq 0$.

$A(\Phi, q') = 0$ at points q' near p where (204) holds.

If q' has permissible parameters satisfying (205), then

$$
\begin{aligned}
u &= y_1^{bk} \\
v &= P(y_1^b) + \alpha^{c - \frac{da}{b}} y_1^d + y_1^d \overline{x}_1
\end{aligned}
$$

(u, v) is invertible at q' Since b ord$(P) < d$ by assumption.

At points q' near p where (205) holds, we have $A(\Phi, q') = d - b$ ord $(P) > 0$.

Let $\pi_1 : X_1 \to X$ be the blow-up of $C = V(x, y)$. Then $\Phi_1 = \Phi \circ \pi_1$ is strongly prepared above p. If $q \in \pi_1^{-1}(p)$ is a 1 point, then q has regular parameters (x, y_1, z) defined by

$$x = x_1, y = x_1(y_1 + \alpha)$$

with $\alpha \neq 0$. There are permissible parameters $(\overline{x}_1, y, z)$ at q where $\overline{x}_1$ is defined by

$$x_1 = \overline{x}_1(y_1 + \alpha)^{-\frac{b}{a+b}}$$

Thus

$$u = \overline{x}_1^{(a+b)k}$$
$$v = P(\overline{x}_1^{a+b}) + \overline{x}_1^{c+d}(y_1 + \alpha)^{d - \frac{(c+d)b}{a+b}}$$
$$= P(\overline{x}_1^{a+b}) + \overline{x}_1^{c+d}\alpha^{d - \frac{(c+d)b}{a+b}} + \overline{x}_1^{c+d}\overline{y}_1$$

If $(a + b)\text{ord}\,(P) \geq c + d$ then $A(\Phi_1, q) = 0$. Assume that $(a + b)\text{ord}(P) < c + d$. Since $\text{ord}(P) > \frac{c}{a}$, we have that

$$c + d - (a + b)\text{ord}(P) = (d - b\,\text{ord}(P)) + (c - a\,\text{ord}(P)) < d - b\,\text{ord}(P)$$

Thus

$$A(\Phi_1, q) \leq c + d - (a + b)\,\text{ord}\,(P) < d - b\,\text{ord}\,(P) \leq A(\Phi)$$

Suppose that $p \in X$ is a 2 point such that $m_q \mathcal{O}_{X,p}$ is not invertible, and (187) holds at p. After possibly interchanging x and y, we may assume that $ad - bc > 0$. We may assume that there exists an open neighborhood U of p such that (x, y, z) are uniformizing parameters on an étale cover of U. We will determine the locus of points in U where $m_q \mathcal{O}_U$ is not invertible. If q' is a 2 point on the curve $x = y = 0$, then q' has the form of (187), so that (u, v) is not invertible along the curve $x = y = 0$.

Suppose that q' is a 1 point near p. q' has permissible parameters (x_1, y_1, z_1) where either

$$x = x_1(y_1 + \alpha)^{-\frac{b}{a}}, y = y_1 + \alpha, z = z_1 + \beta \qquad (206)$$

with $\alpha \neq 0$ or

$$x = x_1 + \alpha, y = y_1(x_1 + \alpha)^{-\frac{a}{b}}, z = z_1 + \beta \qquad (207)$$

with $\alpha \neq 0$. If q' has permissible parameters satisfying (206), then

$$u = x_1^{ak}$$
$$v = x_1^c(y_1 + \alpha)^{d - \frac{bc}{a}}$$
$$= \alpha^{d - \frac{bc}{a}} x_1^c + x_1^c \overline{y}_1$$

(u, v) is thus invertible at q'.

$A(\Phi, q') = 0$ at points q' near p where (206) holds.

If q' has permissible parameters satisfying (207), then

$$u = y_1^{bk}$$
$$v = (x_1 + \alpha)^{c - \frac{da}{b}} y_1^d$$
$$= \alpha^{c - \frac{da}{b}} y_1^d + y_1^d \overline{x}_1$$

Thus (u, v) is invertible at q'.

$A(\Phi, q') = 0$ at points q' near p where (207) holds.

The locus of points where (u, v) is not invertible near p is $V(x, y)$.

Let $\pi_1 : X_1 \to X$ be the blow-up of $C = V(x, y)$. $\Phi_1 = \Phi \circ \pi_1$ is strongly prepared above p. If $q \in \pi_1^{-1}(p)$ is a 1 point, then q has regular parameters (x_1, y_1, z) defined by

$$x = x_1, y = x_1(y_1 + \alpha)$$

with $\alpha \neq 0$. There are permissible parameters $(\overline{x}_1, y, z)$ at q where $\overline{x}_1$ is defined by

$$x_1 = \overline{x}_1(y_1 + \alpha)^{-\frac{b}{a+b}}.$$

Thus

$$\begin{aligned} u &= \overline{x}_1^{(a+b)k} \\ v &= \overline{x}_1^{c+d}(y_1 + \alpha)^{d - \frac{(c+d)b}{a+b}} \\ &= \overline{x}_1^{c+d}\alpha^{d - \frac{(c+d)b}{a+b}} + \overline{x}_1^{c+d}\overline{y}_1 \end{aligned}$$

and $A(\Phi_1, q) = 0$.

Suppose that $p \in X$ is a 1 point such that $m_q\mathcal{O}_{X,p}$ is not invertible, so that (185) holds at p. We may assume that here exists an open neighborhood U of p such that (x, y, z) are uniformizing parameters on an étale cover of U. We will determine the locus of points in U where $m_q\mathcal{O}_U$ is not invertible.

Suppose that q' is a 1 point near p. q' has permissible parameters (x, y_1, z_1) where

$$\begin{aligned} y = y_1 + \alpha, z = z_1 + \beta \\ u &= x^k \\ v &= \alpha x^c + x^c y_1 \end{aligned}$$

(u, v) is thus only not invertible on the curve $V(x, y)$.

Let $\pi : X_1 \to X$ be the blow-up of $V(x, y)$. $\Phi_1 = \Phi \circ \pi_1$ is strongly prepared above p. If $q \in \pi^{-1}(p)$ is a 1 point, then q has permissible parameters (x, y_1, z) defined by

$$x = x_1, y = x_1(y_1 + \alpha)$$

with $\alpha \neq 0$.

$$\begin{aligned} u &= x_1^k \\ v &= x_1^{c+1}(y_1 + \alpha) \end{aligned}$$

and $A(\Phi_1, q) = 0$.

Suppose that $p \in X$ is a 3 point such that $m_p\mathcal{O}_{X,p}$ is not invertible and (194) holds at p. We may assume that there exists an open neighborhood U of p such that (x, y, z) are uniformizing parameters on an étale cover of U. The locus of points in U where $m_q\mathcal{O}_U$ is not invertible is the union of the 2 curves $V(x, z)$ and $V(y, z)$.

Let $\pi : X_1 \to X$ be the blow-up of $C = V(x, z)$, $\Phi_1 = \Phi \circ \pi$. If $q' \in \pi^{-1}(p)$ is a 2 point, then q' has permissible parameters (x_1, y_1, z) where

$$x = x_1, z = x_1(z_1 + \alpha)$$

with $\alpha \neq 0$.

$$\begin{aligned} u &= x_1^a y^b = \overline{x}_1^a \overline{y}_1^b \\ v &= x_1^c(z_1 + \alpha)^c = \overline{x}_1^c \end{aligned}$$

so that Φ_1 is strongly prepared at q'.

Suppose that $q' \in \pi^{-1}(p)$ is a 3 point and q' has permissible parameters (x_1, y_1, z) where $x = x_1, z = x_1 z_1$. Then

$$
\begin{aligned}
v &= x_1^a y^b \\
u &= x_1^c z_1^c
\end{aligned}
$$

so that Φ_1 is strongly prepared at q'. Suppose that $q' \in \pi^{-1}(p)$ is a 3 point and q' has permissible parameters (x_1, y_1, z) where

$$
x = x_1 z_1, z = z_1.
$$

Then

$$
\begin{aligned}
u &= x_1^a y^b z_1^a \\
v &= z_1^c
\end{aligned}
$$

so that Φ_1 is strongly prepared at q'. A similar analysis shows that the blow-up of $V(y, z)$ composed with Φ is strongly prepared.

Suppose that $p \in X$ is a 3 point such that $m_q \mathcal{O}_{X,p}$ is not invertible and (195) holds at p.

We may assume that there exists an open neighborhood U of p such that (x, y, z) are uniformizing on an étale cover of U. The locus of points in U where $m_q \mathcal{O}_U$ is not invertible is the union of the 2 curves $V(x, z)$, $V(x, y,)$ (if $c > b$) and $V(y, z)$ (if $b > c$). If $\pi : X_1 \to X$ is the blow-up of a 2 curve through C, and $\Phi_1 = \Phi \circ \pi$, then Φ_1 is strongly prepared at points $q \in \pi^{-1}(p)$.

Suppose that $p \in X$ is a 2 point such that $m_p \mathcal{O}_{X,p}$ is not invertible, and (191) holds at p. We may assume that there exists an open neighborhood U of p such that (x, y, z) are uniformizing parameters on an étale cover of U. The locus of points in U where $m_q \mathcal{O}_U$ is not invertible is the 2 curve $V(x, y)$.

Let $\pi : X_1 \to X$ be the blow-up of $C = V(x, y)$, $\Phi_1 = \Phi \circ \pi$. If $q' \in \pi^{-1}(p)$ is a 1 point, then q' has permissible parameters

$$
x = x_1, y = x_1(y_1 + \alpha)
$$

with $\alpha \neq 0$.

$$
\begin{aligned}
u &= x_1^a \\
v &= x_1^b(y_1 + \alpha)^b.
\end{aligned}
$$

Set $\bar{y}_1 = (y_1 + \alpha)^b - \alpha^b$ to get

$$
\begin{aligned}
u &= x_1^a \\
v &= \alpha^b x_1^b + x_1^b \bar{y}_1
\end{aligned}
$$

so that $A(\Phi_1, q') = 0$. Φ_1 is strongly prepared at points of $\pi^{-1}(p)$. $\square$

If α, β are real numbers, define

$$
S(\alpha, \beta) = \max \{(\alpha, \beta), (\beta, \alpha)\}
$$

where the maximum is in the Lexicographic ordering.

Suppose that $\Phi : X \to S$ is strongly prepared. Suppose that $q \in D_S$ and $C \subset X$ is a 2 curve such that $m_q \mathcal{O}_X$ is not invertible along C. At a generic point p of C (186), (187) or (191) holds.

If (186) holds, then $\Phi(C)$ is a 1 point $q \in S$. Suppose that $P(t) = \sum a_i t^i$. Since q is a 1 point, we can, after possibly replacing v with $v - \sum a_{ik} u^k$, assume that $k \nmid \text{ord}(P)$ in (186). With this restriction, define

$$\sigma(C) = \begin{cases} S(|\, c - a \text{ ord } (P)\, |, |\, d - b \text{ ord } (P)\, |) \\ \quad \text{if } c - \text{ ord } (P), d - \text{ ord } (P) \text{ have opposite signs,} \\ -\infty \text{ if they have the same sign.} \end{cases}$$

If (187) or (191) holds, define

$$\sigma(C) = -\infty.$$

$\sigma(C)$ is well defined (independent of choice of permissible parameters (u, v) at q with the restriction that $k \nmid \text{ord}(P)$ in (186)). This follows from Lemma 18.8.

If $m_q \mathcal{O}_X$ is invertible, define

$$\overline{\sigma}_q(\Phi) = -\infty.$$

If $m_q \mathcal{O}_X$ is not invertible, define

$$\overline{\sigma}_q(\Phi) = \max \left\{ \begin{array}{ll} \sigma(C) & |\, C \subset X \text{ is a 2 curve} \\ & \text{such that } m_q \mathcal{O}_X \text{ is not invertible along } C \end{array} \right\}.$$

Lemma 18.16. *Suppose that X is strongly prepared, $q \in S$ is such that $m_q \mathcal{O}_X$ is not invertible. Then there exists a sequence of blow-ups of 2 curves $X_1 \to X$ such that the induced map $\Phi_1 : X_1 \to S$ is strongly prepared, $A(\Phi_1, E) < A(\Phi_1) = A(\Phi)$ if E is an exceptional component of E_{X_1} for $X_1 \to X$, and the forms (186), (188) and (192) do not hold at any point $p \in X$ where $m_q \mathcal{O}_{X_1,p}$ is not invertible.*

Proof. Set $\overline{\sigma}(\Phi) = \overline{\sigma}_q(\Phi)$. $\overline{\sigma}(\Phi) \geq 0$ if and only if there exists a point $p \in X$ such that $m_q \mathcal{O}_{X_1,p}$ is not invertible, and a form (186), (188) or (192) holds at p.

Suppose that $\overline{\sigma}(\Phi) \geq 0$. Let C be a 2 curve such that $\sigma(C) = \overline{\sigma}(\Phi)$. Let $\pi : X_1 \to X$ be the blow-up of C. By Theorem 18.15, we need only verify that if $C_1 \subset \pi^{-1}(C)$ is a 2 curve such that $m_q \mathcal{O}_{X_1}$ is not invertible along C_1 then $\sigma(C_1) < \overline{\sigma}(\Phi)$.

First suppose that C_1 is a section over C. Let $p_1 \in C_1$ be a generic point. Then $p = \pi(p_1)$ is a generic point of C. There exist permissible parameters (x, y, z) at p such that

$$\begin{aligned} u &= (x^a y^b)^k \\ v &= P(x^a y^b) + x^c y^d \end{aligned}$$

with

$$\min\{\frac{c}{a}, \frac{d}{b}\} < \text{ord } (P) < \max \{\frac{c}{a}, \frac{d}{b}\}$$

and

$$\min\{\frac{c}{a}, \frac{d}{b}\} < k, k \nmid \text{ord}(P).$$

We may assume, after possibly interchanging x and y that

$$\overline{\sigma}(\Phi) = \sigma(C) = (|\, c - a \text{ ord } (P)\, |, |\, d - b \text{ ord } (P)\, |).$$

Assume that p_1 has permissible parameters (x_1, y_1, z) such that

$$x = x_1, y = x_1 y_1$$

and $x_1 = y_1 = 0$ are local equations of C_1 at p_1.

$$u = (x_1^{a+b} y_1^b)^k$$
$$v = P(x_1^{a+b} y_1^b) + x_1^{c+d} y_1^d$$

$k \nmid \text{ord}(P)$ implies by Lemma 18.8 that

$$\sigma(C_1) \le S(|\, (c+d) - (a+b) \text{ ord } (P)\, |, |\, d - b \text{ ord } (P)\, |).$$

If $c - a$ ord $(P) > 0$ and $d - b$ ord $(P) < 0$ then

$$0 \le (c - a \text{ ord } (P) + (d - b \text{ ord } (P)) < c - \text{ ord } (P)$$

so that $\sigma(C_1) < \sigma(C)$.

If $c - a$ ord $(P) < 0$ and $d - b$ ord $(P) > 0$ then

$$0 \ge (c - a \text{ ord } (P) + (d - b \text{ ord } (P)) > c - \text{ ord } (P)$$

so that $\sigma(C_1) < \sigma(C)$.

Assume that p_1 has permissible parameters (x_1, y_1, z) such that

$$x = x_1 y_1, y = y_1$$

and $x_1 = y_1 = 0$ are local equations of C_1 at p_1.

$$u = (x_1^a y_1^{a+b})^k$$
$$v = P(x_1^a y_1^{a+b}) + x_1^c y_1^{c+d}$$

$k \nmid \text{ord}(P)$ implies

$$\sigma(C_1) \le S(|\, (c+d) - (a+b) \text{ ord } (P)\, |, |\, c - a \text{ ord } (P)\, |).$$

If $c - a$ ord $(P) > 0$ and $d - b$ ord $(P) < 0$ then

$$0 \le (c - a \text{ ord } (P) + (d - b \text{ ord } (P)) < c - \text{ ord } (P)$$

so that $\sigma(C_1) = -\infty$.

If $c - a$ ord $(P) < 0$ and $d - b$ ord $(P) > 0$ then

$$0 \ge (c - a \text{ ord } (P) + (d - b \text{ ord } (P)) > c - \text{ ord } (P)$$

so that $\sigma(C_1) = -\infty$.

Now suppose that $C_1 \subset \pi^{-1}(C)$ is an exceptional 2 curve. Then $p = \pi(C_1)$ satisfies (192). Let $q = \Phi(p)$. q is a 1 point. If E_1, E_2, E_3 are the components of E_X containing p, then $\Phi(E_1) = \Phi(E_2) = \Phi(E_3) = q$. Suppose that $P(t) = \sum a_i t^i$. Since q is a 1 point, we may replace v with $v - \sum a_{ik} u^i$ so that $k \nmid \text{ord}(P)$. We may also assume, after possibly interchanging (x, y, z) that

$$|\, f - c \text{ ord } (P)\, | \ge |\, e - b \text{ ord } (P)\, | \ge |\, d - a \text{ ord } (P)\, |.$$

If C has local equations $x = z = 0$, then

$$\sigma(C) = (|\, f - c \text{ ord}(P)\, |, |\, d - a \text{ ord}(P)\, |)$$

and a generic point of C_1 has permissible parameters (x_1, y, z_1) where

$$x = x_1, z = x_1(z_1 + \alpha)$$

with $\alpha \neq 0$, and $x_1 = y = 0$ are local equations of C_1. Set $\bar{x}_1 = x_1(z_1 + \alpha)^{-\frac{c}{a+c}}$.

$$u = (\bar{x}_1^{\bar{a}} y^{\bar{b}})^{\lambda k}$$
$$v = P((\bar{x}_1^{\bar{a}} y^{\bar{b}})^\lambda) + \bar{x}_1^{d+f} y^e (z_1 + \alpha)^{f - \frac{(d+f)c}{a+c}}$$

where $\lambda = (a + c, b)$, $a + c = \bar{a}\lambda$, $b = \bar{b}\lambda$.

If $\sigma(C_1) \geq 0$, then $\lambda k \nmid \lambda \mathrm{ord}(P)$ implies

$$\sigma(C_1) = S(|\,(d + f) - (a + c) \; \mathrm{ord} \; (P)\,|, |\, e - b \; \mathrm{ord} \; (P)\,|).$$

Similarily, if C has local equations $y = z = 0$,

$$\sigma(C) = (|\, f - c \; \mathrm{ord}(P)\,|, |\, e - b \; \mathrm{ord}(P)\,|)$$

and if $\sigma(C_1) \geq 0$, $k \nmid \mathrm{ord}(P)$ implies

$$\sigma(C_1) = S(|\, d - a \; \mathrm{ord} \; (P)\,|, |\,(e + f) - (b + c) \; \mathrm{ord} \; (P)\,|).$$

If C has local equations $x = y = 0$, then

$$\sigma(C) = (|\, e - b \; \mathrm{ord}(P)\,|, |\, d - a \; \mathrm{ord}(P)\,|)$$

and if $\sigma(C_1) \geq 0$, then

$$\sigma(C_1) = S(|\, f - c \; \mathrm{ord}(P)\,|, |\,(d + e) - (a + b)\mathrm{ord} \; (P)\,|).$$

If one of $f - c \; \mathrm{ord}(P), e - b \; \mathrm{ord}(P), d - a \; \mathrm{ord}(P)$ is zero, then $\sigma(C_1) = -\infty$.

Case 1 Suppose that $f - c \; \mathrm{ord} \; (P) > 0$, $e - b \; \mathrm{ord} \; (P) > 0$, $d - a \; \mathrm{ord} \; (P) < 0$. Then

$$\bar{\sigma}(\Phi) = \sigma(C) = (|\, f - c \; \mathrm{ord} \; (P)\,|, |\, d - a \; \mathrm{ord} \; (P)\,|)$$

and $x = z = 0$ are local equations of C.

$$0 \leq (d - a \; \mathrm{ord} \; (P)) + (f - c \; \mathrm{ord} \; (P)) < f - c \; \mathrm{ord} \; (P)$$

implies $\sigma(C_1) = -\infty$.

Case 2 Suppose that $f - c \; \mathrm{ord} \; (P) > 0$, $e - b \; \mathrm{ord} \; (P) < 0$, $d - a \; \mathrm{ord} \; (P) > 0$. Then

$$\bar{\sigma}(\Phi) = \sigma(C) = (|\, f - c \; \mathrm{ord} \; (P)\,|, |\, e - b \; \mathrm{ord} \; (P)\,|)$$

and $y = z = 0$ are local equations of C.

$$0 \leq (e - b \; \mathrm{ord} \; (P)) + (f - c \; \mathrm{ord} \; (P)) < f - c \; \mathrm{ord} \; (P)$$

implies $\sigma(C_1) = -\infty$.

Case 3 Suppose that $f - c \; \mathrm{ord} \; (P) > 0$, $e - b \; \mathrm{ord} \; (P) < 0$, $d - a \; \mathrm{ord} \; (P) < 0$. Then

$$\bar{\sigma}(\Phi) = \sigma(C) = (|\, f - c \; \mathrm{ord} \; (P)\,|, |\, e - b \; \mathrm{ord} \; (P)\,|)$$

and $y = z = 0$ are local equations of C.

$$0 \leq (e - b \; \mathrm{ord} \; (P)) + (f - c \; \mathrm{ord} \; (P)) < f - c \; \mathrm{ord} \; (P)$$

implies $\sigma(C_1) < \sigma(C)$.

Case 4 Suppose that $f - c \; \mathrm{ord} \; (P) < 0$, $e - b \; \mathrm{ord} \; (P) > 0$, $d - a \; \mathrm{ord} \; (P) > 0$. Then

$$\bar{\sigma}(\Phi) = \sigma(C) = (|\, f - c \; \mathrm{ord} \; (P)\,|, |\, e - b \; \mathrm{ord} \; (P)\,|)$$

and $y = z = 0$ are local equations of C.

$$0 \geq (e - b \; \mathrm{ord} \; (P)) + (f - c \; \mathrm{ord} \; (P)) > f - c \; \mathrm{ord} \; (P)$$

implies $\sigma(C_1) < \sigma(C)$.

Case 5 Suppose that $f - c$ ord $(P) < 0$, $e - b$ ord $(P) > 0$, $d - a$ ord $(P) < 0$. Then

$$\overline{\sigma}(\Phi) = \sigma(C) = (|\, f - c \text{ ord } (P) \,|, |\, e - b \text{ ord } (P) \,|)$$

and $y = z = 0$ are local equations of C.

$$0 \geq (e - b \text{ ord } (P)) + (f - c \text{ ord } (P)) > f - c \text{ ord } (P)$$

implies $\sigma(C_1) = -\infty$.

Case 6 Suppose that $f - c$ ord $(P) < 0$, $e - b$ ord $(P) < 0$, $d - a$ ord $(P) > 0$. Then

$$\overline{\sigma}(\Phi) = \sigma(C) = (|\, f - c \text{ ord } (P) \,|, |\, d - a \text{ ord } (P) \,|)$$

and $x = z = 0$ are local equations of C.

$$0 \geq (d - a \text{ ord } (P)) + (f - c \text{ ord } (P)) > f - c \text{ ord } (P)$$

implies $\sigma(C_1) = -\infty$.

We conclude that if $C_1 \subset \pi^{-1}(C)$ is a 2 curve such that $m_q \mathcal{O}_{X_1}$ is not invertible along C_1, then $\sigma(C_1) < \overline{\sigma}(\Phi)$.

By Theorem 18.15, induction on the number of 2 curves $C \subset X$ such that $\sigma(C) = \overline{\sigma}(\Phi)$, and induction on $\overline{\sigma}(\Phi)$, we achieve the conclusions of the Lemma.

$\square$

Lemma 18.17. *Suppose that $\Phi : X \to S$ is strongly prepared, $q \in S$ is such that $m_q \mathcal{O}_X$ is not invertible and the forms (186), (188) and (192) do not hold at any point $p \in X$ where $m_q \mathcal{O}_{X,p}$ is not invertible.*

Then there exists a sequence of blow-ups of nonsingular curves $X_1 \to X$ which are not 2 curves such that the induced map $\Phi_1 : X_1 \to S$ is strongly prepared, $A(\Phi_1, E) < A(\Phi_1) = A(\Phi)$ if E is an exceptional component of E_{X_1} for $X_1 \to X$, the forms (186), (188) and (192) do not hold at any point $p \in X_1$ where $m_q \mathcal{O}_{X_1,p}$ is not invertible, and if $C \subset X_1$ is a curve such that $m_q \mathcal{O}_{X_1}$ is not invertible along C, then C is a 2 curve.

Proof. Suppose that C is a curve such that $m_q \mathcal{O}_X$ is not invertible along C and C is not a 2 curve. Suppose that $p \in C$. Then one of the following holds:

1. (185) holds at p, $x = y = 0$ are local equations of C at p.
2. (189) holds at p and $x = z = 0$ with $\frac{c}{a} < k$ or $y = z = 0$ with $\frac{d}{b} < k$ are local equations of C at p.
3. (190) holds at p, $x = z = 0$ or $y = z = 0$ are local equations of C at p.

At a generic point $p \in C$ (185) holds. Define

$$\Omega(C) = k - c > 0.$$

Let

$$\overline{\Omega}(\Phi) = \max \left\{ \begin{array}{l|l} \Omega(C) & C \text{ is not a 2 curve} \\ & \text{and } m_q \mathcal{O}_X \text{ is not invertible along } C. \end{array} \right\}$$

Suppose that $\Omega(C) = \overline{\Omega}(\Phi)$. Let $\pi : X_1 \to X$ be the blow-up of C. The forms (186), (188) and (192) cannot hold at points of X_1. By Theorem 18.15, we need only verify that $\Omega(C_1) < \overline{\Omega}(\Phi)$ if C_1 is a curve in $\pi^{-1}(C)$ such that $m_q\mathcal{O}_{X_1}$ is not invertible along C_1 and C_1 is not a 2 curve. We then have $\pi(C_1) = C$.

Let p_1 be a generic point of C_1, $p = \pi(p_1)$. (185) holds at p since p is a generic point of C. $p_1 \in \pi^{-1}(p)$ is a 1 point. Then p_1 has permissible parameters (x_1, y_1, z_1) such that

$$x = x_1, y = x_1(y_1 + \alpha).$$

$$\begin{aligned} u &= x_1^k \\ v &= x_1^{c+1}(y_1 + \alpha). \end{aligned}$$

$m_q\mathcal{O}_{X_1,p_1}$ is invertible if $\alpha \neq 0$. If $\alpha = 0$ and $m_q\mathcal{O}_{X_1,p_1}$ is not invertible, then $x_1 = y_1 = 0$ are local equations of the curve $C_1 \subset X_1$ through p on which $m_q\mathcal{O}_{X_1}$ is not invertible.

$$0 < \Omega(C_1) = k - (c + 1) < \Omega(C) = \overline{\Omega}(\Phi).$$

By induction on the number of curves C on X such that $\Omega(C) = \overline{\Omega}(\Phi)$, we achieve the conclusions of the Lemma. $\qquad\square$

Lemma 18.18. *Suppose that $\Phi : X \to S$ is strongly prepared, $q \in S$ is such that $m_q\mathcal{O}_X$ is not invertible, the forms (186), (188) and (192) do not hold at any point $p \in X$ where $m_q\mathcal{O}_{X,p}$ is not invertible, and if $C \subset X$ is a curve such that $m_q\mathcal{O}_X$ is not invertible along C, then C is a 2 curve.*

Then there exists a sequence of blow-ups of 2 curves $X_1 \to X$ such that the induced map $\Phi_1 : X_1 \to S$ is strongly prepared, $A(\Phi_1, E) < A(\Phi_1) = A(\Phi)$ if E is an exceptional component of E_{X_1} for $X_1 \to X$ and $m_q\mathcal{O}_{X_1}$ is invertible.

Proof. Suppose that $C \subset X$ is a 2 curve such that $m_q\mathcal{O}_X$ is not invertible along C. Suppose that $p \in C$. Then (187), (193), (191), (194) or (195) hold at p. At a generic point $p \in C$ (187) or (191) holds.

If C is a 2 curve such that at a generic point of C (187) holds, define

$$\omega(C) = \begin{cases} S(|ka - c|, |kb - d|) & \text{if } ka - c, kb - d \text{ have opposite signs} \\ & \text{and } m_q\mathcal{O}_X \text{ is not invertible along } C \\ -\infty & \text{otherwise} \end{cases}$$

If C is a 2 curve such that at a generic point of C (191) holds, define

$$\omega(C) = \begin{cases} S(a, b) & \text{if } m_q\mathcal{O}_X \text{ is not invertible along } C \\ -\infty & \text{otherwise} \end{cases}$$

$m_q\mathcal{O}_X$ is not invertible along C if and only if $\omega(C) > 0$. Set

$$\overline{\omega}(\Phi) = \max \left\{ \omega(C) \;\middle|\; \begin{array}{l} C \text{ is a 2 curve such that} \\ m_q\mathcal{O}_X \text{ is not invertible along } C \end{array} \right\}$$

Suppose that $\omega(C) = \overline{\omega}(\Phi)$. Let $\pi : X_1 \to X$ be the blow-up of C. By Theorem 18.15, we need only verify that $\omega(C_1) < \overline{\omega}(\Phi)$ if $C_1 \subset \pi^{-1}(C)$ is a curve such that $m_q\mathcal{O}_{X_1}$ is not invertible along C_1. We must have that C_1 is a 2 curve.

Suppose that C_1 is a section over C. Let $p_1 \in C_1$ be a generic point. Then $p = \pi(p_1)$ is a generic point on C.

Suppose that there exist permissible parameters (x, y, z) at p such that (187) holds.

$$u = (x^a y^b)^k$$
$$v = x^c y^d$$

with

$$\min\{\frac{c}{a}, \frac{d}{b}\} < k < \max\{\frac{c}{a}, \frac{d}{b}\}.$$

After possibly interchanging x and y, we may assume that

$$w(C) = (|\, c - ak\,|, |\, d - bk\,|).$$

Assume that p_1 has permissible parameters (x_1, y_1, z) such that

$$x = x_1, y = x_1 y_1$$

and $x_1 = y_1 = 0$ are local equations of C_1 at p_1.

$$u = (x_1^{a+b} y_1^b)^k$$
$$v = x_1^{c+d} y_1^d$$

$$w(C_1) = \begin{cases} S(|\, (c+d) - (a+b)k\,|, |\, bk - d\,|) & \text{if } (c+d) - (a+b)k, d - bk \\ & \text{have opposite signs} \\ & \text{and } m_q \mathcal{O}_{X_1} \text{is not invertible} \\ & \text{along } C \\ -\infty & \text{otherwise} \end{cases}$$

Suppose that $c - ak > 0$ and $d - bk < 0$.

$$0 \le (c - ak) + (d - bk) < c - ak$$

implies $w(C_1) < w(C)$.

Suppose that $c - ak < 0$ and $d - bk > 0$.

$$0 \ge (c - ak) + (d - bk) > c - ak$$

implies $w(C_1) < w(C)$.

Assume that p_1 has permissible parameters (x_1, y_1, z) such that

$$x = x_1 y_1, y = y_1$$

and $x_1 = y_1 = 0$ are local equations of C_1 at p_1.

$$u = (x_1^a y_1^{a+b})^k$$
$$v = x_1^c y_1^{c+d}$$

$$w(C_1) = \begin{cases} S(|\, (c+d) - (a+b)k\,|, |\, ak - c\,|) & \text{if } (c+d) - (a+b)k, c - ak \\ & \text{have opposite signs} \\ & \text{and } m_q \mathcal{O}_{X_1} \text{is not invertible} \\ & \text{along } C \\ -\infty & \text{otherwise} \end{cases}$$

Suppose that $c - ak > 0$ and $d - bk < 0$.

$$0 \le (c - ak) + (d - bk) < c - ak$$

implies $w(C_1) = -\infty$.

Suppose that $c - ak < 0$ and $d - bk > 0$.

$$0 \geq (c - ak) + (d - bk) > c - ak$$

implies $\omega(C_1) = -\infty$.

Suppose that C_1 is a section over C, $p_1 \in C_1$ is a generic point and $p \in \pi(p_1)$ is a generic point on C such that p satisfies (191). Then a similar argument shows that $\omega(C_1) < \omega(C)$.

Suppose that $C_1 \subset \pi^{-1}(C)$ is an exceptional 2 curve. Suppose that $p = \pi(C_1)$ satisfies (193). Without loss of generality,

$$\mid f - ck \mid \geq \mid e - bk \mid \geq \mid d - ak \mid .$$

If C has local equations $x = z = 0$ then a generic point of C_1 has regular parameters (x_1, y, z_1) such that

$$x = x_1, z = x_1(z_1 + \alpha)$$

(with $\alpha \neq 0$) and $x_1 = y = 0$ are local equations of C_1. Set $\bar{x}_1 = x_1(z_1 + \alpha)^{-\frac{c}{a+c}}$.

$$
\begin{aligned}
u &= (\bar{x}_1^{\bar{a}} y^{\bar{b}})^{\lambda k} \\
v &= \bar{x}_1^{d+f} y^e (z_1 + \alpha)^{f - \frac{(d+f)c}{a+c}}
\end{aligned}
$$

where $\lambda = (a + c, b)$, $a + c = \bar{a}\lambda$, $b = \bar{b}\lambda$. $\omega(C_1) \geq 0$ implies

$$\omega(C_1) = S(\mid (d + f) - (a + c)k \mid, \mid e - bk \mid).$$

Similarily, if C_1 has local equations $y = z = 0$ then $\omega(C_1) \geq 0$ implies

$$\omega(C_1) = S(\mid (e + f) - (b + c)k \mid, \mid d - ak \mid).$$

If C_1 has local equations $x = y = 0$ then $\omega(C_1) \geq 0$ implies

$$\omega(C_1) = S(\mid f - ck \mid, \mid (d + e) - (a + b)k \mid).$$

If one of $d - ak$, $e - bk$, $f - ck$ is zero, then $\omega(C_1) = -\infty$.

The analysis of Cases 1 - 6 of Lemma 18.16 (with ord (P) changed to k and σ to ω) shows that $\omega(C_1) < \omega(C)$.

A similar argument shows that $\omega(C_1) < \omega(C)$ if $p = \pi(C)$ satisfies (194) or (195).

We achieve the conclusions of the Lemma by Theorem 18.15, induction on the number of 2 curves $C \subset X$ such that $\omega(C) = \bar{\omega}(\Phi)$, and by induction on $\bar{\omega}(\Phi)$.   □

Theorem 18.19. *Suppose that $\Phi : X \to S$ is strongly prepared with respect to D_S. Then there exists a finite sequence of quadratic transforms $\pi_1 : S_1 \to S$ and monoidal transforms centered at nonsingular curves $\pi_2 : X_1 \to X$ such that the induced morphism $\Phi_1 : X_1 \to S_1$ is strongly prepared with respect to $D_{S_1} = \pi_1^{-1}(D_S)_{red}$, and all points of X_1 are good for Φ_1.*

Proof. By Remark 18.9, $A(\Phi) = 0$ if and only if all points of X are good. Suppose that $A(\Phi) > 0$ and E is a component of E_X such that $C(\Phi, E) = C(\Phi)$. $A(\Phi, E) > 0$ implies $\Phi(E)$ is a point q.

Let $\pi_1 : S_1 \to S$ be the blow-up of q. By Lemmas 18.16, 18.17, 18.18 there exists a sequence of blow-ups of curves $X_1 \to X$ such that $\Phi_1 : X_1 \to S$ is strongly prepared, $C(\Phi_1) = C(\Phi)$, $A(\Phi_1, \overline{E}) < A(\Phi_1)$ if $\overline{E}$ is exceptional for Φ_1 and $\Phi_2 : X_1 \to S_1$ is a strongly prepared morphism.

By Theorem 18.14, $C(\Phi_2, \tilde{E}) < C(\Phi)$, where $\tilde{E}$ is the strict transform of E on X_1.

By induction on the number of components E of E_X such that $C(\Phi, E) = C(\Phi)$, and induction on $C(\Phi)$, we get the conclusions of the Theorem. $\qquad\square$

Definition 18.20. *Suppose that $\Phi : X \to Y$ is a dominant morphism of k-varieties, (where k is a field of characteristic zero). Φ is a monomial morphism if for all $p \in X$ there exists an étale neighborhood U of p, uniformizing parameters $(x_1, \dots, x_n)$ on U, regular parameters $(y_1, \dots, y_m)$ in $\mathcal{O}_{Y,\Phi(p)}$, and a matrix (a_{ij}) of nonnegative integers such that*

$$y_1 = x_1^{a_{11}} \cdots x_n^{a_{1n}}$$
$$\vdots$$
$$y_m = x_1^{a_{m1}} \cdots x_n^{a_{mn}}$$

Theorem 18.21. *Suppose that $\Phi : X \to S$ is a dominant morphism from a 3 fold X to a surface S (over an algebraically closed field k of characteristic zero). Then there exist sequences of blow-ups of nonsingular subvarieties $X_1 \to X$ and $S_1 \to S$ such that the induced map $\Phi_1 : X_1 \to S_1$ is a monomial morphism.*

Proof. This follows from Theorem 17.3, the fact that prepared implies strongly prepared, Theorem 18.19 and Remark 18.6. $\qquad\square$

Throughout this section we will assume that $\Phi : X \to S$ is strongly prepared with respect to D_S, and all points of X are good.

Definition 19.1. *([24] and [6]) A normal variety X with a SNC divisor E_X on X is called toroidal if for every point $p \in X$ there exists an affine toric variety X_σ, a point $p' \in X_\sigma$ and an isomorphism of k algebras*

$$\hat{\mathcal{O}}_{X,p} \cong \hat{\mathcal{O}}_{X_\sigma,p'}$$

such that the ideal of E_X corresponds to the ideal of $X_\sigma - T$ (where T is the torus in X_σ). Such a pair (X_σ, p') is called a local model at $p \in X$.

A dominant morphism $\Phi : X \to Y$ of toroidal varieties with SNC divisors D_Y, E_X on X, Y and $\Phi^{-1}(D_Y) \subset E_X$ is called toroidal at p, and we will say that p is a toroidal point of Φ, if with $q = \Phi(p)$, there exist local models (X_σ, p') at p, (Y_τ, q') at q and a toric morphism $\Psi : X_\sigma \to Y_\tau$ such that the following diagram commutes

$$\begin{array}{ccc}
\hat{\mathcal{O}}_{X,p} & \xleftarrow{\sim} & \hat{\mathcal{O}}_{X_\sigma,p'} \\
\hat{\Phi}^* \uparrow & & \hat{\Psi}^* \uparrow \\
\mathcal{O}_{Y,q} & \xleftarrow{\sim} & \hat{\mathcal{O}}_{Y_\tau,q'}
\end{array}$$

$\Phi : X \to Y$ *is called toroidal (with respect to D_Y and E_X) if Φ is toroidal at all $p \in X$.*

Remark 19.2.   1. *If one of the forms (177), (178) or (182) holds at $p \in X$ then $q = \Phi(p)$ is a 2 point.*

2. *If $q = \Phi(p)$ is a 2 point, then (181) cannot hold at p, and if (180) or (183) hold at p, we must have $\alpha \neq 0$, since $uv = 0$ is a local equation of E_X.*

Lemma 19.3. *Suppose that $\Phi : X \to S$ is a morphism from a nonsingular 3 fold X to a nonsingular surface S, D_S is a SNC divisor on S such that $E_X = \Phi^{-1}(D_S)$ is a SNC divisor on X. Then Φ is a toroidal morphism if and only if for all $p \in E_X$ there exist regular parameters (x, y, z) in $\hat{\mathcal{O}}_{X,p}$ (u, v) in $\mathcal{O}_{S,p}$ such that one of the following forms hold:*

1. $u = 0$ *is a local equation for D_S.*
   (a) $xy = 0$ *is a local equation for E_X and*

$$\begin{aligned}
u &= x^a y^b \\
v &= z
\end{aligned}$$

   (b) $x = 0$ *is a local equation for E_X and*

$$\begin{aligned}
u &= x^a \\
v &= y
\end{aligned}$$

2. $uv = 0$ *is a local equation for D_S.*
   (a) $xyz = 0$ *is a local equation for E_X and*

$$\begin{aligned}
u &= x^a y^b z^c \\
v &= x^d y^e z^f
\end{aligned}$$

with

$$rank\begin{pmatrix} a & b & c \\ d & e & f \end{pmatrix} = 2.$$

(b) $xy = 0$ is a local equation for E_X and

$$\begin{aligned} u &= x^a y^b \\ v &= x^c y^d \end{aligned}$$

with $ad - bc \neq 0$.

(c) $xy = 0$ is a local equation of E_X and

$$\begin{aligned} u &= (x^a y^b)^k \\ v &= \alpha(x^a y^b)^t + (x^a y^b)^t z \end{aligned}$$

with $a, b > 0$, $k, t > 0$, $0 \neq \alpha \in k$.

(d) $x = 0$ is a local equation for E_X and

$$\begin{aligned} u &= x^a \\ v &= x^c(y + \alpha) \end{aligned}$$

with $0 \neq \alpha \in k$.

Proof. We will first determine the toroidal forms obtainable from a monomial mapping $\Lambda : \mathbf{A}^3 \to \mathbf{A}^2$ defined by

$$\begin{aligned} u &= x^a y^b z^c \\ v &= x^d y^e z^f \end{aligned}$$

with

$$rank\begin{pmatrix} a & b & c \\ d & e & f \end{pmatrix} = 2.$$

First suppose that that no column of

$$\begin{pmatrix} a & b & c \\ d & e & f \end{pmatrix}$$

is zero. Then $\Lambda^{-1}(D) = \dot{E}$, where $xyz = 0$ is an equation of E, $uv = 0$ is an equation of D.

Suppose that $p \in \mathbf{A}^3$ is a 2 point on $y = z = 0$. Then there exists $0 \neq \beta \in k$ and regular parameters $(\overline{x}, y, z)$ at p such that

$$\begin{aligned} u &= (\overline{x} + \beta)^a y^b z^c \\ v &= (\overline{x} + \beta)^d y^e z^f. \end{aligned}$$

If

$$Det\begin{pmatrix} b & c \\ e & f \end{pmatrix} \neq 0,$$

we can make a permissible change of parameters to get 2.(b).

Suppose that

$$Det\begin{pmatrix} b & c \\ e & f \end{pmatrix} = 0.$$

There exist natural numbers $\overline{b}, \overline{c}$ such that $\overline{b}, \overline{c} > 0$, $(\overline{b}, \overline{c}) = 1$,

$$
\begin{aligned}
u &= (\overline{x} + \alpha)^a (y^{\overline{b}} z^{\overline{c}})^k \\
v &= (\overline{x} + \alpha)^d (y^{\overline{b}} z^{\overline{c}})^t.
\end{aligned}
$$

After possibly interchanging u and v, we can assume that $k > 0$ and $t \geq 0$. If $t > 0$ we get the form 2.(c). If $t = 0$, we get the form 1.(a).

Suppose that $p \in \mathbf{A}^3$ is a 1 point on $z = 0$. Then there exist $0 \neq \alpha, \beta \in k$, and regular parameters $(\overline{x}, \overline{y}, z)$ at p such that

$$
\begin{aligned}
u &= (\overline{x} + \alpha)^a (\overline{y} + \beta)^b z^c \\
v &= (\overline{x} + \alpha)^d (\overline{y} + \beta)^e z^f.
\end{aligned}
$$

After possibly interchanging u and v we may assume that $c > 0$. If $f > 0$ we get the form 2.(d). If $f = 0$ we get the form 1.(b).

Now suppose that a column of

$$
\begin{pmatrix}
a & b & c \\
d & e & f
\end{pmatrix}
$$

is zero.

After possibly interchanging (x, y, z), we may assume that $c = f = 0$. Then $\Lambda^{-1}(D) = E$ where $xy = 0$ is an equation of E, $uv = 0$ is an equation of D. We get the forms 2.(b), 2.(d) or 1.(b).

Conversely, suppose that the forms 1. and 2. hold at all points of E_X and $p \in E_X$. By Lemma 18.3, there exists an étale neighborhood U of p and uniformizing parameters (x, y, z) on U such that a form 1. or 2. holds at p. Working backwards through the above proof, we see that Φ is toroidal at p. □

We will call a point $p \in X$ a non toroidal point if Φ is not toroidal at p.

Lemma 19.4. *The locus of non toroidal points is Zariski closed of pure codimension 1 in X, and is a SNC divisor. The image of the non toroidal points in S is a finite set of points.*

Proof. Suppose that $p \in X$ is a non toroidal point. Then $q = \Phi(p)$ is a 1 point, and thus one of the forms (176), (179), (180) with $t > 0$, (181) or (183) with $c > 0$ hold at p.

1. First suppose that p is of the form of (183). We have $c > 0$, and all points nearby on $x = 0$ are non toroidal.
2. Suppose that p has the form (179). Φ is non toroidal on the line $x = y = 0$.
   (a) Suppose that $c > 0$. Consider the point with regular parameters $(x, \tilde{y} + \alpha, \tilde{z} + \beta)$ with $\alpha \neq 0$. Set $x = \overline{x}(\tilde{y} + \alpha)^{-\frac{b}{a}}$. Then

   $$
   \begin{aligned}
   u &= \overline{x}^a \\
   v &= \overline{x}^c (\tilde{y} + \alpha)^{d - \frac{cb}{a}} = \overline{x}^c (\gamma + \overline{y})
   \end{aligned}
   $$

   which is non toroidal. Thus Φ is non toroidal on the surface $x = 0$.
   (b) Suppose that $d > 0$. Then a similar analysis shows that Φ is non toroidal on the surface $y = 0$.

3. Suppose that p has the form (181). A point on $x = y = 0$ with regular parameters $(x, y, \tilde{z} + \beta)$ with $\beta \neq 0$ has the form of (179), and is thus not toroidal.

(a) Suppose that $a, c > 0$. Consider the point with regular parameters $(x, \tilde{y} + \alpha, \tilde{z} + \beta)$ with $\alpha \neq 0$. Set $x = \overline{x}(\tilde{y} + \alpha)^{-\frac{b}{a}}$.

$$
\begin{aligned}
u &= \overline{x}^a \\
v &= \overline{x}^c (\tilde{y} + \alpha)^{d - \frac{bc}{a}} (\tilde{z} + \beta) = \overline{x}^c (\gamma + \overline{y}).
\end{aligned}
$$

Thus Φ is non toroidal on the surface $x = 0$.

(b) Suppose that $b, d > 0$. Then Φ is non toroidal on $y = 0$.

Since $a, b > 0$ (by assumption) one of the cases (a) or (b) must hold.

4. Suppose that p has the form (180). We have $t > 0$. Consider a nearby point with regular parameters $(x, \tilde{y} + \overline{\alpha}, \tilde{z} + \overline{\beta})$ with $\overline{\alpha} \neq 0$. Set $x = \overline{x}(y + \overline{\alpha})^{-\frac{b}{a}}$.

$$
\begin{aligned}
u &= \overline{x}^{am} \\
v &= \overline{x}^{at} (\alpha + \overline{\beta} + \tilde{z}) = \overline{x}^{at} (\tilde{c} + \tilde{z}).
\end{aligned}
$$

The non toroidal locus locally contains $x = 0$ (and $y = 0$).

5. Suppose that p has the form (176). Since $a, b, c > 0$, after possibly interchanging (x, y, z), we may assume that $a, d > 0$. Suppose that $(x, \tilde{y} + \alpha, \tilde{z} + \beta)$ are regular parameters at a nearby point (with $\alpha, \beta \neq 0$). Set $x = \overline{x}(\tilde{y} + \alpha)^{-\frac{b}{a}}(\tilde{z} + \beta)^{-\frac{c}{a}}$.

$$
\begin{aligned}
u &= \overline{x}^a \\
v &= \overline{x}^d (\tilde{y} + \alpha)^{e - \frac{db}{a}} (\tilde{z} + \beta)^{f - \frac{dc}{a}} = \overline{x}^d (\gamma + \overline{y})
\end{aligned}
$$

In a similar way, we see that nearby 2 points on $x = 0$ are non toroidal. Thus the non toroidal locus locally contains $x = 0$.

□

Suppose that $p \in X$ is a 1 point such that $\Phi(p) = q$ is a 1 point. A form (183) holds at p. $c - a$ is independent of permissible parameters (u, v) at q and (x, y, z) at p of the form of (183) (since $u = 0$ must be a local equation of D_S). Define

$$
I(\Phi, p) = c - a.
$$

$I(\Phi, p)$ is locally constant. Thus if E is a component of E_X and p_1, p_2 are two 1 points in E such that $\Phi(p_1)$ and $\Phi(p_2)$ are 1 points, then $I(\Phi, p_1) = I(\Phi, p_2)$. We can thus define

$$
I(\Phi, E) = I(\Phi, p)
$$

if $p \in E$ is a 1 point such that $\Phi(p)$ is a 1 point. Let

$$
B_\Phi = \{ q \in S \mid q \text{ is the image of a non toroidal point by } \Phi \}.
$$

Define

$$
I(\Phi) = \max\{ I(\Phi, p) \mid p \in \Phi^{-1}(B_\Phi) \text{ is a 1 point} \}.
$$

Remark 19.5. If $p \in \Phi^{-1}(B_\Phi)$ is a toroidal point then $I(\Phi, p) < 0$.

Lemma 19.6. *Suppose that $q \in B_\Phi$. Let $\pi : S_1 \to S$ be the blow-up of q. Let U be the largest open set of X such that the rational map $X \to S_1$ is a morphism $\Phi_1 : U \to S_1$. Then Φ_1 is strongly prepared, and all points of U are good for Φ_1.*

Suppose that $p \in U \cap \Phi^{-1}(q)$ is a 1 point. If $I(\Phi, p) \leq 0$, then Φ_1 is toroidal at p. If $I(\Phi, p) > 0$, then $I(\Phi_1, p) < I(\Phi, p)$.

The locus of points where $m_q \mathcal{O}_X$ is not invertible is a union of curves which make SNCs with $\overline{B}_2(X)$. These points have one of the forms (187), (193), (185), (190) or (189) of Lemma 18.12.

Proof. Φ_1 is strongly prepared by Lemma 18.13. All points of U are good for Φ_1, as follows by the analysis in Lemma 18.12. The locus of points where $m_q \mathcal{O}_X$ is not invertible is a union of curves which make SNCs with $\overline{B}_2(X)$ by Theorem 18.15.

Suppose that $p \in X$ is such that $m_q \mathcal{O}_X$ is not invertible at p. p is a good point and $\Phi(p) = q$ a 1 point implies p has one of the forms (176), (179), (180), (181) or (183). By Lemma 18.12, p must have one of the forms (187), (193), (185), (190), or (189).

Suppose that $p \in \Phi^{-1}(q) \cap U$ is a 1 point. Then

$$
\begin{aligned}
u &= x^a \\
v &= x^c(\alpha + y)
\end{aligned}
$$

where $u = 0$ is a local equation of D_S at q, with either $a \leq c$ or $c < a$ and $\alpha \neq 0$. Suppose that $I(\Phi, p) = c - a \leq 0$.

If $c < a$, $\alpha \neq 0$ and we have permissible parameters (u_1, v_1) at $q_1 = \Phi_1(p)$ such that

$$
u = u_1 v_1, v = v_1
$$

so that q_1 is a 2 point. There exists regular parameters $(\overline{x}, \overline{y}, z)$ in $\hat{\mathcal{O}}_{X,p}$ and $0 \neq \overline{\alpha} \in k$ such that

$$
\begin{aligned}
u &= \overline{x}^a(\overline{\alpha} + \overline{y}) \\
v &= \overline{x}^c,
\end{aligned}
$$

$$
\begin{aligned}
u_1 &= \overline{x}^{a-c}(\overline{\alpha} + \overline{y}) \\
v_1 &= \overline{x}^c
\end{aligned}
$$

$(\overline{x}, \overline{y}, z)$ are thus permissible parameters for (v_1, u_1) at p, and p is a toroidal point for Φ_1.

If $c = a$,

$$
u_1 = u, v_1 = \frac{v}{u} - \alpha,
$$

(u_1, v_1) are permissible parameters for Φ_1 at $q_1 = \Phi_1(p)$, and

$$
\begin{aligned}
u_1 &= x^a \\
v_1 &= \tfrac{v}{u} - \alpha = y
\end{aligned}
$$

so that p is a toroidal point for Φ_1.

Suppose that $I(\Phi, p) > 0$. Then (u_1, v_1) are permissible parameters at $q_1 = \Phi_1(p)$, where

$$
u = u_1, v = u_1 v_1.
$$

q_1 is a 1 point.

$$u_1 = x^a$$
$$v_1 = x^{c-a}(\alpha + y).$$

Thus $I(\Phi_1, p) = c - 2a < I(\Phi, p)$.   □

Lemma 19.7. *Suppose that $C \subset X$ is a 2 curve such that $q = \Phi(C)$ is a 1 point, if $p \in C$ then p satisfies (187) or (193) and $m_q \mathcal{O}_X$ is not invertible along C. Let $\pi : X_1 \to X$ be the blow-up of C, $\Phi_1 = \Phi \circ \pi$. Then $\Phi_1 : X_1 \to S$ is strongly prepared, all points of X_1 are good points for Φ_1, and if $m_q \mathcal{O}_{X_1, p_1}$ is not invertible at a point $p_1 \in \pi^{-1}(C)$, then p_1 satisfies (187) or (193). If $p_1 \in \pi^{-1}(C)$ is a 1 point then $I(\Phi_1, p_1) < I(\Phi)$.*

Proof. Suppose that $p \in C$ satisfies (187). Then all points of $\pi^{-1}(p)$ are strongly prepared and are good points for Φ_1. Let $p_1 \in \pi^{-1}(p)$ be a 1 point. $\hat{\mathcal{O}}_{X_1, p_1}$ has regular parameters (x_1, y_1, z) such that

$$x = x_1, y = x_1(y_1 + \alpha)$$

with $\alpha \neq 0$.

$$u = (x_1^{a+b}(y_1 + \alpha)^b)^k = \bar{x}_1^{(a+b)k}$$
$$v = x_1^{c+d}(y_1 + \alpha)^d = \bar{x}_1^{c+d}(\bar{\alpha} + \bar{y})$$

By (187) $c - ak$, $d - bk$ have opposite signs.

$$I(\Phi_1, p_1) = (c + d) - (a + b)k$$
$$= (c - ak) + (d - bk) < \max\{c - ak, d - bk\}$$
$$\leq I(\Phi).$$

If p satisfies (193), then all points of $\pi^{-1}(p)$ are strongly prepared good points.   □

Lemma 19.8. *Suppose that $C \subset X$ is a curve such that $q = \Phi(C)$ is a 1 point, $m_q \mathcal{O}_X$ is not invertible along C, and C is not a 2 curve.*
   Let $\pi : X_1 \to X$ be the blow-up of C, $\Phi_1 = \Phi \circ \pi$. Then $\Phi_1 : X_1 \to S$ is strongly prepared and all points of X_1 are good points for Φ_1. If $p_1 \in \pi^{-1}(C)$ is a 1 point, then

$$I(\Phi, p) < I(\Phi_1, p_1) \leq 0.$$

Proof. Suppose that $p \in C$. p satisfies (185), (190) or (189). Φ_1 is strongly prepared, and all points of X_1 are good points for Φ_1. A generic point $p \in C$ satisfies (185), and $x = y = 0$ is a local equation of C at p. Suppose that $p_1 \in \pi^{-1}(p)$ is a 1 point. Then p_1 has permissible parameters (x_1, y_1, z) such that

$$x = x_1, y = x_1(y_1 + \alpha),$$

$$u = x_1^k$$
$$v = x_1^{c+1}(y_1 + \alpha).$$

$I(\Phi_1, p_1) = (c + 1) - k \leq 0$ since $c < k$ by (185).   □

Theorem 19.9. *Suppose that* $\Phi : X \rightarrow S$ *is strongly prepared and all points* $p \in X$ *are good points for* Φ. *Then there exists a sequence of quadratic transforms* $S_1 \rightarrow S$ *and monoidal transforms centered at nonsingular curves,* $X_1 \rightarrow X$, *such that the induced map* $\Phi_1 : X_1 \rightarrow S_1$ *is strongly prepared, all points of* X_1 *are good for* Φ_1 *and* $I(\Phi_1) \leq 0$.

Proof. Suppose that $I(\Phi) > 0$. Suppose that E is a component of E_X such that $I(\Phi, E) = I(\Phi)$. Then $\Phi(E)$ is a single 1 point q. Let $\pi_1 : S_1 \rightarrow S$ be the blow-up of q.

By Lemmas 18.17, 19.6 and 19.8, there exists a sequence of blow-ups of curves C (which are not 2 curves) $X_1 \rightarrow X$ such that if $\Phi_1 : X_1 \rightarrow S$ is the induced map, Φ_1 is strongly prepared, all points of X_1 are good for Φ_1, if $m_q \mathcal{O}_{X_1, p}$ is not invertible then (187) or (193) holds at p, and all curves in X_1 along which $m_q \mathcal{O}_{X_1}$ are not invertible are 2 curves. We further have that $I(\Phi_1, \overline{E}) \leq 0$ if $\overline{E}$ is exceptional for $X_1 \rightarrow X$.

By Lemmas 18.18 and 19.7, there exists a sequence of blow-ups of 2 curves C, $X_2 \rightarrow X_1$ such that if $\Phi_2 : X_2 \rightarrow S$ is the induced map, Φ_2 is strongly prepared, all points of X_2 are good for Φ_2, $m_q \mathcal{O}_{X_2}$ is invertible, and if $\overline{E}$ is exceptional for $X_2 \rightarrow X$, then $I(\Phi_2, \overline{E}) < I(\Phi)$.

Let $\overline{\Phi} : X_2 \rightarrow S_1$ be the induced map. By Lemma 19.6, $\overline{\Phi}$ is strongly prepared, all points of X_2 are good for $\overline{\Phi}$, $I(\overline{\Phi}) \leq I(\Phi)$, and if $\overline{E}$ is a component of E_{X_2} which contains a 1 point q such that $\overline{\Phi}(p) \in \pi_1^{-1}(q)$, then $I(\overline{\Phi}, \overline{E}) < I(\Phi)$.

The Theorem now follows by induction on the number of components E of X such that $I(\Phi, E) = I(\Phi)$, and induction on $I(\Phi)$. $\qquad\square$

Theorem 19.10. *Suppose that* $\Phi : X \rightarrow S$ *is strongly prepared with respect to* D_S, $E_X = \Phi^{-1}(D_S)_{red}$, *all points* $p \in X$ *are good points for* Φ *and* $I(\Phi) \leq 0$. *Then there exist sequences of quadratic transforms* $\pi_1 : S_1 \rightarrow S$ *and monoidal transforms centered at nonsingular curves* $\pi_2 : X_1 \rightarrow X$ *such that the induced map* $\Phi_1 : X_1 \rightarrow S_1$ *is toroidal with respect to* $D_{S_1} = \pi^{-1}(D_S)_{red}$ *and* $E_{X_1} = \pi_2^{-1}(E_X)_{red}$.

Proof. Suppose that E is a component of E_X such that Φ is not toroidal along E. If $p \in E$ is a generic point, then at p we have an expression

$$u = x^a$$
$$v = x^c(\alpha + y)$$

with $c > 0$. Thus there exists a point $q \in S$ such that $\Phi(E) = q$. q is necessarily a 1 point.

Let $\pi : S_1 \rightarrow S$ be the blow-up of q. By Lemmas 18.17 and 19.8, there exists a sequence of blow-ups of nonsingular curves (which are not 2 curves) $X_1 \rightarrow X$ such that if $\Phi_1 : X_1 \rightarrow S$ is the induced morphism, then Φ_1 is strongly prepared, all points of X_1 are good for Φ_1, $I(\Phi_1) \leq 0$, the locus of points p_1 of X_1 such that $m_q \mathcal{O}_{X_1, p_1}$ is not invertible is a union of 2 curves, and if $m_q \mathcal{O}_{X_1, p_1}$ is not invertible, then p_1 satisfies (187) or (193).

Suppose that C is a 2 curve on X_1 such that $m_q \mathcal{O}_{X_1}$ is not invertible along C. A generic point p of C satisfies (187). Let E_1 be the component of E_{X_1}

with local equation $x = 0$ at p, E_2 be the component of E_{X_1} with local equation $y = 0$ at p.

$$I(\Phi_1, E_1) = c - ak, \quad I(\Phi_1, E_2) = d - bk.$$

Since $m_q \mathcal{O}_{X_1,p}$ is not invertible, either $0 < d - bk$ or $0 < c - ak$, a contradiction since $I(\Phi_1) \leq 0$. Thus $m_q \mathcal{O}_{X_1}$ is invertible and $\Phi_1 : X_1 \to S$ induces a morphism $\overline{\Phi} : X_1 \to S_1$. By Lemma 19.6, $\overline{\Phi}$ is strongly prepared, all points of X_1 are good for $\overline{\Phi}$, and $I(\overline{\Phi}) \leq 0$. Further, if $p \in X_1$ is a 1 point such that $p \in \Phi_1^{-1}(q)$, then $\overline{\Phi}$ is toroidal at p.

By induction on the number of components of E_X along which Φ is not toroidal, we achieve the conclusions of the Theorem. $\qquad\square$

Theorem 19.11. *Suppose that* $\Phi : X \to S$ *is a dominant morphism from a 3 fold* X *to a surface* S *(over an algebraically closed field* k *of characteristic 0) and* D_S *is a reduced 1 cycle on* S *such that* $E_X = \Phi^{-1}(D_S)_{red}$ *contains* $sing(X)$ *and* $sing(\Phi)$. *Then there exist sequences of blow-ups of nonsingular subvarieties* $\pi_1 : X_1 \to X$ *and* $\pi_2 : S_1 \to S$ *such that the induced morphism* $X_1 \to S_1$ *is a toroidal morphism with respect to* $\pi_2^{-1}(D_S)_{red}$ *and* $\pi_1^{-1}(E_X)_{red}$.

Proof. This follows from Theorem 17.3, the fact that prepared implies strongly prepared and Theorems 18.19, 19.9 and 19.10. $\qquad\square$

20. Glossary of Notations and Definitions

$\nu(p)$, Definition 6.8.

$\gamma(p)$, Definition 6.8.

$\tau(p)$, Definition 6.8.

$S_r(X)$, $\overline{S}_r(X)$, After Definition 6.8.

$B_2(X)$, $\overline{B}_2(X)$, $B_3(X)$, Before Definition 6.17.

SNCs with $\overline{B}_2(X)$, Definition 6.17.

r small, Definition 8.3.

r big, Definition 8.3.

1 point, 2 point, 3 point, Definition 6.4 and before Definition 18.1.

1-resolved, Definition 9.6.

$\overline{\nu}(p)$, After Definition 9.6.

$\sigma(p)$, Before Lemma 9.9.

$\delta(p)$, before Lemma 9.13.

Inv(p), Before Theorem 9.15.

$A_r(X)$, Definition 10.2.

$\overline{A}_r(X)$, Definition 10.1.

$C_r(X)$, Definition 14.1.

(E), Definition 15.5.

$*$-permissible parameters, After Definition 18.1.

$\nu_E(f)$, Before Definition 18.7.

$A(\Phi, p)$, Definition 18.7.

$C(\Phi, p)$, Definition 18.7.

$A(\Phi, E)$, After Definition 18.7.

$C(\Phi, E)$, After Definition 18.7.

$A(\Phi)$, Before Lemma 18.10.

$C(\Phi)$, Before Lemma 18.10.

$I(\Phi, p)$, $I(\Phi, E)$, $I(\Phi)$, Before Remark 19.5.

B_Φ, Before Remark 19.5.

SNC divisor, Definition 5.1.

$P_t(x)$, After Definition 5.1.

bad point, Definition 18.5.

étale neighborhood, Definition 6.18.

good point, Definition 18.5.

monoidal transform, After Definition 5.1.

monomial mapping, Definition 18.20.

non toroidal point, Before Lemma 19.4.

permissible monoidal transform, Definition 10.4.

permissible parameters, Before Definition 6.4 and 6.4.

permissible parameters for C at p, After Lemma 6.16.

prepared, Definition 6.5.

resolved, Definition 6.9.

strongly prepared, Definition 18.1.

toroidal mapping, 19.1.

REFERENCES

[1] ABHYANKAR, S., Local uniformization on algebraic surfaces over ground fields of characteristic $p \neq 0$, Annals of Math, 63 (1956), 491-526.

[2] ABHYANKAR, S., Simultaneous resolution for algebraic surfaces, Amer. J. Math 78 (1956), 761-790.

[3] ABHYANKAR, S., Resolution of singularities of embedded algebraic surfaces, second edition, Springer Verlag, 1998.

[4] ABHYANKAR, S., Good points of a hypersurface, Advances in Math. 68 (1988), 87-256.

[5] ABHYANKAR, S., Resolution of Singularities and Modular Galois Theory, Bulletin of the AMS 38 (2001), 131-171.

[6] ABRAMOVICH D., KARU, K., Weak semistable reduction in characteristic 0, preprint.

[7] ABRAMOVICH, D., KARU, K., MATSUKI, K., WLODARCZYK, J., Torification and Factorization of Birational Maps, preprint.

[8] AKBULUT, S. AND KING, H., Topology of algebraic sets, MSRI publications 25, Springer-Verlag Berlin.

[9] BARTLET, D., MAIRE, H.M., Asymptotique des intgrales-fibres, Ann. Inst. Fourier 43, 1267-1299 (1993).

[10] CANO TANO, F., Desingularization strategies for three-dimensional vector fields, LNM 1259 (1987).

[11] CHRISTENSEN, C., Strong domination/ weak factorization of three dimensional regular local rings, Journal of the Indian Math Soc., 45 (1981), 21-47.

[12] COSSART, V., Polyedre caracteristique d'une singularite, Thesis, Universite de Paris-Sud, Centre d'Orsay (1987).

[13] CUTKOSKY, S.D., Local Factorization of Birational Maps, Advances in Math. 132, (1997), 167-315.

[14] CUTKOSKY, S.D., Local Monomialization and Factorization of Morphisms, Astérisque 260, (1999).

[15] CUTKOSKY, S.D., Simultaneous resolution of singularities, Proc. American Math. Soc. 128, (2000), 1905-1910.

[16] CUTKOSKY, S.D., Generically Finite Morphisms, preprint.

[17] CUTKOSKY, S.D., Resolution of Singularities, Lectures on resolution, preprint.

[18] CUTKOSKY, S.D. AND PILTANT, O., Monomial Resolutions of Morphisms of Algebraic Surfaces, Communications in Algebra 28 (Hartshorne Volume), (2000), 5935-5960.

[19] CUTKOSKY, S.D. AND PILTANT, O., Ramification of valuations, preprint.

[20] CUTKOSKY, S.D. AND SRINIVASAN, H., An Intrinsic Criterion for isomorphism of singularities, American. Journal of Math. 115, (1993), 789-821.

[21] GROTHENDIECK, A., Revêtements étales et Groupe Fondemental, Lecture Notes in Math. 224, (1971) Springer-Verlag Heidelberg (1971).

[22] DE JONG, A.J., Smoothness, semistability and Alterations, Publ. Math. I.H.E.S. 83, 1996, 51-93.

[23] HIRONAKA, H., Resolution of singularities of an algebraic variety over a field of characteristic zero, Annals of Math, 79 (1964), 109-326.

[24] KEMPF, G., KNUDSEN, F., MUMFORD, D., SAINT-DONAT, B., Toroidal embeddings I, LNM 339, Springer Verlag (1973).

[25] KUHLMANN, FV, Valuation theoretic and model theoretic aspects of local uniformization, preprint.

[26] LIPMAN, J., Introduction to Resolution of Singularities, in Algebraic Geometry, Arcata 1974, Amer. Math. Soc. proc. Symp. Pure Math. 29 (1975) 187-230.

[27] MATSUMURA, H., Commutative Algebra 2nd edition, W.A. Benjamin Co., N.Y.

[28] MOH, TT., Quasi-Canonical uniformization of hypersurface singularities of cahracteristic zero, Comm. Algebra 20, 3207-3249 (1992).

[29] MUMFORD, D., Red Book, Lecture Notes in Math. 1358, (1988) Springer Verlag.

[30] SALLY, J., Regular overrings of regular local rings, Trans. Amer. Math. Soc. 171 (1972) 291-300.

[31] SHANNON, D.L., monoidal transforms, Amer. J. Math, 45 (1973), 284-320.

[32] SPIVAKOVSKY, M., Sandwiched singularities and desingularization of surfaces by normalized Nash transforms, Ann. of Math. 131, 1990, 441-491.

[33] TEISSIER, B., Valuations, Deformations and Toric Geometry, preprint.

[34] VILLAMAYOR, O., Constructiveness of Hironaka's resolution, Ann. Scient. Ecole Norm Sup 22, 1-32, 1989

[35] WLODARCZYK, J., Birational Cobordisms and factorization of birational maps, to appear in Journal of Alg. Geometry.

[36] ZARISKI, O., The reduction of the singularities of an algebraic surface, Annals of Math., 40 (1939) 639-689.

[37] ZARISKI, O., Local uniformization of algebraic varieties, Annals of Math., 41, (1940), 852-896.

[38] ZARISKI, O. AND SAMUEL, P., Commutative Algebra II, Van Nostrand, Princeton (1960).

Druck: Strauss Offsetdruck, Mörlenbach
Verarbeitung: Schäffer, Grünstadt

Vol. 1701: Ti-Jun Xiao, J. Liang, The Cauchy Problem of Higher Order Abstract Differential Equations, XII, 302 pages. 1998.

Vol. 1702: J. Ma, J. Yong, Forward-Backward Stochastic Differential Equations and Their Applications. XIII, 270 pages. 1999.

Vol. 1703: R. M. Dudley, R. Norvaiša, Differentiability of Six Operators on Nonsmooth Functions and p-Variation. VIII, 272 pages. 1999.

Vol. 1704: H. Tamanoi, Elliptic Genera and Vertex Operator Super-Algebras. VI, 390 pages. 1999.

Vol. 1705: I. Nikolaev, E. Zhuzhoma, Flows in 2-dimensional Manifolds. XIX, 294 pages. 1999.

Vol. 1706: S. Yu. Pilyugin, Shadowing in Dynamical Systems. XVII, 271 pages. 1999.

Vol. 1707: R. Pytlak, Numerical Methods for Optimal Control Problems with State Constraints. XV, 215 pages. 1999.

Vol. 1708: K. Zuo, Representations of Fundamental Groups of Algebraic Varieties. VII, 139 pages. 1999.

Vol. 1709: J. Azéma, M. Émery, M. Ledoux, M. Yor (Eds), Séminaire de Probabilités XXXIII. VIII, 418 pages. 1999.

Vol. 1710: M. Koecher, The Minnesota Notes on Jordan Algebras and Their Applications. IX, 173 pages. 1999.

Vol. 1711: W. Ricker, Operator Algebras Generated by Commuting Projections: A Vector Measure Approach. XVII, 159 pages. 1999.

Vol. 1712: N. Schwartz, J. J. Madden, Semi-algebraic Function Rings and Reflectors of Partially Ordered Rings. XI, 279 pages. 1999.

Vol. 1713: F. Bethuel, G. Huisken, S. Müller, K. Steffen, Calculus of Variations and Geometric Evolution Problems. Cetraro, 1996. Editors: S. Hildebrandt, M. Struwe. VII, 293 pages. 1999.

Vol. 1714: O. Diekmann, R. Durrett, K. P. Hadeler, P. K. Maini, H. L. Smith, Mathematics Inspired by Biology. Martina Franca, 1997. Editors: V. Capasso, O. Diekmann. VII, 268 pages. 1999.

Vol. 1715: N. V. Krylov, M. Röckner, J. Zabczyk, Stochastic PDE's and Kolmogorov Equations in Infinite Dimensions. Cetraro, 1998. Editor: G. Da Prato. VIII, 239 pages. 1999.

Vol. 1716: J. Coates, R. Greenberg, K. A. Ribet, K. Rubin, Arithmetic Theory of Elliptic Curves. Cetraro, 1997. Editor: C. Viola. VIII, 260 pages. 1999.

Vol. 1717: J. Bertoin, F. Martinelli, Y. Peres, Lectures on Probability Theory and Statistics. Saint-Flour, 1997. Editor: P. Bernard. IX, 291 pages. 1999.

Vol. 1718: A. Eberle, Uniqueness and Non-Uniqueness of Semigroups Generated by Singular Diffusion Operators. VIII, 262 pages. 1999.

Vol. 1719: K. R. Meyer, Periodic Solutions of the N-Body Problem. IX, 144 pages. 1999.

Vol. 1720: D. Elworthy, Y. Le Jan, X-M. Li, On the Geometry of Diffusion Operators and Stochastic Flows. IV, 118 pages. 1999.

Vol. 1721: A. Iarrobino, V. Kanev, Power Sums, Gorenstein Algebras, and Determinantal Loci. XXVII, 345 pages. 1999.

Vol. 1722: R. McCutcheon, Elemental Methods in Ergodic Ramsey Theory. VI, 160 pages. 1999.

Vol. 1723: J. P. Croisille, C. Lebeau, Diffraction by an Immersed Elastic Wedge. VI, 134 pages. 1999.

Vol. 1724: V. N. Kolokoltsov, Semiclassical Analysis for Diffusions and Stochastic Processes. VIII, 347 pages. 2000.

Vol. 1725: D. A. Wolf-Gladrow, Lattice-Gas Cellular Automata and Lattice Boltzmann Models. IX, 308 pages. 2000.

Vol. 1726: V. Marić, Regular Variation and Differential Equations. X, 127 pages. 2000.

Vol. 1727: P. Kravanja M. Van Barel, Computing the Zeros of Analytic Functions. VII, 111 pages. 2000.

Vol. 1728: K. Gatermann Computer Algebra Methods for Equivariant Dynamical Systems. XV, 153 pages. 2000.

Vol. 1729: J. Azéma, M. Émery, M. Ledoux, M. Yor Séminaire de Probabilités XXXIV. VI, 431 pages. 2000.

Vol. 1730: S. Graf, H. Luschgy, Foundations of Quantization for Probability Distributions. X, 230 pages. 2000.

Vol. 1731: T. Hsu, Quilts: Central Extensions, Braid Actions, and Finite Groups. XII, 185 pages. 2000.

Vol. 1732: K. Keller, Invariant Factors, Julia Equivalences and the (Abstract) Mandelbrot Set. X, 206 pages. 2000.

Vol. 1733: K. Ritter, Average-Case Analysis of Numerical Problems. IX, 254 pages. 2000.

Vol. 1734: M. Espedal, A. Fasano, A. Mikelić, Filtration in Porous Media and Industrial Applications. Cetraro 1998. Editor: A. Fasano. 2000.

Vol. 1735: D. Yafaev, Scattering Theory: Some Old and New Problems. XVI, 169 pages. 2000.

Vol. 1736: B. O. Turesson, Nonlinear Potential Theory and Weighted Sobolev Spaces. XIV, 173 pages. 2000.

Vol. 1737: S. Wakabayashi, Classical Microlocal Analysis in the Space of Hyperfunctions. VIII, 367 pages. 2000.

Vol. 1738: M. Émery, A. Nemirovski, D. Voiculescu, Lectures on Probability Theory and Statistics. XI, 356 pages. 2000.

Vol. 1739: R. Burkard, P. Deuflhard, A. Jameson, J.-L. Lions, G. Strang, Computational Mathematics Driven by Industrial Problems. Martina Franca, 1999. Editors: V. Capasso, H. Engl, J. Periaux. VII, 418 pages. 2000.

Vol. 1740: B. Kawohl, O. Pironneau, L. Tartar, J.-P. Zolesio, Optimal Shape Design. Tróia, Portugal 1999. Editors: A. Cellina, A. Ornelas. IX, 388 pages. 2000.

Vol. 1741: E. Lombardi, Oscillatory Integrals and Phenomena Beyond all Algebraic Orders. XV, 413 pages. 2000.

Vol. 1742: A. Unterberger, Quantization and Non-holomorphic Modular Forms.VIII, 253 pages. 2000.

Vol. 1743: L. Habermann, Riemannian Metrics of Constant Mass and Moduli Spaces of Conformal Structures. XII, 116 pages. 2000.

Vol. 1744: M. Kunze, Non-Smooth Dynamical Systems. X, 228 pages. 2000.

Vol. 1745: V. D. Milman, G. Schechtman, Geometric Aspects of Functional Analysis. VIII, 289 pages. 2000.

Vol. 1746: A. Degtyarev, I. Itenberg, V. Kharlamov, Real Enriques Surfaces. XVI, 259 pages. 2000.

Vol. 1747: L. W. Christensen, Gorenstein Dimensions. VIII, 204 pages. 2000.

Vol. 1748: M. Ruzicka, Electrorheological Fluids: Modeling and Mathematical Theory. XV, 176 pages. 2001.

Vol. 1749: M. Fuchs, G. Seregin, Variational Methods for Problems from Plasticity Theory and for Generalized Newtonian Fluids. VI, 269 pages. 2001.

Vol. 1750: B. Conrad, Grothendieck Duality and Base Change. X, 296 pages. 2001.

Vol. 1751: N. J. Cutland, Loeb Measures in Practice: Recent Advances. XI, 111 pages. 2001.

Vol. 1752: Y. V. Nesterenko, P. Philippon, Introduction to Algebraic Independence Theory. XIII, 256 pages. 2001.

Vol. 1753: A. I. Bobenko, U. Eitner, Painlevé Equations in the Differential Geometry of Surfaces. VI, 120 pages. 2001.

Vol. 1754: W. Bertram, The Geometry of Jordan and Lie Structures. XVI, 269 pages. 2001.

Vol. 1755: J. Azéma, M. Émery, M. Ledoux, M. Yor, Séminaire de Probabilités XXXV. VI, 427 pages. 2001.

Vol. 1756: P. E. Zhidkov, Korteweg de Vries and Nonlinear Schrödinger Equations: Qualitative Theory. VII, 147 pages. 2001.

Vol. 1757: R. R. Phelps, Lectures on Choquet's Theorem. VII, 124 pages. 2001.

Vol. 1758: N. Monod, Continuous Bounded Cohomology of Locally Compact Groups. X, 214 pages. 2001.

Vol. 1759: Y. Abe, K. Kopfermann, Toroidal Groups. VIII, 133 pages. 2001.

Vol. 1760: D. Filipović, Consistency Problems for Heath-Jarrow-Morton Interest Rate Models. VIII, 134 pages. 2001.

Vol. 1761: C. Adelmann, The Decomposition of Primes in Torsion Point Fields. VI, 142 pages. 2001.

Vol. 1762: S. Cerrai, Second Order PDE's in Finite and Infinite Dimension. IX, 330 pages. 2001.

Vol. 1763: J.-L. Loday, A. Frabetti, F. Chapoton, F. Goichot, Dialgebras and Related Operads. IV, 132 pages. 2001.

Vol. 1764: A. Cannas da Silva, Lectures on Symplectic Geometry. XII, 217 pages. 2001.

Vol. 1765: T. Kerler, V. V. Lyubashenko, Non-Semisimple Topological Quantum Field Theories for 3-Manifolds with Corners. VI, 379 pages. 2001.

Vol. 1766: H. Hennion, L. Hervé, Limit Theorems for Markov Chains and Stochastic Properties of Dynamical Systems by Quasi-Compactness. VIII, 145 pages. 2001.

Vol. 1767: J. Xiao, Holomorphic Q Classes. VIII, 112 pages. 2001.

Vol. 1768: M.J. Pflaum, Analytic and Geometric Study of Stratified Spaces. VIII, 230 pages. 2001.

Vol. 1769: M. Alberich-Carramiñana, Geometry of the Plane Cremona Maps. XVI, 257 pages. 2002.

Vol. 1770: H. Gluesing-Luerssen, Linear Delay-Differential Systems with Commensurate Delays: An Algebraic Approach. VIII, 176 pages. 2002.

Vol. 1771: M. Émery, M. Yor, Séminaire de Probabilités 1967-1980. A Selection in Martingale Theory. IX, 553 pages. 2002.

Vol. 1772: F. Burstall, D. Ferus, K. Leschke, F. Pedit, U. Pinkall, Conformal Geometry of Surfaces in S^4. VII, 89 pages. 2002.

Vol. 1773: Z. Arad, M. Muzychuk, Standard Integral Table Algebras Generated by a Non-real Element of Small Degree. VI, 126 pages. 2002.

Vol. 1774: V. Runde, Lectures on Amenability. XIV, 296 pages. 2002.

Vol. 1775: W. H. Meeks, A. Ros, H. Rosenberg, The Global Theory of Minimal Surfaces in Flat Spaces. Martina Franca 1999. Editor: G. P. Pirola. X, 117 pages. 2002.

Vol. 1776: K. Behrend, C. Gomez, V. Tarasov, G. Tian, Quantum Comohology. Cetraro 1997. Editors: P. de Bartolomeis, B. Dubrovin, C. Reina. VIII, 319 pages. 2002.

Vol. 1777: E. García-Río, D. N. Kupeli, R. Vázquez-Lorenzo, Osserman Manifolds in Semi-Riemannian Geometry. XII, 166 pages. 2002.

Vol. 1778: H. Kiechle, Theory of K-Loops. X, 186 pages. 2002.

Vol. 1779: I. Chueshov, Monotone Random Systems. VIII, 234 pages. 2002.

Vol. 1780: J. H. Bruinier, Borcherds Products on O(2,1) and Chern Classes of Heegner Divisors. VIII, 152 pages. 2002.

Vol. 1781: E. Bolthausen, E. Perkins, A. van der Vaart, Lectures on Probability Theory and Statistics. Ecole d' Eté de Probabilités de Saint-Flour XXIX-1999. Editor: P. Bernard. VII, 480 pages. 2002.

Vol. 1782: C.-H. Chu, A. T.-M. Lau, Harmonic Functions on Groups and Fourier Algebras. VII, 100 pages. 2002.

Vol. 1783: L. Grüne, Asymptotic Behavior of Dynamical and Control Systems under Perturbation and Discretization. IX, 231 pages. 2002.

Vol. 1784: L.H. Eliasson, S. B. Kuksin, S. Marmi, J.-C. Yoccoz, Dynamical Systems and Small Divisors. Cetraro, Italy 1998. Editors: S. Marmi, J.-C. Yoccoz. VIII, 199 pages. 2002.

Vol. 1785: J. Arias de Reyna, Pointwise Convergence of Fourier Series. XVIII, 175 pages. 2002.

Vol. 1786: S. D. Cutkosky, Monomialization of Morphisms from 3-Folds to Surfaces. V, 235 pages. 2002.

Vol. 1787: S. Caenepeel, G. Militaru, S. Zhu, Frobenius and Separable Functors for Generalized Module Categories and Nonlinear Equations. XIV, 354 pages. 2002.

Vol. 1788: A. Vasil'ev, Moduli of Families of Curves for Conformal and Quasiconformal Mappings. IX, 211 pages. 2002.

Vol. 1789: Y. Sommerhäuser, Yetter-Drinfel'd Hopf algebras over groups of prime order. V, 157 pages. 2002.

Vol. 1790: X. Zhan, Matrix Inequalities. VII, 116 pages. 2002.

Vol. 1791: M. Knebusch, D. Zhang, Manis Valuations and Prüfer Extensions I: A new Chapter in Commutative Algebra. VII, 264 pages. 2002.

Recent Reprints and New Editions

Vol. 1200: V. D. Milman, G. Schechtman, Asymptotic Theory of Finite Dimensional Normed Spaces. 1986. – Corrected Second Printing. X, 156 pages. 2001.

Vol. 1618: G. Pisier, Similarity Problems and Completely Bounded Maps. 1995 – Second, Expanded Edition VII, 198 pages. 2001.

Vol. 1629: J. D. Moore, Lectures on Seiberg-Witten Invariants. 1997 – Second Edition. VIII, 121 pages. 2001.

Vol. 1638: P. Vanhaecke, Integrable Systems in the realm of Algebraic Geometry. 1996 – Second Edition. X, 256 pages. 2001.

Vol. 1702: J. Ma, J. Yong, Forward-Backward Stochastic Differential Equations and Their Applications. 1999. – Corrected Second Printing. XIII, 270 pages. 2000.

4. Lecture Notes are printed by photo-offset from the master-copy delivered in camera-ready form by the authors. Springer-Verlag provides technical instructions for the preparation of manuscripts. Macro packages in $\mathrm{T_EX}$, $\mathrm{L^AT_EX2e}$, $\mathrm{L^AT_EX2.09}$ are available from Springer's web-pages at

http://www.springer.de/math/authors/b-tex.html.

Careful preparation of the manuscripts will help keep production time short and ensure satisfactory appearance of the finished book.

The actual production of a Lecture Notes volume takes approximately 12 weeks.

5. Authors receive a total of 50 free copies of their volume, but no royalties. They are entitled to a discount of 33.3 % on the price of Springer books purchase for their personal use, if ordering directly from Springer-Verlag.

Commitment to publish is made by letter of intent rather than by signing a formal contract. Springer-Verlag secures the copyright for each volume. Authors are free to reuse material contained in their LNM volumes in later publications: A brief written (or e-mail) request for formal permission is sufficient.

Addresses:

Professor Jean-Michel Morel
CMLA, École Normale Supérieure de Cachan
61 Avenue du Président Wilson
94235 Cachan Cedex France
e-mail: Jean-Michel.Morel@cmla.ens-cachan.fr

Professor Bernard Teissier
Institut de Mathématiques de Jussieu
Equipe "Géométrie et Dynamique"
175 rue du Chevaleret
75013 PARIS
e-mail: Teissier@ens.fr

Professor F. Takens, Mathematisch Instituut
Rijksuniversiteit Groningen, Postbus 800
9700 AV Groningen, The Netherlands
e-mail: F.Takens@math.rug.nl

Springer-Verlag, Mathematics Editorial, Tiergartenstr. 17
D-69121 Heidelberg, Germany
Tel.: +49 (6221) 487-701
Fax: +49 (6221) 487-355
e-mail: lnm@Springer.de